EINSTEIN'S THEORY OF RELATIVITY

by Max Born

Revised Edition

prepared with the collaboration of

Günther Leibfried *and* Walter Biem

DOVER PUBLICATIONS, INC.
NEW YORK

TO MY WIFE

Published in Canada by General Publishing Company, Ltd., 30 Lesmill Road, Don Mills, Toronto, Ontario.
Published in the United Kingdom by Constable and Company, Ltd.

This Dover edition, first published in 1962, is a revised and enlarged version of the work published by Methuen Company in 1924.

Standard Book Number: 486-60769-0
Library of Congress Catalog Card Number: 62-5801

Manufactured in the United States of America
Dover Publications, Inc.
180 Varick Street
New York, N. Y. 10014

PREFACE

The first German edition of this book appeared in 1920 and an English translation appeared in 1924. When the publishing firm of Dover Publications, Inc., approached me with the suggestion of reprinting the book, I was at first unable to agree, since large sections of it were by then completely out of date. Yet there was a peculiar feature of the book which seemed to me worth preserving or renewing. This text was originally an elaboration of a series of lectures given at Frankfurt am Main to a large audience when a wave of popular interest in the theory of relativity and in Einstein's personality had spread around the world, following the first confirmation by a British solar-eclipse expedition of Einstein's prediction that a beam of light should be bent by the gravitational action of the sun. Though sensationalism was probably the main cause of this interest, there was also a considerable and genuine desire to understand. I set myself the task of satisfying this desire as far as possible. The main obstacle was the low standard of knowledge of physics and mathematics in the audience. In the lectures I followed a quasi-historical method of presentation and illustrated each step by simple experiments and diagrams. In the printed text the experiments could be described only with the help of figures. I restricted myself in the use of mathematical formulae, to the simplest algebra, using nothing more than linear equations and square roots (avoiding even quadratic equations and trigonometric functions)—in short, to matter contained in the syllabus of the lower forms of high school. Limiting processes could not quite be avoided but were presented in a form which could be understood by applying common sense.

Since then, many books on relativity have appeared, both scientific and popular ones. The latter, including some presentations by Einstein himself, avoid, in general, all mathematical formulae and diagrams and give only descriptions of the facts and ideas in ordinary language or in philosophical terms—a way in which, I fear, only a very superficial acquaintance with relativity can be obtained. Nevertheless, in our time science, and physics in particular, has become a fundamental part of our civilization, and the number of people who wish to grasp its essence has grown immensely. Now, rereading my old book, I get the impression that its way of presentation should have an appeal to a considerable number of people, particularly to those who, without knowing higher mathematics and modern physics, remember something of what they learned at school and are willing to do a little thinking. I believe that they could gather from a book of this kind more than a vague feeling about grand, but dark and abstruse, mysteries of nature; they might really obtain an understanding of modern scientific thinking.

Therefore I suggested to the publishers a thoroughly revised and modernized edition, provided a younger collaborator could be found. Professor Günther Leibfried, then at Göttingen, now at Aachen, accepted my proposal to help me, and as the work of scanning the literature, formulating new sections, revising the text, and the like, was more than he could do, he found a collaborator in Dr. Walter Biem, one of the members of his department. We have used the English edition of 1924 as a skeleton, but have revised the whole text, rewriting many sections and adding a few. The main changes are in Chapters VI (Special Relativity) and VII (General Relativity). For instance, the derivation of the law connecting mass and energy and of the formula expressing the velocity dependence of these quantities has been much improved by applying the conservation laws of energy and momentum to the case of an inelastic collision. The account of the empirical tests of general relativity has been brought up to date, and the prospects of future developments are indicated. The section "Macrocosm and Microcosm" of the old edition has been replaced by one on "Cosmology," giving a very brief survey of the modern situation. I wish to mention two short and impressive papers which have been helpful, one by O. Heckmann (*Von Erde und Weltall*; A. Kröner Verlag, Stuttgart, p. 149), the other by W. L. Ginsburg (*Fortschritte der Physik*, Vol. V, 1957, p. 16). In spite of these modernizations the text reflects at many places the situation of physical knowledge forty years ago. A complete adaptation to modern standards would have meant writing a new book, and this was not our purpose.

As in the original edition, no references are given in the text.

The vexatious question of units has been solved by retaining the classical Gaussian system used in the original book. I am still convinced that it is the most satisfactory one from the standpoint of logic and epistemology—though perhaps not from that of the practical physicist and engineer—and that it is therefore preferable for teaching.

I wish to thank my collaborators for their untiring efforts, and Dr. Chr. Lehmann for redrawing and improving all the figures, my grandson John Pryce for checking the English, Mrs. Felicitas Ludwig for clerical help, and Dover Publications, Inc., for the excellent printing.

MAX BORN

Bad Pyrmont, 1962

CONTENTS

INTRODUCTION

The development of science is a continuous and steady process. Nevertheless, definite periods are recognizable, marked by outstanding empirical discoveries or theoretical ideas. One of these turning points happened about 1600 and is connected with the name of Galileo, who laid the foundations of the empirical method through his researches in mechanics and produced convincing evidence for the Copernican system of the universe, published fifty years earlier. That meant the end of the scholastic philosophy of nature based on the teaching of Aristotle, and the beginning of modern science.

Another turning point came about 1900, when a flood of new experimental discoveries—x-rays, radioactivity, the electron, etc.—occurred, and two fundamental theories—quantum theory and relativity—were developed. The quantum theory dates from the year 1900, when Max Planck announced his revolutionary concept of energy atoms, or "quanta." This event was so decisive for the development of science that it is usually considered as the dividing point between *classical physics* and *modern* or *quantum physics*. Relativity actually ought not to be connected with a single name or with a single date. It was in the air about 1900 and several great mathematicians and physicists—Larmor, Fitzgerald, Lorentz, Poincaré, to mention a few—were in possession of many of its contents. In 1905 Albert Einstein based the theory on very general principles of a philosophical character, and a few years later Hermann Minkowski gave it final logical and mathematical expression. The reason Einstein's name alone is usually connected with relativity is that his work of 1905 was only the initial step to a still more fundamental "general relativity," which included a new theory of gravitation and opened new vistas in our understanding of the structure of the universe.

The special theory of relativity of 1905 can be justifiably considered the end of the classical period or the beginning of a new era. For it uses the well-established classical ideas of matter spread continuously in space and time, and of causal or, more precisely, deterministic laws of nature. But it introduces revolutionary notions of space and time, resolutely criticizing the traditional concepts as formulated by Newton. Thus it opens a new way of thinking about natural phenomena. This seems today Einstein's most remarkable feat, the one which distinguishes his work from that of his predecessors, and modern science from classical science.

Even before Einstein the investigation of the physical world had led to a trespassing on the limits of the domain of the human senses. Scientists knew about invisible (ultraviolet, infrared) light, inaudible sound; they operated with electromagnetic fields in empty space which were imperceptible to the senses and only indirectly open to observation through their action on matter, and so on. These generalizations were possible and necessary when the restricted value of the direct sense impressions was recognized. To give a simple example: the feeling of hot and cold was not sufficiently precise for a theory of heat to be built upon it; it was replaced by thermometers, therefore, where a thermal difference could be observed as the length of a mercury column, or by some similar device. There are innumerable cases where one of the senses has been replaced, or at least checked, by another one. In fact, the whole of science is a maze of such cross-connections whereby the purely geometric structures, as given by vision or touch, are preferred because they are the most reliable ones. This process is the essence of *objectivization*, which aims at making observations as independent of the individual observer as possible. In this way electromagnetic fields, for instance, which are not directly accessible to any human sense, could be introduced by reducing them to mechanical quantities measurable in space and time.

Another general feature of science was the principle of *relativization*. One famous example is connected with the discovery of the spherical shape of the earth. As long as the earth was regarded as a flat disk, the "up-down," or vertical, direction at a place on the earth was something absolute. Now it became the direction towards the center of the globe and thus was defined only relative to the standpoint of the observer. The general question as to whether a direction or a point in space and an instant in the flux of time was something absolute was answered for science by Newton's celebrated axioms. Their wording leaves no doubt that Newton's answer is affirmative. But his equations of motion contradict this in a way: there exist certain equivalent systems of reference in relative motion each of which can be regarded with equal justification as absolutely at rest. Newton's space is therefore absolute only in a restricted sense. Later research, particularly in electromagnetism and optics, revealed other and more severe difficulties in the Newtonian position.

Einstein broke through this barrier by a critical assessment of the current ideas of space and time. He found them unsatisfactory and replaced them by better ones. Thereby he followed the leading principles of scientific research, objectivization and relativization, and in addition used another principle which certainly had been known before but which was used mainly for logical criticism and not for scientific construction—for instance by Ernst Mach, the physicist and philosopher whose work had made a strong impression on Einstein. This principle said that concepts and statements which are not empirically verifiable should have no place in a physical theory. Einstein analyzed the simultaneity of two events happening at different places in space and found it to be such a nonverifiable notion. This discovery led him, in 1905, to a new formulation of the fundamental properties of space and time. About ten years later the same principle, applied to motion under gravitational forces, guided him in the establishment of his theory of general relativity.

This principle, demanding the elimination of the unobservable, has been the object of much philosophical discussion. It was called positivistic, and it is certainly in the line of the philosophy of which Mach was a prominent partisan. But positivism accepts only the immediate sense impressions as real, everything else as a construct of the mind; it leads to a skeptical attitude towards the existence of an external world. Nothing was more remote from Einstein's convictions; in later years he emphatically declared himself opposed to positivism.

One should regard this method, used with such success by Einstein, as a heuristic principle pointing to weak spots in a traditional theory which has turned out to be empirically unsatisfactory. It has become the outstanding method of fundamental research in modern physics, particularly in the development of quantum theory; and because of this fact Einstein's way of thinking has not only led to the summit of the classical period but has opened a new age of physics.

CHAPTER I

GEOMETRY AND COSMOLOGY

1. The Origin of the Art of Measuring Space and Time

The physical problem presented by space and time consists in fixing numerically a place and a point of time for every physical event, thus enabling us to single it out, as it were, from the chaos of the coexistence and succession of things.

The first problem of man was to find his way about on the earth. Hence the art of measuring the earth (geodesy) became the source of the science of space, which derived its name "geometry" from the Greek word for earth. From the very outset, however, the measure of time arose from the regular changes of night and day, the phases of the moon, and the seasons. These phenomena forced themselves on man's attention and caused him to direct his gaze toward the stars, which were the source of the science of the universe, cosmology. Astronomic technique applied to the heavenly regions the teachings of geometry that had been tested on the earth, allowing distances and orbits to be defined, and gave the inhabitants of the earth the celestial (astronomic) measure of time which taught them to distinguish between past, present, and future, and to assign to each event its place in time.

2. Units of Length and Time

The foundation of every space and time measurement is laid by fixing the unit. The phrase "a length of so and so many meters" denotes the ratio of the length to be measured to the length of a meter. The phrase "a time of so many seconds" denotes the ratio of the time to be measured to the duration of a second. Thus we are always dealing with ratios, relative data concerning units which are themselves to a high degree arbitrary, and are chosen for reasons of their being easily reproduced, easily transported, durable, and so forth.

In physics the measure of length is the *centimeter* (cm.), the hundredth part of a meter rod that is preserved in Paris. This was originally intended to bear a simple ratio to the circumference of the earth—namely, to be the ten-millionth part of a quadrant—but more recent measurements have disclosed that this statement is not accurate.

The unit of time in physics is the *second* (sec.), which bears the well-known relation to the time of rotation of the earth on its axis.

These definitions of the units derived from the circumference and rotation of the earth have turned out to be inconvenient. Today we use more readily reproducible units based on the atomic properties of matter. Thus the meter is now defined by saying that it contains a certain number of wave lengths of a certain, well-defined electromagnetic radiation sent out by a cadmium atom. The second can be defined as a given multiple of the oscillation time of certain molecules.

3. Origin and Coordinate System

If we wish not only to determine lengths and periods of time but also to designate places and points of time, further conventions must be made. In the case of time, which we regard as a one-dimensional configuration, it is sufficient to specify an *origin* (or zero point). Historians reckon dates by counting the years from the birth of Christ. Astronomers choose other origins or initial points, according to the objects of their researches; these they call epochs. If the unit and the origin are fixed, every event may be singled out by assigning a number to it.

In geometry in the narrower sense, in order to determine position on the earth, two data must be given to fix a point. To say "My house is on Baker Street" is not sufficient to locate it. The house number must also be given. In many American towns the streets themselves are numbered. The address 25 13th Street thus consists of two number data. It is exactly what mathematicians call a "coordinate determination." The earth's surface is covered with a network of intersecting lines which are numbered, or whose position is determined by a number, distance, or angle (made with respect to a fixed initial line or zero-line).

Geographers generally use geographic longitude (east or west of Greenwich) and latitude (north or south of the equator) (Fig. 1).

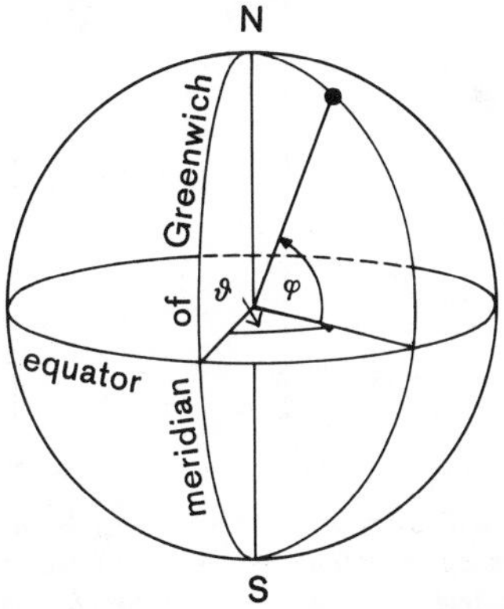

Fig. 1 *Geographic longitude ϑ and latitude φ of a point* P *on the earth's surface. ϑ is counted from the meridian of Greenwich, φ from the equator.* N and S *are the North and the South Poles.*

These determinations at the same time fix the zero lines from which the coordinates are to be reckoned—for geographical longitude, the meridian of Greenwich, and for the latitude, the equator. In investigations of plane geometry we generally use *rectangular* (*Cartesian*)

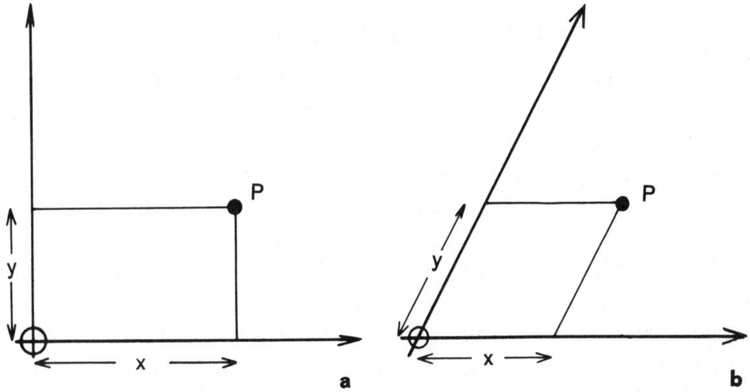

Fig. 2 *A point* P *in the plane is defined by the projections on the axis* x *and* y *in a rectangular coordinate system* (*2a*) *or in an oblique coordinate system* (*2b*).

coordinates (Fig. 2*a*) *x*, *y*, which signify the distances from two mutually perpendicular *coordinate axes*; or occasionally we also use *oblique coordinates* (Fig. 2*b*), *polar coordinates* (Fig. 3), and others.

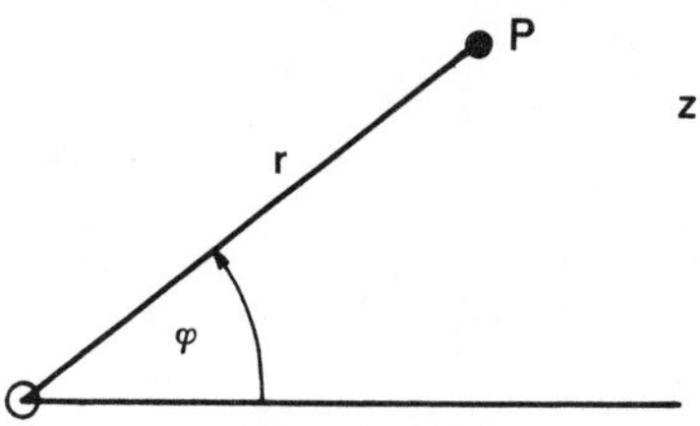

Fig. 3 *Definition of* P *in polar coordinates by the distance* r *from the origin* O *and the angle* φ *from an axis through the origin.*

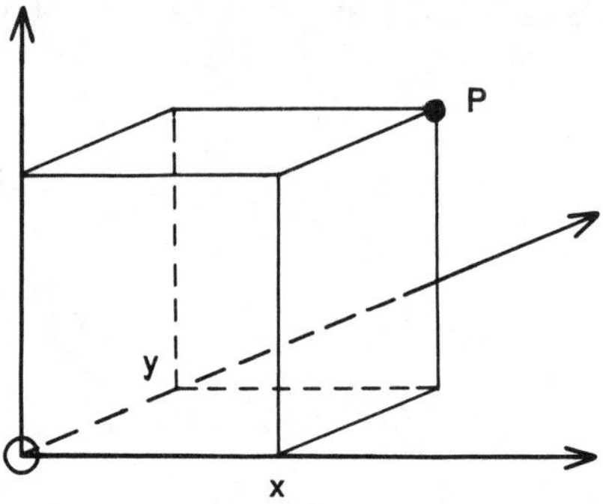

Fig. 4 *A point* P *in space is defined by three axis intercepts* x, y, z *of a rectangular coordinate system.*

When the coordinate system has been specified, we can determine each point by giving it two numbers.

In precisely the same way we require three coordinates to fix points in space. Mutually perpendicular rectilinear coordinates are again the simplest choice; we denote them by *x*, *y*, *z* (Fig. 4).

4. The Axioms of Geometry

Ancient geometry, regarded as a science, was less concerned with the question of fixing positions on the earth's surface than with determining the size and form of areas, volumes of figures in space, and the laws governing these figures. Geometry originated in the arts of surveying and architecture. Thus it managed without the concept of coordinates. First and foremost, geometric theorems assert properties of things that are called points, straight lines, and planes. In the classic canon of Greek geometry, the work of Euclid (300 B.C.), these things are not defined further but are only named or described. Thus an appeal to intuition is made. You must already know what a straight line is if you wish to take up the study of geometry. Picture the edge of a house, or a stretched string; abstract what is material and you will get your straight line. Next, laws are set up that are to hold between configurations of such abstract things. It is to the credit of the Greeks to have made the great discovery that we need assume only a small number of these statements to derive all others correctly with logical inevitability. These statements which are used as the foundation are called *axioms*. Their correctness cannot be proved. They do not arise from logic but from other sources of knowledge. What these sources are has formed a subject for the philosophical speculations of succeeding centuries. The science of geometry itself, up to the end of the eighteenth century, accepted these axioms as given, and built up its purely deductive system of theorems on them.

Later we shall have to discuss the question of the meaning of the elementary configurations called point, straight line, and so forth, and the sources of our knowledge of the geometric axioms. For the present, however, we shall assume that we are clear about these things and shall thus operate with the geometric concepts in the way we learned at school. The intuitive truth of numerous geometric theorems and the utility of the whole system in giving us bearings in our ordinary real world are sufficient for the present as our justification for using them.

5. The Ptolemaic System

To the eye the sky appears as a more or less flat dome to which stars are attached. In the course of a day the whole dome turns about an axis whose position on the sky is very close to the polestar. So long as this visual appearance was regarded as reality, an application of geometry to astronomic space was superfluous and was, as a matter of fact, not carried out. There were no lengths and distances measurable with terrestrial units. To determine the position of a star one had only to know the pair of angles formed by the observer's line of vision to the star with respect to the horizon and with respect to another appropriately chosen plane. At this stage of knowledge the earth's surface was considered at rest and was the eternal basis of the universe. The words "above" and "below" had an absolute meaning, and when poetic fancy or philosophical speculation undertook to estimate the height of the heavens or the depth of

Tartarus, the meaning of these terms required no explanation. Scientific concepts were still being drawn from the abundance of subjective data. The world system named after Ptolemy (A.D. 150) is the scientific formulation of this frame of mind. It was already aware of a great number of facts concerning the motion of the sun, the moon, and the planets and provided theoretical methods to predict them, but it retained the notion that the earth is at rest and that the stars are revolving about it at immeasurable distances. Their orbits were assumed to be circles and epicycles according to the laws of terrestrial geometry, yet astronomic space was not actually considered as an object for geometrical consideration, for the orbits were fastened like rings to crystal spheres, which, arranged in shells, formed the sky.

6. The Copernican System

It is known that Greek thinkers had already discovered the *spherical* shape of the earth and had ventured to take the first steps from the geocentric world systems (Aristarchus, third century B.C.) to higher abstractions. But only long after Greek civilization and culture had died did the peoples of other countries accept the spherical shape of the earth as a physical reality. This is the first truly great departure from the evidence of our eyes, and at the same time the first truly great step towards relativization. Centuries have passed since that first turning point, and what was at that time an unprecedented discovery has now become a platitude for school children. This makes it difficult to convey an impression of what it meant to people of that time to see the concepts "above" and "below" lose their absolute meaning and to recognize the right of the inhabitants of the antipodes to call "above" in their regions what we call "below" in ours. But after the earth had once been circumnavigated, all dissident voices became silent. Thus the discovery of the spherical shape of the earth offered no reason for strife between the objective and the subjective view of the world, between scientific research and the Church. This strife broke out only after Copernicus (1543) displaced the earth from its central position in the universe and created the *heliocentric world system.*

The importance of this discovery for the development of the human mind lay in the fact that the earth, mankind, and the individual ego became dethroned. The earth became a satellite of the sun which carried around in space the peoples swarming on it. Similar planets of equal importance accompanied it, describing orbits about the sun. Man was no longer important in the universe, except to himself. None of these amazing facts arose from ordinary experience (such as with a circumnavigation of the globe), but from observations which were, for the time in question, very delicate and subtle and from accurate calculations of planetary orbits. At any rate, the evidence was such as was neither accessible to all men nor of importance to everyday life. Visual evidence, intuitive perception, sacred and pagan tradition alike spoke against the new doctrine. In place of the visible disk of the sun the new doctrine put a ball of fire, gigantic beyond imagination; in place of the friendly lights of the sky, similar balls of fire at inconceivable distances, or spheres like the earth, that reflected light from other sources; and all immediate sense impressions were to be regarded as deception, whereas immeasurable distances and incredible velocities were to represent the true state of affairs. Yet this new doctrine was destined to be victorious. For it drew its power from the burning desire of all thinking minds to comprehend all things in the material world—be they ever so unimportant for human existence—by simple, unambiguous, though abstract, concepts. In this process, which constitutes the essence of scientific research, the human spirit neither hesitates nor fears to doubt the most self-evident facts of visual perception and to declare them to be illusions, but prefers to resort to the most extreme abstractions rather than exclude from the scientific description of nature one established fact, however insignificant it might seem.

The great relativizing achievement of Copernicus was the root of the many similar but lesser relativizations of a growing natural science until the time when Einstein's discovery was to stand alongside that of its great predecessor.

But now we must sketch in a few words the cosmos as mapped out by Copernicus. We have first to remark that the concepts and laws of geometry were directly applied to astronomic space. In place of the cycles of the Ptolemaic world, which were supposed to be fixed to the surfaces of crystal spheres, we now have real plane orbits in space, the planes of which could have different positions. The center of the world system is the sun. The planets describe their circles about it, and one of them is the earth, which rotates about its own axis, while the moon in its turn revolves in its orbit about the earth. Beyond, at enormous distances, the fixed stars are suns like our own, at rest in space. Copernicus' constructive achievement was that his system explained in a simpler way the phenomena which the traditional world system was able to explain only by means of complicated and artificial hypotheses. The alternation of day and night, the seasons, the moon's phases, the planetary orbits, all these things became at a single stroke clear, intelligible, and open to simple calculations.

7. The Elaboration of the Copernican Doctrine

Soon, however, the circular orbits of Copernicus no longer sufficed to account for the observations. The real orbits turned out to be considerably more complicated than believed. Now, an important point for the new view of the world was whether artificial constructions, such as the epicycles of the Ptolemaic system, were necessary, or whether an improvement in the calculations of the orbits could be carried out without introducing complications. It was the immortal achievement of Kepler (1618) to discover the simple and striking laws of the planetary orbits and hence save the Copernican system at a critical period. The orbits are not circles about the sun but curves closely related to circles, namely ellipses, in one focus of which the sun is situated. Just as this law describes the form of the orbits in a very simple manner, so the other two laws of Kepler determine the velocities with which they are traversed and the relation of the periods of revolution to the dimensions of the ellipses.

Kepler's contemporary, Galileo (1610), directed the recently invented telescope at the sky and discovered the moons of Jupiter. In them he recognized a model of the planetary system in a smaller scale and saw Copernicus' ideas as optical realities. But it is Galileo's greater merit to have developed the principles of mechanics, which were applied by Newton (1687) to planetary orbits, thus bringing about the completion of the Copernican world system.

Copernicus' circles and Kepler's ellipses are what modern science calls a *kinematic* or *phoronomic* description of the orbits—a mathematical formulation of the motions which does not contain the conditions and causes that bring about these motions. The causal expression of the laws of motion is the content of *dynamics* or *kinetics*, founded by Galileo. Newton applied this doctrine to the motions of the heavenly bodies and by interpreting Kepler's laws in a very ingenious way introduced the causal conception of *mechanical force* into astronomy. Newton's law of gravitation proved its superiority over the older theories by accounting for all the deviations from Kepler's laws, the so-called perturbations of orbits which had been brought to light by later refinements in the methods of observation.

This dynamical view of the phenomena of motion in astronomic space, however, demanded at the same time a more precise formulation of the assumptions concerning *space* and *time*. These axioms occur in Newton's work for the first time as explicit definitions. It is therefore justifiable to regard the theory that was accepted until the advent of Einstein's theory as an expression of Newton's doctrine of space and time. To understand these ideas it is essential to have a clear notion of the fundamental laws of mechanics, and to have it from a point of view which places the question of relativity in the foreground, an approach usually neglected in the elementary textbooks. We shall therefore have to discuss next the simplest facts, definitions, and laws of mechanics.

CHAPTER II

THE FUNDAMENTAL LAWS OF CLASSICAL MECHANICS

1. Equilibrium and the Concept of Force

Historically, mechanics took its start from the *doctrine of equilibrium*, or *statics*; the development from this point is also the most natural one logically.

The fundamental concept of statics is *force*. It is derived from the subjective feeling of exertion experienced when we perform work with our bodies. The stronger of two men, we say, is the one who can lift the heavier stone or stretch the stiffer bow. This measure of force, with which Ulysses established his right among the suitors, and which indeed plays a great part in the stories of ancient heroes, already contains the germ of the objectivization of the subjective feeling of exertion. The next step was the choice of a unit of force and the measurement of all forces in terms of their ratios to this unit, that is, the relativization of the concept of force. Weight, being the most evident manifestation of force, and making all things tend downwards, offered the unit of force in a convenient form—a piece of metal which was chosen as the unit of weight through some decree of the state or of the Church. Nowadays it is an international congress that fixes the units. In technical matters, the unit of weight today is the weight of a definite piece of platinum which is maintained in Paris. This unit, called the *pond* (p) will be used in our discussion till otherwise stated. The instrument used to compare the weights of different bodies is the *balance*.

Two bodies have the same weight, or are equally heavy, when on being placed in the two scales of the balance they do not disturb its equilibrium. If we place these two bodies in one pan of the balance, and in the other a body such that the equilibrium is again not disturbed, this new body has twice the weight of either of the other two. Continuing in this way, we get, starting from the unit of weight, a set of weights by means of which the weight of every body can be conveniently determined.

It is not our task here to show how these means enabled man to find and interpret the simple laws of the statics of rigid bodies, such as the laws of levers. We here introduce only those concepts that are indispensable for an understanding of the theory of relativity.

Besides the forces that occur in man's body or in those of his domestic animals he encounters others, above all in the events that we nowadays call *elastic*. The force necessary to stretch a bow or a crossbow belongs to this category. Now, these forces can easily be compared with weights. If, for example, we wish to measure the force that is necessary to stretch a spiral spring a certain distance (Fig. 5), then we find by trial what weight must be suspended from it

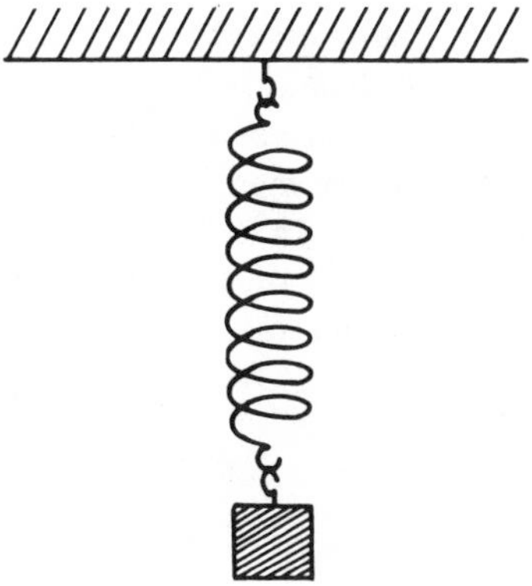

Fig. 5 *Comparison of an elastic force with a weight.*

to effect equilibrium for just this extension. The force of the spring is then equal to that of the weight, except that the former exerts a pull upwards but the latter downwards. The principle used here is that in equilibrium all forces cancel; this is Newton's *principle of the equality of action and reaction.*

If such a state of equilibrium be disturbed by weakening or removing one of the forces, *motion* occurs. The raised weight falls when it is released by the hand supporting it, that is, furnishing the reacting force. The arrow shoots forth when the archer releases the string of the stretched bow. The spring in Fig. 5 moves back if the weight is taken off. Force tends to produce motion. This is the starting point of *dynamics*, which seeks to discover the laws of this process.

2. The Study of Motion—Rectilinear Motion

It is first necessary to subject the concept of motion itself to analysis. The exact mathematical description of the motion of a point consists of specifying at what place relative to the previously selected coordinate system the point is situated from moment to moment. Mathematicians use formulae to express this. We shall as much as possible avoid this method of representing laws and relationships, which is not familiar to everyone, and shall instead make use of a graphical method of representation.

Let us illustrate with the simplest case, the motion of a point in a straight line. Let the unit of length be the centimeter, as is usual in physics, and let the moving point be at the distance $x=1$ cm. from the zero point, or origin, at the moment when we start our considerations, the moment we call $t=0$. In the course of 1 sec. suppose the point to have moved a distance of $\frac{1}{2}$ cm. to the right, so that for $t=1$ sec. the distance from the origin amounts to 1.5 cm. In the next

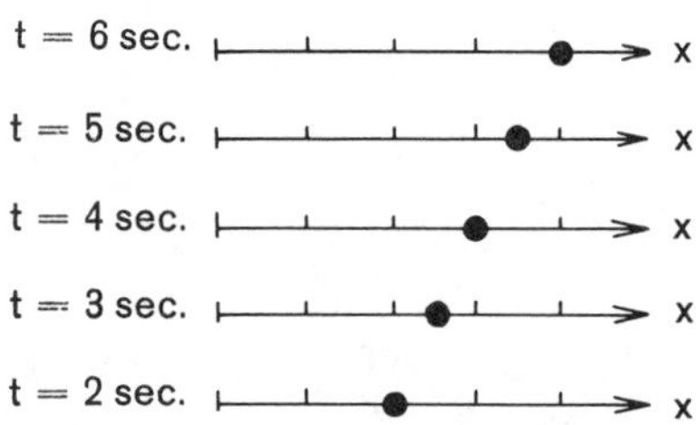

Fig. 6 *Motion of a point on the* x-*axis with constant velocity* $v=\frac{1}{2}$ *cm./sec.*

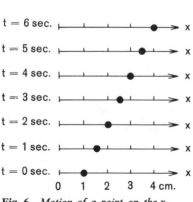

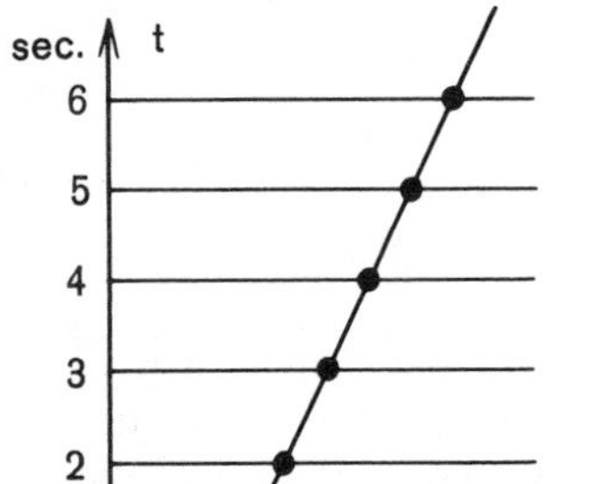

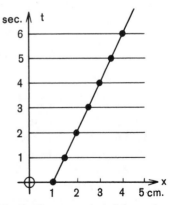

Fig. 7 *Representation of the motion of a point (Fig. 6) in an* xt-*coordinate system.*

second let it move by the same amount to $x=2$ cm., and so forth. The following table gives the distances x corresponding to the times t:

t	0	1	2	3	4	5	6	7	8	sec.
x	1	1.5	2	2.5	3	3.5	4	4.5	5	cm.

We see the same relationship pictured in the successive lines of Fig. 6, in which the moving point is indicated as a small point on the scale of distances. Now, instead of drawing a number of small diagrams, one above the other, we may also draw a single figure in which the x's and the t's occur as *coordinates* (Fig. 7). This has the advantage of allowing the place of the point to be depicted not only at the beginning of each full second but also at all intermediate times. We need only connect the positions marked in Fig. 6 by a continuous curve. In our case this is obviously a straight line, for the point advances equal distances in equal times; the coordinates x, t thus change in the same ratio (or proportionally), and it is evident that the graph of this law is a straight line. Such a motion is called *uniform*. The term *velocity* v of the motion designates the ratio of the path traversed to the time required in doing so:

$$v = \frac{\text{path traversed}}{\text{time required}}. \tag{1}$$

In our example the point traverses $\frac{1}{2}$ cm. of path per sec. The velocity remains the same throughout and amounts to $\frac{1}{2}$ cm./sec.

The unit of velocity is already fixed by this definition; it is the velocity which the point would have if it traversed 1 cm./sec. It is said to be a *derived* unit, and, without introducing a new word, we call it cm. per sec. or cm./sec. To express the fact that the measurement of velocities may be referred back to measurements of lengths and times in accordance with formula (1) we also say that velocity has the *dimension* length divided by time, written thus: $[v]=\left[\frac{l}{t}\right]$ or $[l \cdot t^{-1}]$. In the same way we assign definite dimensions to every quantity that can be derived from the fundamental quantities length l, time t, and weight G. When the latter are known, the unit of the quantity may at once be expressed by means of those of length, time, and weight—say, cm., sec., and p.

In the case of great velocities the path traversed in one second is great, thus the graph line has only a small inclination to the x-axis: the smaller the velocity, the steeper the graph. A point that is at

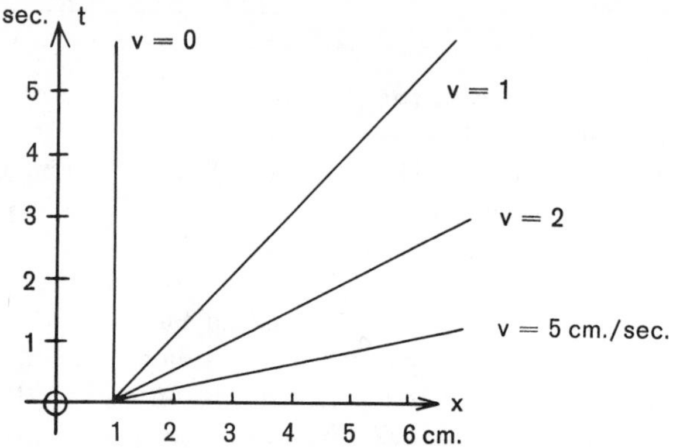

Fig. 8 *Uniform motion with different velocities* v=*0, 1, 2, 5 cm./sec.*

rest has zero velocity and is represented in our diagram by a straight line parallel to the t-axis, for the points of this straight line have the same value of x for all times t (Fig. 8).

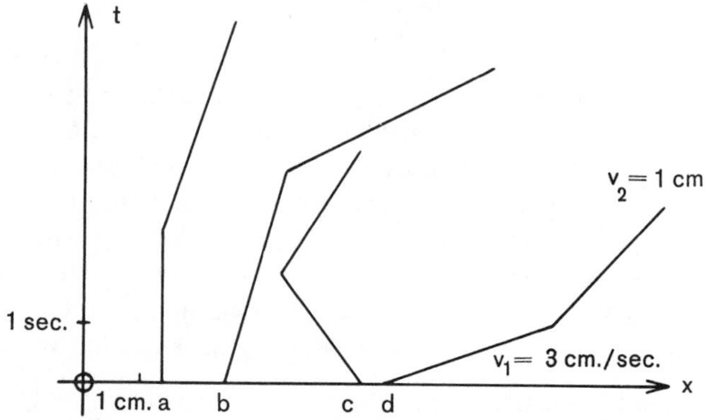

Fig. 9 *Uniform motions with sudden changes of velocity.*

If a point starts at rest and then suddenly acquires a velocity and moves on with this velocity, the graph is a broken line one part of which is inclined, the other being vertical (Fig. 9*a*). Similarly broken lines represent cases in which a point that is initially moving uniformly to the right or to the left suddenly changes its velocity (Figs. 9*b*, *c*, *d*).

If the velocity before the sudden change is v_1 (say, 3 cm./sec.), and afterwards v_2 (say, 5 cm./sec.), then the increase of velocity is v_2-v_1 (i.e., $5-3=2$ cm./sec.). If v_2 is less than v_1 (say, $v_2=1$ cm./sec.), then v_2-v_1 is negative (namely, $1-3=-2$ cm./sec.), and this clearly denotes that the moving point is suddenly retarded (Fig. 9*d*).

If a point experiences a series of sudden changes of velocity, then the graph of its motion is a succession of straight lines joined together (polygon) as in Fig. 10.

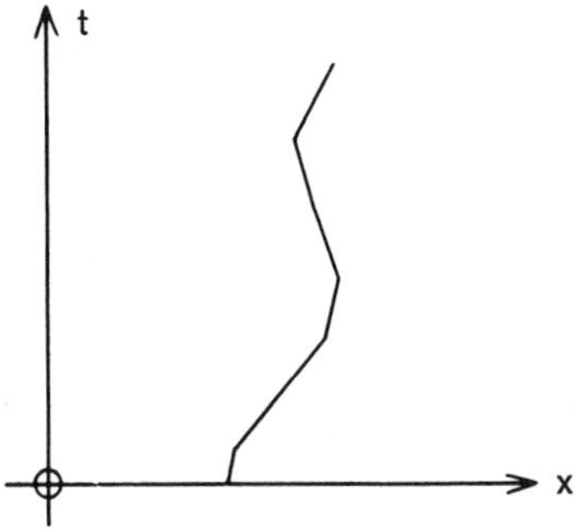

Fig. 10 *Motion of a point with a series of sudden changes of velocity.*

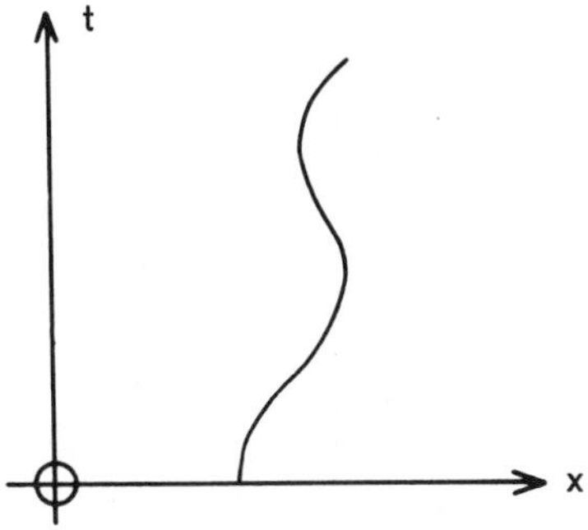

Fig. 11 *Continually changing velocity.*

If the changes of velocity occur more and more frequently and are sufficiently small, the polygon will no longer be distinguishable from a curved line. It then represents a motion whose velocity is continually changing, that is, one which is nonuniform, accelerated or retarded (Fig. 11).

An exact measure of the velocity and its rate of change, acceleration, can be obtained in this case only with the aid of the methods of differential calculus. It is sufficient for us to imagine the continuous curve replaced by a polygon whose straight sides represent uniform motions with definite velocities. The bends of the polygon (that is, the sudden changes of velocity) may be supposed to succeed each other at equal intervals of time, say, $\tau=\frac{1}{n}$ sec.

If, in addition, these changes are equal in size, the motion is said to be uniformly accelerated. Let each such change of velocity have the value w; then if there are n per sec. the total change of velocity per sec. is

$$nw \text{ per sec.} = \frac{w}{\tau} = b. \tag{2}$$

For example, in Fig. 12:

$$\tau = \tfrac{1}{10} \text{ sec.}, \qquad n = 10,$$
$$w = 10 \text{ cm./sec.},$$
$$v_0 = 5, \qquad v_1 = 15, \qquad v_2 = 25 \text{ cm./sec.} \ldots$$
$$b = \frac{w}{\tau} = 100 \text{ cm./sec.}^2.$$

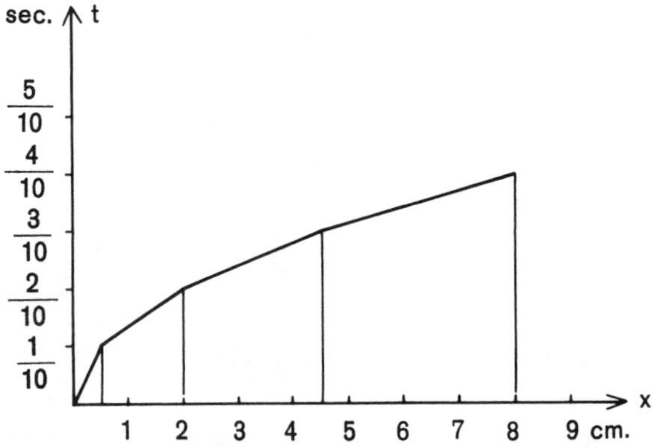

Fig. 12 *A point starts at the time* t=0 *at* x=0 *with the velocity 5 cm./sec. and undergoes a change of 10 cm./sec. after each tenth of a second.*

This quantity b is the measure of the *acceleration*. Its dimensions are clearly $[b]=\left[\frac{v}{t}\right]=\left[\frac{l}{t^2}\right]$, and its unit is that acceleration which causes the velocity to increase by one unit in the unit of time, that is, referred to the physical system of measure cm./sec.^{2}.

If we wish to know how far a uniformly accelerated point moves forward in a time t, we imagine the time t divided into n equal parts,* and suppose the point to receive a sudden increase of velocity w at

* In this case any arbitrary length of time t (and not as before, the unit of time, 1 sec.) is divided into n parts.

the end of each small interval of time $\frac{t}{n}$. This little increase is connected with the acceleration b by formula (2) if we replace the small interval of time τ by $\frac{t}{n}$, thus: $w=b\,\frac{t}{n}$.

If the point starts with zero velocity from $x=0$ at $t=0$, then the velocity

after the first interval of time is: $v_1 = w$,
,, second ,, ,, $v_2 = v_1+w = 2w$,
,, third ,, ,, $v_3 = v_2+w = 3w$,
and so forth.

The point advances

after the first interval of time to: $x_1 = v_1\,\frac{t}{n}$,
,, second ,, ,, $x_2 = x_1+v_2\,\frac{t}{n} = (v_1+v_2)\,\frac{t}{n}$,
,, third ,, ,, $x_3 = x_2+v_3\,\frac{t}{n} = (v_1+v_2+v_3)\,\frac{t}{n}$,
and so forth.

After the nth interval of time, that is, at the end of the time t, the point will have arrived at

$$x = (v_1+v_2+\cdots+v_n)\,\frac{t}{n}.$$

But

$$\begin{aligned} v_1+v_2+\cdots+v_n &= 1w+2w+3w+\cdots+nw \\ &= (1+2+3+\cdots+n)w. \end{aligned}$$

The sum of the numbers from 1 to n can be calculated quite simply by adding the first and the last; the second and the second to last; and so forth. In each case we get for the sum of the two numbers $n+1$, and altogether we have $\frac{n}{2}$ of such pairs. Thus we get $1+2+\cdots+n=\frac{n}{2}(n+1)$. If, further, we replace w by $b\frac{t}{n}$, we get

$$v_1+v_2+\cdots+v_n = \frac{n}{2}(n+1)\,\frac{bt}{n} = \frac{bt}{2}(n+1),$$

thus

$$x = \frac{bt}{2}(n+1)\frac{t}{n} = \frac{bt^2}{2}\left(1+\frac{1}{n}\right).$$

Here we may choose n to be as great as we please. If we let n be arbitrarily large, then $\frac{1}{n}$ becomes arbitrarily small and we get

$$x = \tfrac{1}{2}bt^2.$$

This signifies that the paths traversed are proportional to the squares of the times. If, for example, the acceleration $b = 100$ cm./sec.^{2}, then after 1 sec. the point has moved 50 cm., after 2 sec. $50 \times 2^2 = 200$ cm., after 3 sec. $50 \times 3^2 = 450$ cm., and so forth. If we use smaller time intervals of $\frac{1}{10}$ sec., we see that the point has traversed $\frac{1}{2} \times 100 \times (\frac{1}{10})^2 = \frac{1}{2}$ cm. after the first tenth of a second, $\frac{1}{2} \times 100 \times (\frac{2}{10})^2 = 2$ cm. after the second tenth of a second, and so on. This relationship is represented by a curved line, called a parabola, in the xt-plane

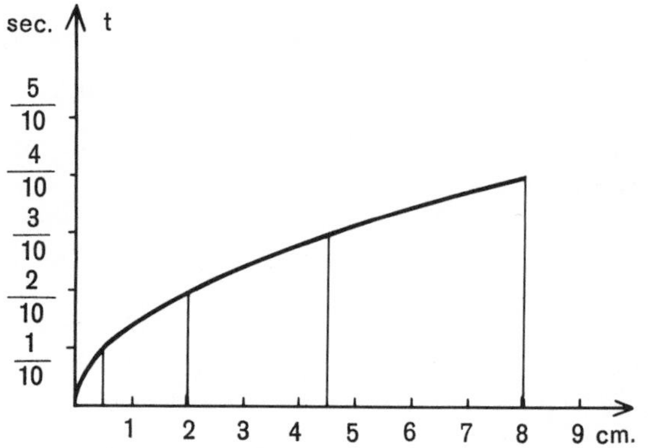

Fig. 13 *A point starts at the time* t=0 *at* x=0 *with an acceleration of 100 cm./sec.*^{2} *Compare with Fig. 12.*

(Fig. 13). Comparing this figure with Fig. 12 we see how the polygon approximately represents the continuously curved parabola. In both figures the acceleration $b = 100$ cm./sec.^{2} has been chosen, and this choice determines the appearance of the curves.

We may also apply the concept of acceleration to nonuniformly accelerated motions by using, instead of 1 sec., so small a time of observation that the motion may be regarded as uniformly accelerated. The acceleration itself then becomes continuously variable.

All these definitions become rigorous, and at the same time convenient to handle, if the process of subdivision into small intervals during which the quantity under consideration may be regarded as constant is carefully studied. This leads to the concept of limiting value which forms the starting point of the differential calculus. Historically, it was when Newton was investigating the problems of motion that he was actually led to invent the differential calculus and its inverse, the integral calculus.

The *theory of motion* (kinematics, phoronomy) is the forerunner of the proper mechanics of forces, or dynamics. It is evidently a sort of geometry of motion. As a matter of fact, in our graphical representation each motion is represented by a geometric configuration in the plane with the coordinates x, t. In this we are concerned with more than a mere analogy. For it is the principle of relativity that attaches fundamental importance to the introduction of time as a coordinate in conjunction with the spatial dimensions.

3. Motion in a Plane

If we wish to study the motion of a point in a plane, we can at once extend our method of representation to this case. We take in

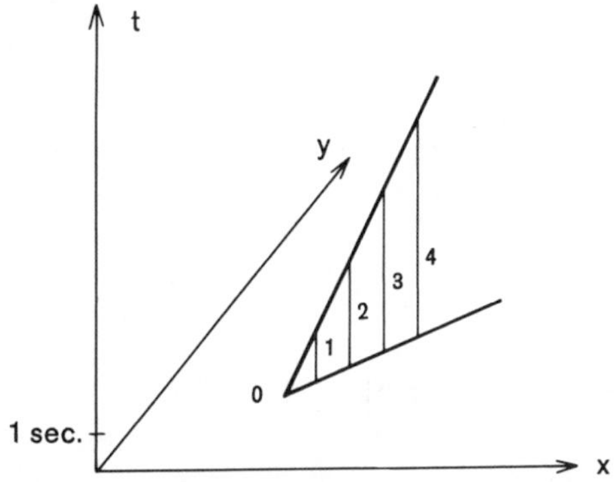

Fig. 14 *Uniform motion in a plane represented in an* x, y, t*-coordinate system After 1, 2, 3, 4, . . . sec. the point in the* x y*-plane has reached the foot of the parallel line which bears the number 1, 2, 3, 4, . . . respectively.*

the plane an xy-coordinate system and erect a t-axis perpendicular to it (Fig. 14). Then a straight line in the xyt-space corresponds to a rectilinear and uniform motion in the xy-plane. For if we project the points of the straight line that correspond to the points of time $t=0, 1, 2, 3, \ldots$ sec. on to the xy-plane, we see that the positional displacement takes place along a straight line and at equal intervals.

Every nonrectilinear motion is said to be *accelerated* even if, for example, a *curved* path is traversed with *constant* speed. For in this case the *direction* of the velocity changes although its numerical value remains constant. An accelerated motion is represented in the xyt-space (Fig. 15) by an arbitrary curve. The projection of this

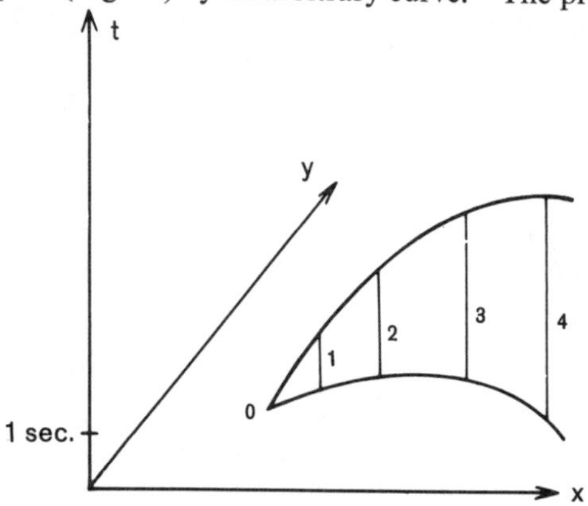

Fig. 15 *Accelerated motion in a plane (compare legend of Fig. 14).*

curve into the xy-plane is the plane orbit. The velocity and the acceleration are again calculated by supposing the curve replaced by a polygon closely wrapped around the curve. At each corner of this polygon not only the amount but also the direction of the velocity alters. A more exact analysis of the concept of acceleration would take us too far; it is sufficient to mention that it is best to project the graph of the moving point onto the coordinate axes x, y, and to follow out the rectilinear motion of these two points, or what is the same, how the coordinates x, y change with time. The concepts defined for rectilinear motions as given above may now be applied to these projected motions. We thus get two *components of velocity* v_x, v_y, and two *components of acceleration* b_x, b_y, that together fix the velocity and the acceleration of the moving point at a given instant.

In the case of a plane motion (and also in one that occurs in space), velocity and acceleration are thus directed magnitudes (vectors). They have a definite direction and a definite magnitude. The latter can be calculated from the components. For example,

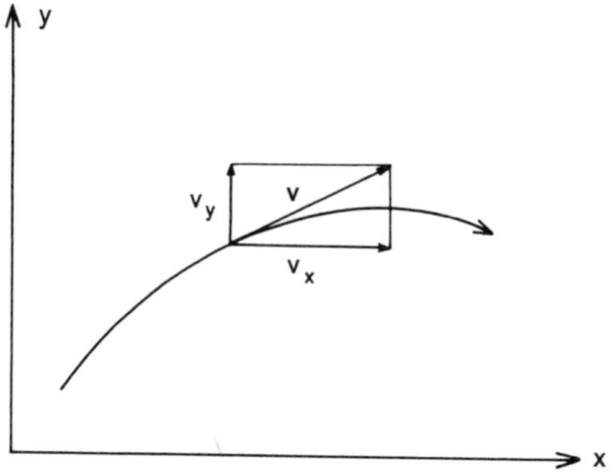

Fig. 16 *The velocity* v *of a motion in a plane has the components* v_x *and* v_y.

we get the direction and magnitude of the velocity from the diagonal of the rectangle with the sides v_x and v_y (Fig. 16). Thus, by Pythagoras' theorem, its magnitude is

$$v = \sqrt{v_x^2 + v_y^2}. \tag{3}$$

A corresponding result holds for the acceleration.

4. Circular Motion

There is only one case which we wish to consider in greater detail, that of the motion of a point in a circular orbit with constant speed (Fig. 17*a*). According to what was already said, it is an accelerated motion, since the direction of the velocity constantly alters. If the motion *were* unaccelerated, the moving point would move forward from *A* in a straight line with the uniform velocity v. But in reality the point is to remain on the circle, and hence must have a supplementary velocity or acceleration that is directed to the central point *M*. This is called the *centripetal acceleration*. It causes the velocity at a neighboring point *B*, which is reached after a short interval τ, to have a direction different from that at the point A. On a separate diagram (Fig. 17*b*) we draw from a point C to the points D and E the velocities at A and B, paying due regard to their magnitude and direction. Their magnitude will be the same, namely v, since the circle is to be traversed with constant speed, but their direction is different. If we connect the end points D and E of the two velocity lines, then the connecting line is clearly the supplementary velocity w, which transforms the first velocity state

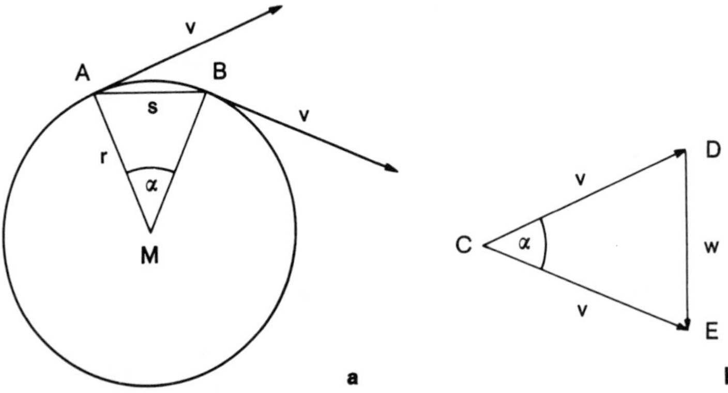

Fig. 17 *Explanation of the centripetal acceleration of a circular motion with constant velocity* v.

into the second. We thus get an isosceles triangle CED, having the base w and the sides v, and we at once see that the angle α at the vertex is equal to the angle subtended by the arc AB, which the point traverses, at the center of the circle. For the velocities at A and B are perpendicular to the radii MA and MB, and hence include the same angle. Consequently the two isosceles triangles MAB and CDE are similar, and we get the proportion

$$\frac{DE}{CD} = \frac{AB}{MA}.$$

Now $DE=w$, $CD=v$, and further, MA is equal to the radius r of the circle, and AB is equal to the arc s except for a small error that can be made as small as we please by choosing the time interval τ sufficiently small.

Hence we have

$$\frac{w}{v} = \frac{s}{r} \quad \text{or} \quad w = \frac{sv}{r}.$$

We now divide by τ and notice that $\frac{s}{\tau}=v, \frac{w}{\tau}=b$. Hence the acceleration

$$b = \frac{v^2}{r}, \tag{4}$$

that is, the centripetal acceleration is equal to the square of the velocity round the circle, divided by the radius.

This theorem, as we shall see, is the basis of one of the first and most important empirical proofs of Newton's theory of gravitation.

Perhaps it is not superfluous to have a clear idea of what this uniform circular motion looks like in the graphical representation in the xyt-space. This is obviously produced by allowing the moving point to move upwards uniformly parallel to the t-axis during the circular motion. We thus get a helix (screw line), which now represents the orbit and the course of the motion in time completely. In Fig. 18 it is drawn on the surface of a cylinder that has its base on the xy-plane.

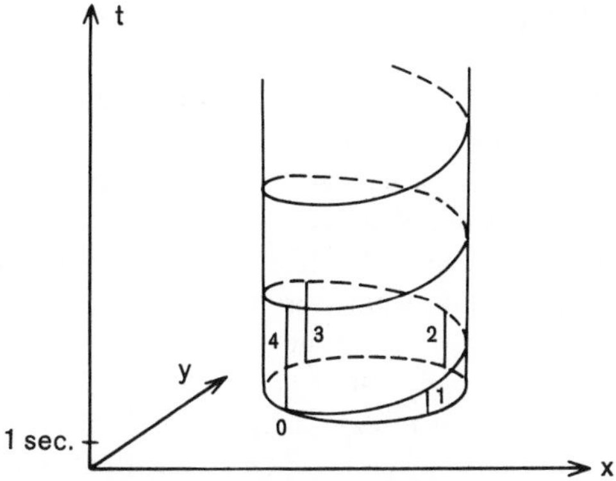

Fig. 18 *Representation of a circular motion with constant velocity. The velocity* v *and the radius of the circle are chosen so that after four seconds the point has traversed the circle just once.*

5. Motion in Space

Our graphical method of representation fails for motions in space, for in this case we have three space coordinates x, y, z, and time has to be added as a fourth coordinate. But unfortunately our visual powers are confined to three-dimensional space. The symbolic language of mathematics must now lend us a helping hand. For the methods of *analytical geometry* allow us to treat the properties and relationships of spatial configurations as matters of pure calculation without requiring us to use our visual power or to sketch figures. Indeed, this process is much more powerful than geometric construction. Above all, it is not confined to the dimensional number three but is immediately applicable to spaces of four or more dimensions. In the language of mathematics the concept of a space of more than three dimensions is not at all mystical but is simply an abbreviated expression of the fact that we are dealing with things that are fully determined by more than three number data. Thus the position of a point at a given moment of time can be fixed only by specifying four number data, the three space coordinates x, y, z and the time t. After we have learned to deal with the xyt-space as a means of depicting plane motion, it will not be difficult also to handle motion in three-dimensional space in the light of curves in the $xyzt$-space. This view of kinematics as geometry in a four-dimensional $xyzt$-space has the advantage of allowing us to apply the well-known laws of geometry to the study of motion. But it has a still deeper significance that will become apparent in the discussion of Einstein's theory. It will be shown that the concepts space and time, which are derived from quite different kinds of experience, cannot be sharply differentiated at all as objects of physical measurement. If physics is to retain its maxim of recognizing as real only what is physically observable, it must combine the concepts space and time into a higher unity, namely, a four-dimensional expanse. Minkowski called this the "*world*" (1908), by which he wished to express that the element of all order of real things is not place or point of time, but the "*event*" or the "*world point*," that is, a place at a definite time. He called the graphical picture of a moving point a "*world line*," an expression that we shall continue to use. Rectilinear uniform motion thus corresponds to a straight world line, accelerated motion to one that is curved.

6. Dynamics—The Law of Inertia

After these preliminaries we revert to the question with which we started: How do forces generate motion?

The simplest case is that in which no forces are present at all. A body at rest will then certainly not be set into motion. The ancients had already made this discovery, but, more than this, they also believed the converse to be true—that wherever there is motion there must be forces that maintain it. This view at once leads to difficulties if we reflect on why a stone or a spear that has been thrown continues to move when it has been released from the hand. It is clearly the latter that has set it into motion, but its influence is at an end as soon as the motion has actually begun. Ancient thinkers were much troubled in trying to discover what forces actually maintain the motion of the thrown stone. Galileo was the first to take the right point of view. He observed that it is a fallacy to assume that wherever there is motion there must always be force. Rather it must be asked what quantitative property of motion is in a regular fixed relation to force, whether it be the position of the moving body, its velocity, its acceleration, or some composite quantity dependent on all of these. No amount of reflection will allow us to derive an answer from philosophy. We must address ourselves directly to nature. The answer which she gives is that force has an influence in effecting *changes* of velocity but that no force is necessary to maintain a motion in which the magnitude and the direction of the velocity remain unaltered. And conversely, where there are no forces, the magnitude and direction of the velocity remain unaltered; thus a body which is at rest remains at rest, and one that is moving uniformly and rectilinearly continues to move uniformly and rectilinearly.

This *law of inertia* (or persistence) is by no means so obvious as its simple expression might lead us to surmise. In our experience we do not know of bodies that are really withdrawn from all external influences; and, if we use our imaginations to picture how they travel in their solitary rectilinear paths with constant velocity through astronomic space, we are at once confronted with the problem of the absolutely straight path in space absolutely at rest, with which we shall have to deal later. For the present, we shall interpret the law of inertia in the restricted sense in which Galileo meant it.

Let us consider a smooth, exactly horizontal table on which a smooth sphere is resting. This is kept pressed against the table by its own weight. Evidently there is no force acting in a horizontal direction on the sphere; otherwise it would not itself remain at rest at any point on the table.

But if we now give the sphere a velocity, it will continue to move in a straight line and will lose only very little of its speed. This retardation was recognized as a secondary effect by Galileo, and is to be ascribed to the friction of the table and the air, even if the frictional forces cannot be proved to be present by the statical methods with which we started. It is just this intuition, which correctly differentiates what is essential in an occurrence from disturbing subsidiary effects, that characterizes the great scientist.

The law of inertia is at any rate confirmed for motion on the table. It has been established that in the absence of forces the velocity remains constant in direction and magnitude.

Consequently the forces will be associated with the change of velocity, the acceleration. In what way they are associated can again be decided only by experiment.

7. Impulses

We have presented the acceleration of a nonuniform motion as a limiting case of sudden changes of velocity of brief uniform motions. Hence we shall first have to inquire how a single sudden change of velocity is produced by the application of a force. For this a force must act for only a short time; it is then what we call an instantaneous or impulsive force. The result of such an impulsive force depends not only on the magnitude of the force but also on the duration of the action, even if this is very short. We therefore define a new quantity, called *impulse* J, with the help of the following consideration: n impulses J, each of which consists of the force K acting during the time $\tau = \frac{1}{n}$ sec., will, if they follow each other without appreciable pauses, have exactly the same effect as if the force K were to continue to act throughout the whole second. Thus we have

$$nJ \text{ per sec.} = \frac{1}{\tau} J = K,$$

or

$$J = \tau K. \tag{5}$$

To visualize this, let us imagine a weight placed on one side of a lever having equal arms (such as a balance), and suppose a hammer to tap very quickly and evenly on the other side with blows just powerful enough to preserve equilibrium except for inappreciable fluctuations (Fig. 19). It is clear that we may tap more weakly but

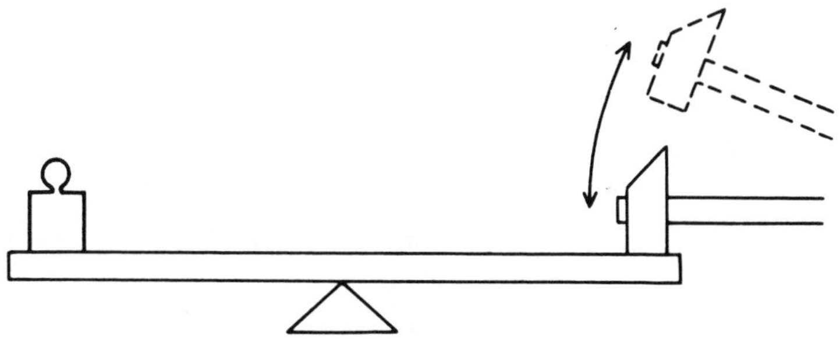

Fig. 19 *Balance of a weight by a series of impulses with a hammer.*

more often, or more strongly and less often, so long as the intensity J of the blow multiplied by the number of blows per second, or divided by the time required by each blow, always remains exactly equal to the weight K. This "impulse balance" enables us to measure the intensity of blows even when we cannot ascertain the duration and the force of each one singly. We need find only the force K that keeps equilibrium with n such equal blows per second (disregarding the inappreciable trembling of the arms); then the magnitude of each blow is the nth part of K.

The dimensions of impulse are $[J]=[t \cdot G]$, where G denotes weight.

8. The Effect of an Impulse

We again consider the sphere on the table and study the action of impulses on it. To do this we require a hammer that may be swung, say, about a horizontal axis. First, we calibrate the power of the blows of our hammer for each drop by means of our "impulse balance." Then we allow it to impinge against the sphere resting on the table and observe the velocity that it acquires through the blow by measuring how many centimeters it rolls in 1 sec. (Fig. 20).

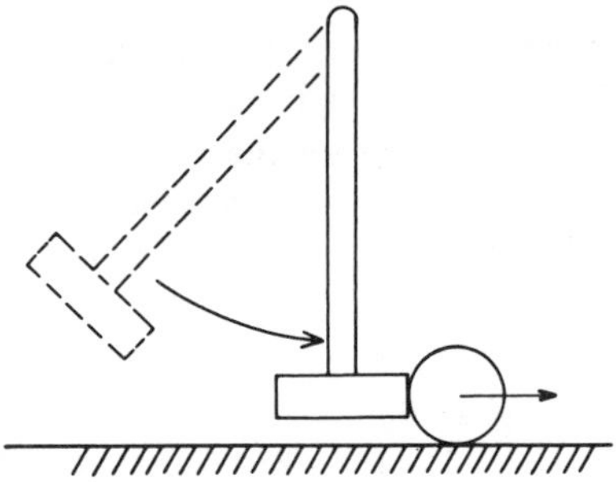

Fig. 20 *A hammer hits a sphere on a table. The impulses are proportional to the velocities of the sphere.*

The result is very simple. The more powerful the blow, the greater the velocity, the relation being such that twice the blow imparts twice the velocity, three times the blow three times the velocity, and so forth, that is, the velocity and the blow bear a constant ratio to each other (they are proportional). If the sphere already has an initial velocity, the blow will increase or decrease it according to whether it strikes the sphere in the rear or in the front. By a strong counterblow it is possible to reverse the direction of motion of the sphere.

The effect of impulses consists of sudden changes of velocity which are in the ratio of the impulses that produce them. The velocities are here considered as positive or negative according to their direction.

9. Mass and Momentum

Hitherto we have dealt with a single sphere. We shall now perform the same impulse experiment with spheres of different kinds, say, of different size or of different material, some being solid and others hollow. Suppose all these spheres are to be set into motion by equal blows. We produce these with the help of a hammer which is very light compared with the spheres. The exchange of impulse of the hammer and the spheres is in consequence of the heaviness of the spheres so fast that these have hardly moved from the beginning to the end of the blow; therefore all spheres obtain the same blow. Experiment shows that they then acquire quite different velocities and indeed it is at once observed that light spheres (still heavy compared with the hammer) are made to travel with greater speed than heavy ones. Thus we find a relationship with weight, which we shall explore later, for it is one of the empirical foundations of the general theory of relativity. From the abstract point of view, however, we must emphasize that the fact that various spheres acquire various velocities after equally strong impacts, has nothing to do with weight. Weight acts downwards and produces the pressure of the spheres on the table but exerts no horizontal force. We now find that *one* sphere offers greater resistance to the blow than *another*; it is a new fact of experience that the former is at the same time the heavier one. But this fact, from the point of view here adopted, cannot be deduced from the concept of weight. What we establish is a difference of resistance of the spheres to impacts. We call it *inertial resistance,* and define it to be the ratio of the impulse J to the velocity v measured from rest. The name *mass* has been chosen for this ratio and it is denoted by the symbol m. Thus we get

$$m = \frac{J}{v}. \tag{6}$$

This formula states that for one and the same body an increase of the impulse J produces a greater velocity v in such a way that their ratio always has the same value m. When mass has been defined in this way its unit can no longer be chosen arbitrarily, because the units of velocity and of impulse have already been fixed. Rather, mass has the dimensions

$$[m] = \left[\frac{t^2G}{l}\right],$$

and its unit in the ordinary system of measures is sec.^{2}p/cm.

In ordinary language the word *mass* denotes something like amount of substance or quantity of matter, these concepts themselves being defined no further. The concept of substance is considered self-evident. In physics, however, as we must very strongly emphasize, the word *mass* has no meaning other than that given by formula (6). It is the measure of the resistance of a body to changes of velocity.

We may write the law of impulses more generally thus:

$$mw = J. \tag{7}$$

It determines the *change* of velocity w that a body in motion experiences as the result of an impulse J.

The product mv of mass and velocity is called *momentum* p. The quantity mw, where w is the change of velocity, is the change of momentum produced by the impulse J.

We have so far supposed that the hammer is light as compared with the spheres. Therefore we could consider the impulse transferred as a given quantity. Actually this is not correct. The hammer puts a sphere into motion already during the impact, a light one more than a heavy one. Therefore the blows transferred and also the momentum received will not be quite equal. It is preferable to consider the two colliding bodies on equal footing and to replace the hammer by another sphere. Let us therefore consider two bodies moving in the same straight line with the velocities v_1 and v_2. If they collide they exert impulsive forces on one another; the impulse exerted on the first body by the second is J_1, that exerted on the second by the first, J_2. Now according to Newton's principle that action is equal to reaction (p. 15), the forces effected by the bodies on each other are equal, (and in opposite directions) and as the short time action is the same for both, the impulses are also equal and opposite, $J_2 = -J_1$. Hence their sum is zero:

$$J_1 + J_2 = m_1 w_1 + m_2 w_2 = 0. \tag{8}$$

From this it follows that

$$w_2 = -\frac{m_1}{m_2} w_1,$$

that is, when one sphere loses velocity (w_1 negative), the other gains velocity (w_2 positive), and vice versa.

If we introduce the velocities of the two spheres before and after the impact, namely, v_1, v_1' for the first sphere, and v_2, v_2' for the second, then the changes of velocity are

$$w_1 = v_1' - v_1, \qquad w_2 = v_2' - v_2$$

and we may also write the equation (8) thus:

$$m_1(v_1'-v_1)+m_2(v_2'-v_2) = 0.$$

If we then collect all the quantities referring to motion before impact on the one side, and all those referring to motion after impact on the other, we get

$$m_1v_1+m_2v_2 = m_1v_1'+m_2v_2'. \tag{9}$$

On the left side we find the total momentum $p=m_1v_1+m_2v_2$ of the spheres before the impact, on the right side we find the total momentum $p'=m_1v_1'+m_2v_2'$ after it. Therefore we have

$$p = p', \tag{9a}$$

and this equation may be interpreted as follows:

The total momentum of two bodies is not changed as a result of the impact. This is the law of *conservation of momentum.*

10. Force and Acceleration

Before pursuing the striking parallel between mass and weight mentioned in the foregoing section we shall apply the laws so far established to the case of forces that act continuously. Obviously, again, the theorems can be rigorously formulated only with the aid of the methods of the infinitesimal calculus, yet the following considerations may serve to give an approximate idea of the relationships involved.

A force that acts continuously generates a motion whose velocity alters continuously. We now suppose the force replaced by a rapid succcession of blows or impulses. At each blow the velocity will then suffer a sudden change, and there will result a world line that is bent many times, as in Fig. 10, which will approximate the true, uniformly curved, world line and may be used in place of the latter in the calculations. Now if n blows per second replace the force K, then by (5) each of them has the value $J=\tau K$, where τ is the short interval occupied by each blow. At each impulse a change of velocity w occurs which, according to (7), is determined by $mw=J=\tau K$. But by (2), $\frac{w}{\tau}=b$, thus we get

$$mb = K. \tag{10}$$

This is the *law of motion of dynamics* for forces that act continuously. It states that *a force produces an acceleration that is proportional to it; the constant ratio* K:b *is the mass.*

We may give this law still a different form which is advantageous for many purposes, particularly for the generalization necessary in the dynamics of Einstein (see VI, 7, p. 274). If the velocity v alters by the amount w, then the momentum p carried along by the moving body, namely $p=mv$, alters by mw. Thus we have $mb=\frac{mw}{\tau}$, the change of the momentum of the body in the time τ required to effect it. Accordingly, we may express the fundamental law expressed in formula (10) thus:

If a force K *acts on a body, then the momentum* p = mv *carried along by the body changes in such a way that its change per unit of time is equal to the force* K.

Expressed in this form the law holds only for motions which take place in a straight line and in which the force acts in the same straight line. If this is not the case, that is, if the force acts obliquely to the momentary direction of motion, the law must be generalized somewhat. Let us suppose the force drawn as an arrow which is then projected onto three mutually perpendicular directions, say, the coordinate axes. In Fig. 21 the force is shown acting in the xy-plane, with its projections drawn on the x- and y-axes. Let us

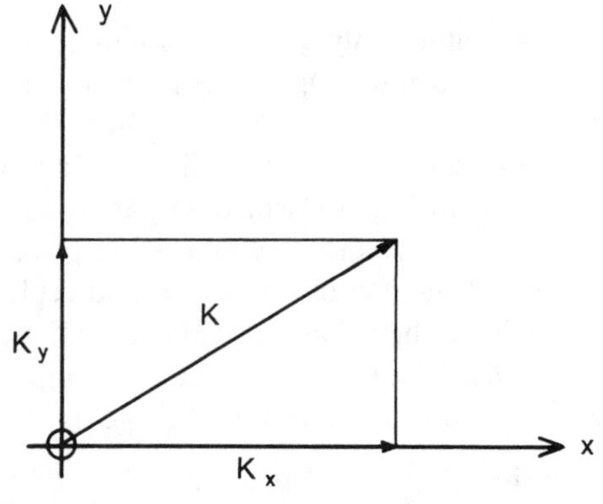

Fig. 21 *The components* K_x *and* K_y *of a force* K *in a plane with coordinates* x *and* y.

imagine the moving point projected on the axes in the same way. Then each of the projected points executes a motion on its axis of projection. The law of motion then states that the accelerations of these projected motions bear the relation $mb = K$ to the corresponding components of force. But we shall not enter more closely into these mathematical generalizations, which involve no new concepts.

11. Example—Elastic Vibrations

As an example of the relation between force, mass, and acceleration we now consider a body that can execute vibrations under the action of elastic forces. Let us take a straight broad steel spring and fasten it at one end so that it lies horizontally in its position of rest (and does not hang downwards). It bears a sphere at the other

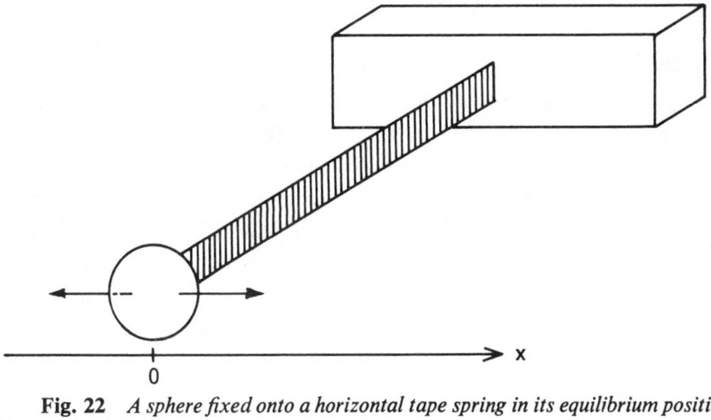

Fig. 22 *A sphere fixed onto a horizontal tape spring in its equilibrium position (spring pendulum).*

end (Fig. 22). The sphere can then swing to and fro in the horizontal plane. Gravity has no influence on its motion, which depends only on the elastic force of the spring. When the displacements are small, the sphere moves almost in a straight line. Let its direction of motion be the x-axis.

If we set the sphere into motion, it executes a periodic vibration, the nature of which we can understand as follows: If we displace the sphere slightly out of the position of equilibrium with our hands, we sense the restoring force of the spring. If we let the sphere go, this force imparts to it an acceleration, which causes it to return to the mean position with increasing velocity. In this process the restoring force, and hence also the acceleration, continuously decreases, and becomes zero when passing through the mean position itself, for here the sphere is in equilibrium and no accelerative force acts on it. At the place, therefore, at which the velocity is greatest, the acceleration is least. In consequence of its inertia the sphere passes rapidly through the position of equilibrium, and then the force of the spring begins to retard it and applies a brake, as it were, to the motion. When the original deflection has been attained on the other side the velocity has decreased to zero and the force has reached its highest value. At the same time the acceleration has reached its greatest value in reversing the direction of the velocity at this moment. From this point onwards it repeats the process in reverse.

If we next replace the sphere by another of different mass, we see that the character of the motion remains the same but the time of vibration is changed. When the mass is greater, the motion is slowed down and the acceleration becomes less; while a decrease of mass increases the number of vibrations per second.

In many cases the restoring force K may be assumed to be exactly proportional to the deflection x. The course of the motion may then be represented geometrically as follows: Consider a point P moving uniformly around the circumference of a circle of radius a, revolving once in time T. The number of revolutions per second is then $\nu=\frac{1}{T}$. The point P traverses the circumference, which is $2\pi a$ (where $\pi=3.14\ldots$), with velocity $\frac{2\pi a}{T}=2\pi a\nu$. If P moves along a small distance s on the circle in the time τ we get the velocity $\frac{s}{\tau}=2\pi a\nu$.

Let us now take the center O of the circle as the origin of a rectangular set of coordinates in which P has the coordinates x, y. Then the point of projection A of the point P on the x-axis will move to and fro during the motion just like the mass fastened to the spring. *This point* A *is to represent the vibrating mass.* If P moves forward a small distance s, then A moves along the x-axis a small distance

ξ, and the velocity of A is $v=\frac{\xi}{\tau}$. Fig. 23 now shows that the displacements ξ and s are the side and the hypotenuse of a small

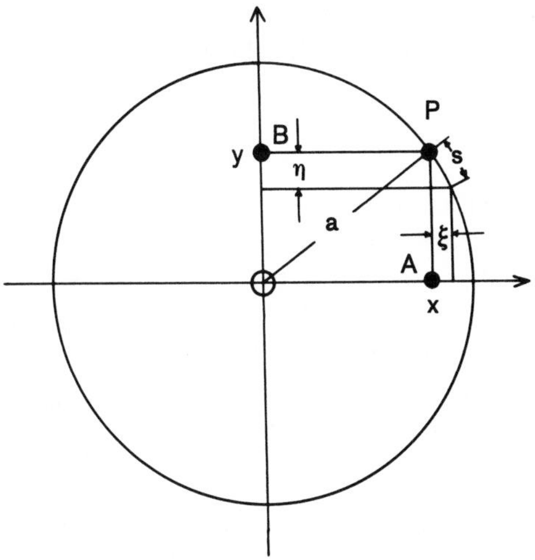

Fig. 23 *Representation of the motion of the spring pendulum of Fig. 22 by the projection* A *of a point* P *on the* x*-axis which moves on a circle with constant velocity.*

right-angled triangle, which is similar to the large right-angled triangle OAP, because corresponding sides are perpendicular to one another. Hence we have the proportion

$$\frac{\xi}{s} = \frac{y}{a} \quad \text{or} \quad \xi = \frac{sy}{a}.$$

Hence the velocity of A becomes

$$v = \frac{\xi}{\tau} = \frac{s}{\tau}\times\frac{y}{a} = 2\pi\nu y.$$

Now, the point of projection B of the point P executes exactly the same pendulum motion on the y-axis. During the small displacement s of P the point B moves backwards a distance η, and just as for ξ, we have

$$\frac{\eta}{s} = -\frac{x}{a} \quad \text{or} \quad \eta = -s\times\frac{x}{a}.$$

η is negative, for there is a decrease of y.

This change η of y corresponds to a change in the velocity $v=2\pi\nu y$ of the point A which is given by

$$w = 2\pi\nu\eta = -2\pi\nu s\frac{x}{a}.$$

The minus sign of w indicates that the velocity decreases. The acceleration is

$$b = \frac{w}{\tau} = -2\pi\nu\frac{s}{\tau}\times\frac{x}{a} = -(2\pi\nu)^2 x$$

and also has the negative sign.

The acceleration in this vibrational motion of the point A is thus actually proportional at each moment to the deflection x. We get for the force

$$K = mb = -m(2\pi\nu)^2 x. \tag{11}$$

The negative sign of K means that the force in any position drives the sphere back to the equilibrium position $x=0$.

By measuring the force corresponding to a deflection x and by counting the vibrations we can thus determine the mass m of the spring pendulum.

The picture of the world line of such a vibration is clearly a wavy line in the xt-plane, if x is the direction of vibration (Fig. 24).

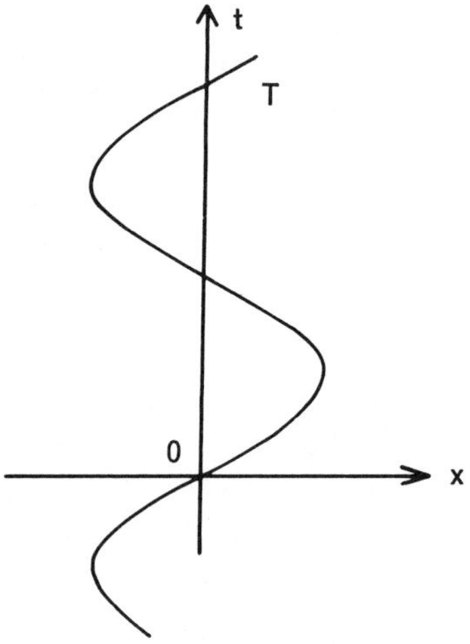

Fig. 24 *The vibration of the spring pendulum in the* xt-*diagram. T is the time of vibration.*

In the figure it has been assumed that at the time $t=0$ the sphere is moving through the middle position $x=0$ towards the right. We see that whenever the sphere passes through the t-axis, that is, for $x=0$, the direction of the curve is most inclined to the x-axis, and this indicates the greatest velocity. Hence the line is not curved at this point, and the change of velocity or the acceleration is zero. The opposite is true of those points that correspond to the extreme deflections.

12. Weight and Mass

At the beginning of this chapter when we introduced the concept of mass, we observed that mass and weight exhibit a remarkable parallelism. Heavy bodies offer a stronger resistance to an accelerating force than light bodies. Is this, then, an exact law? As a matter of fact, it is. To make the facts quite clear, let us again consider the experiment of setting into motion spheres on a smooth horizontal table by means of impacts or impulses. We take two spheres A and B, of which B is twice as heavy as A, that is, on the balance B exactly counterpoises two bodies each exactly like A. We next apply equal blows to A and B on the table and observe the velocity attained. We find that A rolls away twice as quickly as B.

Thus the sphere B, which is twice as heavy as A, opposes a change of velocity exactly twice as strongly as A. We may also express this as follows: Bodies having twice the mass have twice the weight; or, more generally, the masses m are in the ratio of the weights G. The ratio of the weight to the mass is a definite quantity. It is denoted by g, and we write

$$\frac{G}{m} = g \qquad \text{or} \qquad G = mg. \tag{12}$$

Of course, the experiment used to illustrate the law is very rough*. But there are many other phenomena that prove the same fact—above all, the observed phenomenon that all bodies fall equally fast.

* For example, we have neglected the circumstance that in producing the rotation of the rolling sphere a resistance must also be overcome which depends on the distribution of mass in the interior of the sphere (the moment of inertia).

It is here assumed, of course, that no forces other than gravity exert an influence on the motion. This means that the experiment must be carried out *in vacuo* so that the resistance of the air may be eliminated. For purposes of demonstration we may use an inclined plane (Fig. 25)

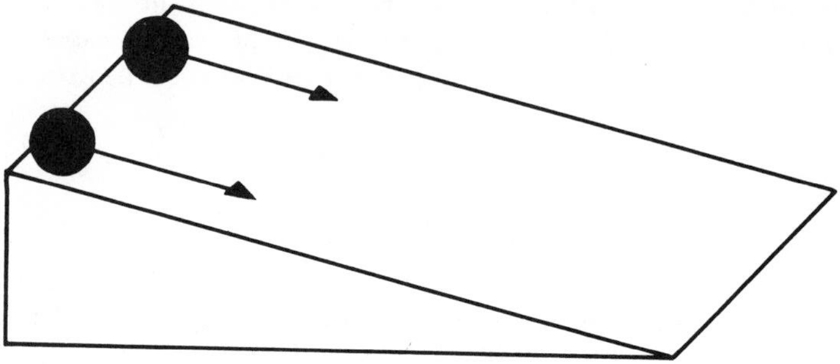

Fig. 25 *Two spheres similar in appearance but with different weights roll down the inclined plane with the same velocity.*

down which two spheres, similar in appearance but of different weight, are rolling. It is observed that they reach the bottom at the same moment.

The weight is the driving force; the mass determines the resistance. If they are proportional to each other, then a heavy body will indeed be driven forward more strongly than a lighter one, but to balance this it resists the impelling force more strongly, and the result is that the heavy and the light body slide down or fall equally fast. We also see this from our formulae. For if in (10) we replace the force by the weight G and assume the latter, by (12), proportional to the mass, we get

$$mb = G = mg,$$

that is,

$$b = g. \tag{13}$$

Thus all bodies have one and the same acceleration vertically downwards, if they move under the influence of gravity alone, whether they fall from rest or are thrown. The quantity g, the acceleration due to gravity, has the value

$$g = 981 \text{ cm./sec.}^2 \text{ (or 32 ft./sec.}^2\text{)}.$$

The most searching experiments for testing this law may be

carried out with the aid of a simple pendulum, a sphere carried by a thin string. Newton even in his time noticed that the times of swing are always the same for pendulums of the same length, whatever the composition of the bob. The process of vibration is exactly the same as that described above for the elastic pendulum, except that now it is not a steel spring but gravity that pulls back the sphere. We must imagine the force of gravity acting on the sphere to be resolved into two components, one acting in the direction of the string, keeping it stretched, the other acting in the direction of motion as the driving force of the sphere or bob.

Fig. 26 exhibits the bob deflected a distance x. We see at once

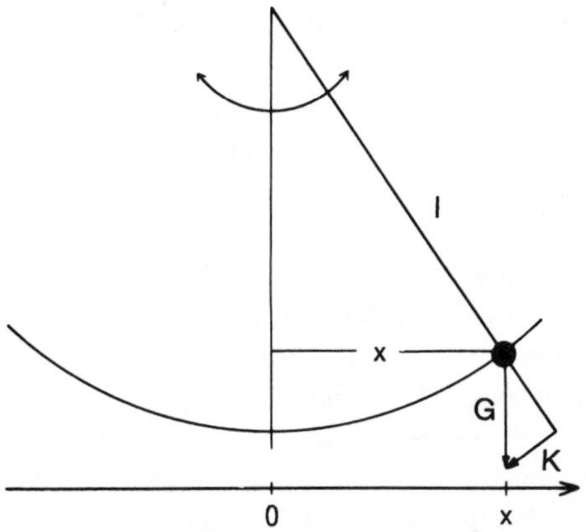

Fig. 26 *Illustration of the forces on a string pendulum.*

the two similar right-angled triangles, the sides of which are in the same proportion:

$$\frac{K}{x} = -\frac{G}{l}.$$

Again the negative sign of K shows that the force is directed towards the equilibrium position $x=0$. Accordingly, formula (11) gives for two pendulums, the weights of which are G_1 and G_2 respectively,

$$(2\pi\nu)^2 m_1 = \frac{G_1}{l}, \qquad (2\pi\nu)^2 m_2 = \frac{G_2}{l},$$

thus

$$\frac{G_1}{m_1} = \frac{G_2}{m_2} = (2\pi\nu)^2 l,$$

that is, the ratio of the weight to the mass is the same for both pendulums. We called this ratio g in formula (12). Hence we get the equation

$$g = (2\pi\nu)^2 l, \tag{14}$$

from which we see that g may be determined by measuring the length l of the pendulum and the vibration number ν.

The law of the proportionality of weight to mass is often expressed as follows:

Gravitational and inertial mass are equal.

Here gravitational mass simply signifies the weight divided by g, and the proper mass is distinguished by prefixing the word "inertial."

The fact that this law holds exactly was already known to Newton. Nowadays it has been confirmed by most delicate measurements carried out by Eötvös (1890).* Hence we are completely justified in using the balance to compare not only weights but also masses.

One might imagine that such a law is firmly embedded in the foundations of mechanics. Yet this is by no means the case, as shown by our account, which follows the ideas of classical mechanics. Rather, it is attached somewhat loosely, as a sort of oddity, to the fabric of the other laws. Probably it has been a source of wonder to many, but no one suspected or sought any deeper relationship that might be hidden in it. For there are many kinds of forces that can act on a mass. Why should there not be one that is exactly proportional to the mass? A question to which no answer is *expected* will receive none. And so the matter rested for centuries. This was possible only because of the overwhelming success of the Galilean-Newtonian mechanics. It controlled not only the movements of terrestrial bodies but also those of the stars, and showed itself to be a trustworthy foundation for the whole realm of exact science. In the middle of the nineteenth century, indeed, it was believed that the aim of all research was the interpretation of physical phenomena in terms of Newtonian mechanics. And thus in building up their imposing edifice physicists forgot to ascertain whether the basis was strong enough to support the whole. Einstein

*New measurements made by Dicke (1961) confirmed the results obtained by Eötvös with still greater accuracy. The relative difference between inertial and gravitational mass is according to Eötvös at most 10^{-9}, according to Dicke at most 10^{-11}.

was the first to recognize the importance of the law of equality of inertial and gravitational mass for the foundations of the physical sciences.

13. Analytical Mechanics

The problem of analytical mechanics is to find from the law of motion

$$mb = K,$$

the motion when the forces K are given. The formula itself gives us only the acceleration, that is, the change of velocity. To get from the latter the velocity, and from this again the varying position of the moving point, is a problem of integral calculus that may be very difficult if the force alters in a complicated way with place and time. An idea of the nature of the problem is given by our derivation of the change of position in a uniformly accelerated motion along a straight line (p. 20–22). The motion becomes more complicated when it takes place in a plane and is due to the action of a constant force of definite

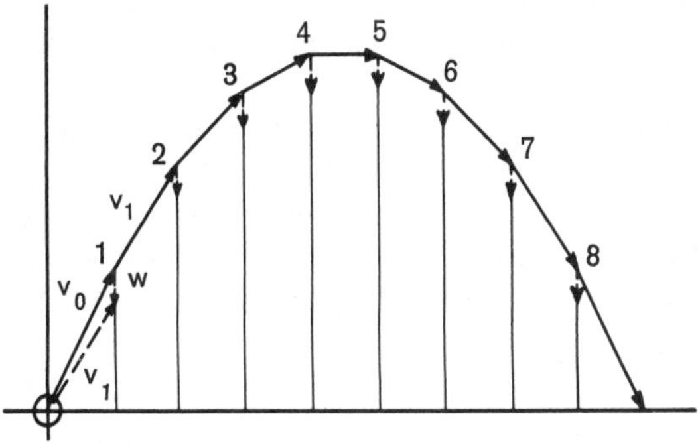

Fig. 27 *Motion of a sphere on a table. At the points 1, 2, . . . 8 blows of the same strength strike the sphere and produce a change* w *of velocity.*

direction, as in the case of a falling or thrown body. Here, too, we may substitute as an approximation to the continuous course of the actual motion a fictional one consisting of a series of uniform motions, each of which is transformed into the next by means of impulses. We again use our horizontal table with a sphere rolling, and give the sphere a blow of constant magnitude and direction at short equal intervals τ. (Fig. 27). Now, if the sphere starts from the point O with an arbitrary initial velocity, it arrives after time τ at a point 1 where the first blow strikes it. From this point it pursues its course in another direction with a different velocity for the time τ until at a point 2 it is struck by the second blow, which again deflects it, and so forth. Each individual deflection may be determined from the law of impulses. Accordingly, we may draw the whole motion, and we see that the initial point, the initial direction, and the initial velocity completely determine the subsequent course of the motion. This jerky motion gives us a rough picture of the motion of a sphere on an inclined plane. The graph coincides more closely with the actual continuous motion as we choose smaller time intervals between the blows.

This approximate construction can be replaced in the case of forces acting continuously by a rigorous treatment with the methods of the integral calculus. In this case, too, the point of departure and the magnitude and direction of the initial velocity remain quite arbitrary. But if these are given, the further course of the motion is fully determined. Thus one and the same law of force may produce an infinity of motions according to the choice of the initial conditions, so that the enormous number of motions of falling or thrown bodies depends on the same law of force (the force being gravity which acts vertically downwards).

In mechanical problems we are usually concerned with the motion not of one body but of several that exert forces on one another. The forces themselves are not given but depend for their part on the unknown motion. It is obvious that the problem of determining the motions of several bodies by calculation becomes highly complicated.

14. The Law of Energy

There is a law, however, which makes these problems much simpler and supplies a general survey of the motion. It is the *law of the conservation of energy*, which has become of very great importance for the development of the physical sciences. Of course, we cannot give its general formulation or prove it here. We shall only illustrate its content with simple examples.

A pendulum which is released after the bob has been raised to a certain point rises on the opposite side of its low point to the same height—except for a small deviation caused by friction and the resistance of the air (Fig. 28). If we replace the circular orbit by some other, e.g., allowing the sphere to run on rails as in a "toy railway" (Fig. 29), then the same result holds: The sphere always rises to the same height as that from which it started.

From this it easily follows that the velocity of the sphere at any point *P* of its path depends only on the depth of this point *P* below the initial point *A*. To see this we imagine the section *AP* of the orbit changed, the rest *PB* remaining unaltered. Now, if the sphere were to arrive at *P* along the one orbit from *A* with a velocity different

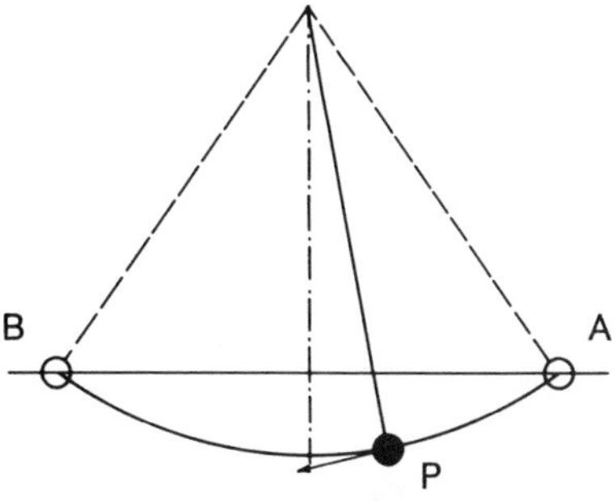

Fig. 28 *If the bob begins its motion at* A, *it rises to a position* B *of equal height.*

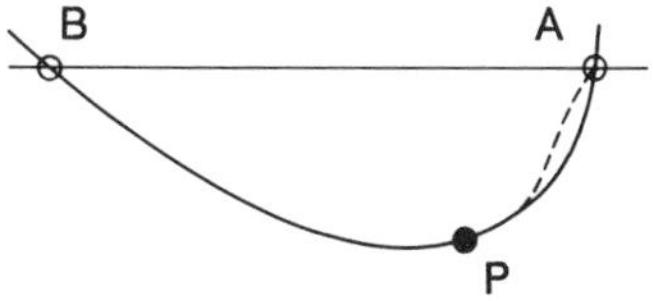

Fig. 29 *If the sphere starts at* A, *the turning point* B *will be at the same level, independent of the shape of its path. The velocity at* P *is determined only by the height of fall from* A *to* P.

from that with which it arrives along the other, then in its further course from *P* to *B* it would not in each case exactly reach its goal *B*. For, to achieve this, a uniquely determined velocity is clearly necessary at *P*. Consequently the velocity at *P* does not depend on the form of the section of orbit traversed, and since *P* is an arbitrary point, this result holds generally. Hence the velocity v must be determined by the height of fall h alone. The validity of this law depends on the assumption that the path (the rails) as such offers no resistance to the motion (i.e., exerts no force on the sphere in its direction of motion), but receives only its perpendicular pressure. If the rails are not present, we have the case of a body falling freely or of one that has been thrown, and the same result holds: The velocity at each point depends only on the height of fall.

This fact may not only be established experimentally but may also be derived from the laws of motion. Thus one can also obtain the law that relates the velocity to the height, namely:

Let x be the altitude above ground (Fig. 30), v the velocity, m the

Fig. 30 *The coordinate* x *measures the altitude from the earth's surface* (x=*0*).

mass, and G the weight of the body. Then the quantity

$$E = \frac{m}{2} v^2 + Gx \tag{15}$$

has the same value during the whole process of falling.

To prove this we first suppose E to stand for any arbitrary quantity that depends on the motion and hence alters from moment to moment. Let E alter by the amount e in a small interval of time τ; then we shall call the ratio $\frac{e}{\tau}$ the rate of change of E, and, exactly as before in defining the orbital velocity v and the acceleration b, we suppose that the time interval τ may be made as small as we please. If the quantity E does not change in the course of time, then its rate of change is, of course, zero, and vice versa. The change e of E can be determined in the following manner: During the time τ the height of fall x decreases by $v\tau$, and the velocity increases by $w = b\tau$. Hence after the time τ the value of E becomes

$$E' = \frac{m}{2}(v+w)^2 + G(x - v\tau).$$

Now,

$$(v+w)^2 = v^2+w^2+2vw.$$

This states that the square erected over v and w, joined together in the same straight line, may be resolved into a square having the side v, one having the side w, and two equal rectangles having the sides v and w (Fig. 31). Hence we get

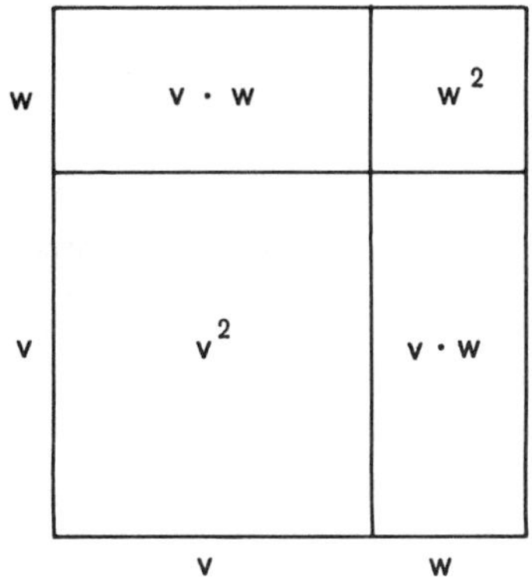

Fig. 31 $(v+w)^2 = v^2+2vw+w^2$.

$$E' = \frac{m}{2}v^2+\frac{m}{2}w^2+mvw+Gx-Gv\tau.$$

If we subtract the old value of E from this, we get

$$e = E'-E = \frac{m}{2}w^2+mvw-Gv\tau,$$

or, since $w = b\tau$,

$$e = \frac{m}{2}b^2\tau^2+mvb\tau-Gv\tau.$$

Hence the rate of change becomes

$$\frac{e}{\tau} = \frac{m}{2}b^2\tau+mvb-Gv.$$

The term involving τ may be neglected since it can be made infinitesimally small by making the time interval infinitesimally small. Hence we get finally for the rate of change of E,

$$\frac{e}{\tau} = v(mb - G).$$

But according to the laws of mechanics this expression has the value zero, for by (13) we have $mb = mg = G$. Hence we have proved that the quantity E defined in (15) remains unchanged with time. If the initial point and the initial velocity of the motion are given, that is, the values of x and v for $t = 0$, then the expression E, according to (15), has a definite value and retains it during the whole motion.

From this it follows that if the body rises, that is, if x increases, v must decrease, and vice versa. One of the two terms of the expression E can increase only at the expense of the other. The first term is a measure of the velocity of the body; the second, of the height that it has attained against the force of gravitation. There are special names for these terms.

$T = \frac{m}{2} v^2$ is called the *vis viva*, or *kinetic energy*.

$U = Gx$ is called the *capacity for work*, or the *potential energy*. Their sum,

$$T + U = E, \tag{16}$$

is called simply the *mechanical energy* of the body; and the law which states that it remains constant during the motion of the body is called the law of *conservation of energy*.

The dimensions of energy are $[E] = [Gl]$. Its unit is p.cm.

The name *capacity for doing work* is of course derived from the work done by the human body in lifting a weight. According to the law of conservation of energy this work becomes transformed into kinetic energy in the process of falling. If, on the other hand, we give a body kinetic energy by throwing it upwards, this energy changes into potential energy or capacity for doing work.

Exactly the same as has been described for the motion of falling bodies holds in the widest sense for systems composed of any number of bodies, so long as two conditions are fulfilled, namely:

1. External influences must not be involved, that is, the system must be self-contained or isolated.

2. Phenomena must not occur in which mechanical energy is transformed into thermal, electrical, or chemical energy, and such like.

If these are fulfilled the law that

$$E = T+U$$

always remains constant holds true, the kinetic energy depending on the velocities, the potential energy on the positions of the moving bodies.

In the mechanics of the heavenly bodies this ideal case is realized in perfection. Here the ideal dynamics whose principles we have developed is strictly valid. But on the earth this is by no means the case. Every motion is subject to friction, whereby its energy is transformed into heat. The machines by means of which we produce motion transform thermal, chemical, electric, and magnetic forces into mechanical forces, and hence the law of energy in its narrow mechanical form does not apply. But it may always be maintained in an extended form. If we call the heat energy Q, the chemical energy C, the electromagnetic energy W, and so forth, then the law that for closed systems the sum

$$E = T+U+Q+C+W\ .\ .\ . \tag{17}$$

is always constant holds.

It would lead us too far to pursue the discovery and logical evolution of this fact by Robert Mayer, Joule (1842), and Helmholtz (1847), or to investigate how the nonmechanical forms of energy are determined quantitatively. But we shall use the concept of energy later when we speak of the intimate relationship that the theory of relativity has disclosed between mass and energy.

15. Dynamical Units of Force and Mass

The method by which we have derived the fundamental laws of mechanics, by experiments on a table or with an inclined plane, a pendulum, and such simple things, apparently restricts their validity. We have abstracted our concepts and laws from experiments in the laboratory. The advantage of this is that we need not trouble about assumptions concerning space and time. The rectilinear motions with which the law of inertia deals may be measured on the table with a ruler. Compasses, rulers, and clocks are assumed to be available for measuring the orbits and the motions.

Our next concern will be to step out of the narrow confines of our rooms into the wider world of astronomic space. The first stage will be a "voyage round the world," meaning not the universe but the globe of the earth. We shall put the question: Do the laws of mechanics apply just as much in a laboratory in Buenos Aires or in Cape Town as in Berlin or New York?

They do, with one exception—the value of the gravitational acceleration g. We have seen that this can be measured precisely by observations of pendulums. It has been found, however, that the same pendulum will swing somewhat more slowly at the equator than in the more southerly or more northerly regions. Fewer vibrations occur in the course of a day, in the course of one rotation of the earth. From this it follows that g has a minimum value at the equator and increases towards the north and the south. This increase is quite regular up to the poles, where g has its greatest value. We shall examine the cause of this later. Here we merely take note of the fact. However, for the system which we have hitherto used for measuring forces and masses, this fact has very awkward consequences.

So long as weights are compared with each other only by means of the scale balance, there are no difficulties. But let us imagine a spring balance here in the laboratory which has been calibrated with weights. If we bring this spring balance into more southerly or more northerly regions, we shall find that when loaded with the same weights it will give different deflections. If, therefore, we identify weight with force as we have hitherto done, there is nothing left for us but to assert that the force of the spring has altered and that it depends on the geographical latitude. But this is obviously not the case. It is not the force of the spring that has altered but the gravitational force. Therefore it is wrong to take the weight of one and the same piece of metal as the unit of force at all points of the earth. We may choose the weight of a definite body at a definite point on the earth as the unit of force, and this may be applied at other points if the acceleration g due to gravity is known by pendulum measurements at both points. This is, indeed, just what technical science does. Its unit of force is the weight of a definite normal body in Paris, the pond. Hitherto we have always used this without taking into account its variability with position. In exact measurements, however, the value must be reduced to that at the normal place (Paris).

Science has departed from this system of measures, at which one place on the earth is favored, and has selected a system that is less arbitrary.

The fundamental law of mechanics itself offers a suitable method of doing this. Instead of referring the mass to the force, we establish the mass as the fundamental quantity; we ascribe to it an independent dimension $[m]$ and choose its unit arbitrarily: a definite piece of metal is to represent the unit of mass. As a matter of fact, the same piece of metal that serves technical science as the unit of weight, the Paris pond, is taken for this purpose, and this unit of mass is called the gram (gm.).

Hereafter we will use the physical system of measure, the fundamental units of which are cm. for length, sec. for time, gm. for mass.

Force now has the derived dimensions

$$[K] = [mb] = \left[\frac{ml}{t^2}\right],$$

and its unit, called the dyne, is mg. cm./sec.^{2}

Weight is defined by $G=mg$; thus the unit of mass has the weight $G=1$ gm.$\times g$. It changes with the geographical latitude. In our own latitude g has the value 981 dynes. This is the technical unit of force. The strength of a spring balance expressed in dynes is, of course, a constant; for its power of accelerating a definite mass is independent of the geographical latitude.

The dimensions of impulse or momentum are now:

$$[J] = [tK] = \left[\frac{ml}{t}\right] = [mv] = [p],$$

and its unit is gm. cm./sec. Finally, the dimensions of energy are

$$[E] = [mv^2] = \left[\frac{ml^2}{t^2}\right],$$

and its unit is gm. cm.^{2}/sec.^{2}

Now that we have cleansed the system of measures of all terrestrial impurities, we can proceed to the mechanics of the stars.

CHAPTER III

THE NEWTONIAN WORLD SYSTEM

1. Absolute Space and Absolute Time

The principles of mechanics, as here developed, were partly suggested to Newton by Galileo's works and partly created by himself. To Newton we owe above all the definitions and laws in such a generalized form that they appear detached from earth-bound experiments and can be applied to events in astronomic space.

In arriving at these laws Newton had to preface the actual mechanical principles by making definite assertions about space and time. Without such determinations even the simplest law of mechanics, that of inertia, makes no sense. According to this law, a body on which no force is acting is to move uniformly in a straight line. Let us consider the table upon which we first experimented with the rolling sphere. If the sphere rolls on the table in a straight line, an observer who follows and measures its path from another planet would have to assert that the path is not a straight line according to his point of view. For the earth itself is rotating, and a motion that appears rectilinear to the observer traveling with the earth, because it leaves the trace of a straight line on his table, must appear curved to another observer who does not participate in the rotation of the earth. This may be roughly illustrated as follows:

A circular disk of white cardboard is mounted on an axis so that it can be turned by means of a handle. A ruler is fixed in front of the disk. Now turn the disk as uniformly as possible and at the same time draw a pencil along the ruler with constant velocity so that the pencil marks its course on the disk. This path will, of course, not be a straight line on the disk, but a curved line, which will even take the form of a loop if the rotary motion is sufficiently rapid. Thus, the same motion which an observer fixed to the ruler would call uniform and rectilinear would be called curvilinear (and nonuniform) by an observer moving with the disk. This motion may be constructed point for point, as is illustrated in the drawing (Fig. 32).

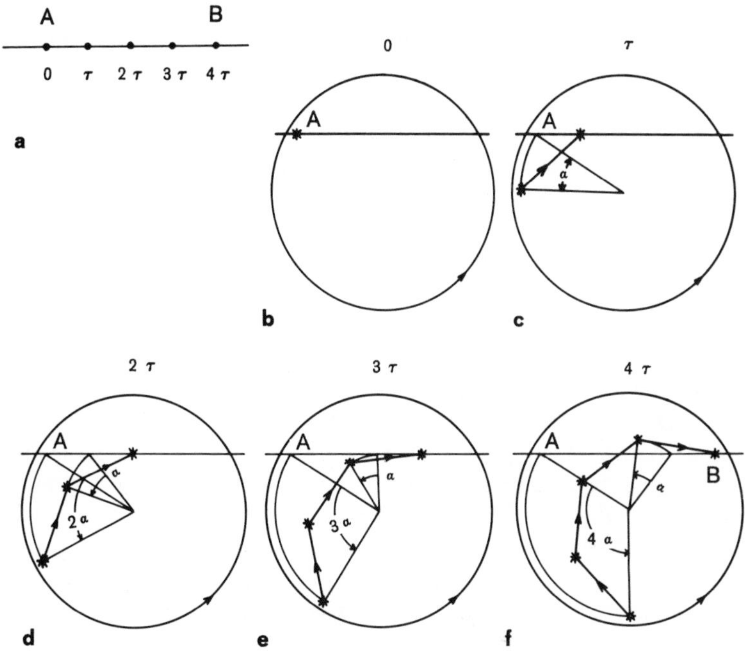

Fig. 32*a* *A body runs uniformly from* A *to* B *in four time intervals τ. The motion is observed by a resting observer.*

b t=*0. The body is at* A*; the observer marks this point by a star.*

c t=τ. *The position of the body is characterized by a dot and also marked by a star. The disk and with it the star marked in 32b has rotated through an angle* α.

d–f *The observer has continued to mark the position of the body in the same way.*
The polygon connecting the stars represents, approximately, the path of the body on the moving disk.

This example shows clearly that the law of inertia makes sense, indeed, only when the space, or rather, the system of reference in which the rectilinear character of the motion is to hold, is exactly specified.

It is in conformity with the Copernican world picture, of course, not to regard the earth as the system of reference for which the law of inertia holds, but one that is somehow fixed in astronomic space. In experiments on the earth—for example, of the sphere rolling on the table—the path of the freely moving body is not actually straight but a little curved. The fact that this escapes our immediate notice is due only to the shortness of the paths used in the experiments compared with the dimensions of the earth. Here, as has often happened in science, the inaccuracy of observation led to the discovery of an important fact. If Galileo had been able to make observations as refined as those of later centuries, the confused accumulation of phenomena would have made the discovery of the laws much more difficult. Perhaps Kepler would never have unravelled the motions of the planets if the orbits had been known to him as accurately as at the present day. For Kepler's ellipses are only approximations from which the real orbits differ considerably in long periods of time. A similar thing happened in the case of modern physics in regard to the regularities of spectra; the discovery of simple relationships was rendered more difficult and considerably delayed by the very abundance of data of observation.

Newton was therefore confronted with the task of finding the system of reference in which the law of inertia and all the other laws of mechanics were to hold. If he had chosen the sun, the question would not have been solved but would only have been postponed, for the sun might one day be discovered also to be in motion, as has actually happened in the meantime.

Probably it was for such reasons that Newton arrived at the conviction that an empirical system of reference fixed by material bodies could never be the foundation of a law involving the idea of inertia. But the law itself, through its close connection with Euclid's doctrine of space, the element of which is the straight line, appears as the natural starting point of the dynamics of astronomic space. It is indeed in the law of inertia that Euclidean space manifests itself outside the narrow limits of the earth. Similar conditions obtain in the case of time, the flow of which receives expression in the uniform motion due to inertia. If one were for instance to take the period of rotation of the earth as the unit of time, the law of inertia would not be exactly valid because there are some irregularities in the motion of the earth.

In this way Newton came to the conclusion that there is an absolute space and an absolute time. It will be best to give the substance of his own words (the quotations are from the translation of the original Latin by Newton's contemporary, Andrew Motte, 1729). Concerning time he says:

> Absolute, True, and Mathematical Time, of itself, and from its own nature flows equably without regard to any thing external, and by another name is called Duration: Relative, Apparent, and Common Time is some sensible and external (whether accurate or unequable) measure of Duration by the means of motion, which is commonly used instead of True time; such as an Hour, a Day, a Month, a Year. . . .
>
> For the natural days are truly unequal, though they are commonly consider'd as equal, and used for a measure of time: Astronomers correct this inequality for their more accurate deducing of the celestial motions. It may be, that there is no such thing as an equable motion, whereby time may be accurately measured. All motions may be accelerated and retarded, but the True, or equable progress, of Absolute time is liable to no change. The duration or perseverance of the existence of things remains the same, whether the motions are swift or slow, or none at all. . . .

Concerning space Newton expresses similar opinions. He says:

> Absolute Space, in its own nature, without regard to any thing external, remains always similar and immoveable. Relative Space is some moveable dimension or measure of the absolute spaces; which our senses determine, by its position to bodies; and which is vulgarly taken for immoveable space. . . .
>
> And so instead of absolute places and motions, we use relative ones; and that without any inconvenience in common affairs: but in Philosophical disquisitions, we ought to abstract from our senses, and consider things themselves, distinct from what are only sensible measures of them. For it may be that there is no body really at rest, to which the places and motions of others may be referred. . . .

The definite statement, both in the definition of absolute time and in that of absolute space, that these two quantities exist "without reference to any external object whatsoever" seems strange from one like Newton, for he often emphasizes that he wishes to investigate only what is actual, what is ascertainable by observation. "*Hypotheses non fingo*" is his brief and definite expression. But what exists "without reference to any external object whatsoever" is not ascertainable and is not a fact. Here we have clearly a case in which the ideas of unanalyzed consciousness are applied without reflection to the objective world. We shall investigate this question in detail later on.

Our next task is to describe how Newton interpreted the laws of the cosmos and how his doctrine advanced the contemporary world view.

2. Newton's Law of Attraction

Newton succeeded in developing a dynamical theory of planetary orbits, or, as we nowadays express it, in founding *celestial mechanics*. To do this it was necessary to apply Galileo's concept of force to the motions of the stars. Yet Newton discovered the law according to which the heavenly bodies act on one another, not by setting up bold hypotheses but by pursuing the systematic and rigorous method of analyzing the known facts of planetary motions. These facts were expressed in the three Kepler laws that compressed all the observations at that period of time into a wonderfully concise form. We must here state Kepler's laws in full. They are:

1. The planets move in ellipses with the sun at one of the foci (Fig. 33).

2. The radius vector drawn from the sun to a planet describes equal areas in equal times.

3. The cubes of the major axes of the ellipses are proportional to the squares of the periods of revolution.

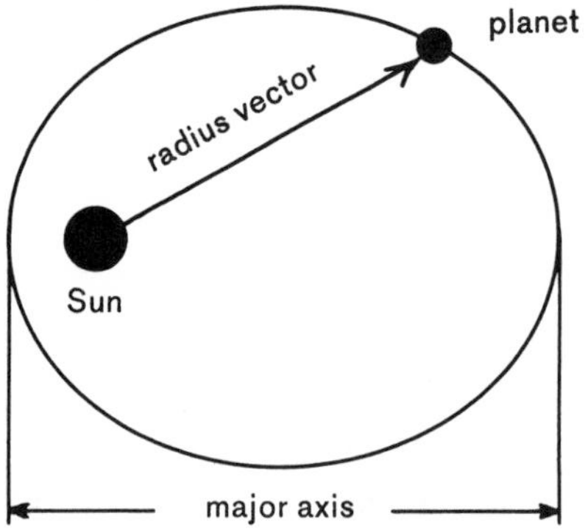

Fig. 33 *The path of a planet around the sun is an ellipse. The sun is situated in one focus.*

Now the fundamental law of mechanics gives a relation between the acceleration b of the motion, and the force K that produces it. The acceleration b is completely determined by the course of the motion, and if this is known, k can be calculated. Newton recognized that the orbit as defined by Kepler's laws just sufficed to allow a calculation of b. Then the law

$$K = mb$$

also allows the acting force to be calculated.

The ordinary mathematics of his time would not have enabled Newton to carry out his project. But he was already in the possession of the mathematical tool necessary for his purpose, differential and integral calculus almost simultaneously discovered by Leibniz (1684) and today one of the roots of modern mathematics. In his fundamental book *Philosophiae Naturalis Principia Mathematica*, however, Newton refrained from using these new methods and represented the matter in the guise of the customary classical geometrical methods.

We follow Newton's method but restrict ourselves to a simple example in order to illustrate his general results.

The orbits of the planets are ellipses of slight eccentricity—that is, almost circular. It is permissible to assume that the planets describe approximate circles about the sun, as was, indeed, supposed by Copernicus. Since a circle is a special kind of ellipse, this assumption certainly fulfills Kepler's first law.

The second law now implies that every planet traverses its circle with constant speed. We already know (II, 4) all about the acceleration in such circular motions. It is directed towards the center and, by formula (4), has the value

$$b = \frac{v^2}{r},$$

where v is the speed in the orbit, and r is the radius of the circle.

If now T is the period of revolution, the velocity is determined as the ratio of the circumference $2\pi r$ to the time T, thus

$$v = \frac{2\pi r}{T} \tag{18}$$

so that

$$b = \frac{4\pi^2 r^2}{rT^2} = \frac{4\pi^2 r}{T^2}.$$

We next consider the third Kepler law which, in the case of a circular orbit, clearly states that the ratio of the cube of the radius, r^3, to the square of the time of revolution, T^2, has the same value C for all planets:

$$\frac{r^3}{T^2} = C \quad \text{or} \quad \frac{r}{T^2} = \frac{C}{r^2}. \tag{19}$$

If we insert this in the value for b above, we get

$$b = \frac{4\pi^2 C}{r^2}. \tag{20}$$

Hence the value of the centripetal acceleration depends only on the distance of the planet from the sun, being inversely proportional to the square of the distance, but it is quite independent of the properties of the planet, such as its mass. For the quantity C is, by Kepler's third law, the same for all planets, and can therefore involve at most the nature of the sun and not that of the planets.

It is a remarkable circumstance that the same law emerges for elliptic orbits—by a rather more laborious calculation. The acceleration is always directed towards the sun situated at a focus, and has the value given by formula (20), where r is the length of the radius vector (see Fig. 33).

3. Universal Gravitation

The law of acceleration found in the preceding section has an important property in common with the gravitational force on the earth (weight): it is quite independent of the nature of the moving body. If we calculate the force from the acceleration, we find it likewise directed towards the sun. It is thus an attraction and has the value

$$K = mb = m\frac{4\pi^2 C}{r^2}. \tag{21}$$

It is proportional to the mass of the moving body, just as the weight

$$G = mg$$

of a body on the earth.

This fact suggested to Newton the idea that both forces may have one and the same origin. Nowadays this circumstance, handed down to us through the centuries, has become such a truism that we can scarcely conceive the boldness and breadth of Newton's innovation. What a prodigious imagination it required to conceive the motion of the planets about the sun or of the moon about the earth as a process of "falling" that takes place according to the same laws and under the action of the same force as the falling of a stone released by my hand. The fact that the planets or the moon do not actually plunge into their central attracting bodies is due to the law of inertia that here results in a centrifugal force. We shall have to deal with this again later.

Newton first tested this idea of *universal gravitation* in the case of the moon, the distance of which from the earth was known by direct measurements. This test is so important that we shall repeat the very simple calculation here as evidence of the fact that all scientific ideas become valid and of worth only when calculated and measured numerical values agree.

The central body is now the earth; the moon takes the place of the planet. The symbol r denotes the radius of the moon's orbit, T the period of revolution of the moon. Let the radius of the earth be a. If the gravitational force on the earth is to have the same origin as the attraction that the moon experiences from the earth, then the acceleration g due to gravity must, by Newton's law (20), have the form

$$g = \frac{4\pi^2 C}{a^2},$$

where C has the same value as for the moon, namely, by (19),

$$C = \frac{r^3}{T^2}.$$

If we insert this value in that for g, we get

$$g = \frac{4\pi^2 r^3}{T^2 a^2}. \tag{22}$$

Now, the "sidereal" period of revolution of the moon, that is, the time between two positions in which the line connecting the earth to the moon has the same direction with respect to the stars, is

$$T = 27 \text{ days } 7 \text{ hours } 43 \text{ minutes } 12 \text{ seconds}$$
$$= 2{,}360{,}592 \text{ seconds.}$$

In physics it is customary only to write down a number to as many places as are required for further calculation. So we write here

$$T = 2.36 \times 10^6 \text{ sec.}$$

The distance of the moon from the center of the earth is about 60 times the earth's radius, or, more exactly,

$$r = 60.1a.$$

The earth's radius itself is easy to remember because the metric system is related to it in a simple way. For 1 m. = 100 cm. = one ten-millionth of the earth's quadrant, that is, the forty-millionth part or (4×10^7)th part of the earth's circumference $2\pi a$

$$100 \text{ cm.} = \frac{2\pi a}{4 \times 10^7} \quad \text{or} \quad a = 6.37 \times 10^8 \text{ cm.} \tag{23}$$

If we insert all these values in (22) we get

$$g = \frac{4\pi^2 \times 60.1^3 \times 6.37 \times 10^8}{2.36^2 \times 10^{12}} = 981 \text{ cm./sec.}^2 \tag{24}$$

This value agrees exactly with that found by pendulum observations on the earth (see II, 12).

The great importance of this result is that it represents the *relativization of the force of weight.* To the ancients weight denoted a pull towards the absolute "below," which is experienced by all terrestrial bodies. The discovery of the spherical shape of the earth brought with it the relativization of the direction of weight; it became a pull towards the center of the earth.

And now the identity of weight with the force of attraction that keeps the moon in her orbit is proved, and since there can be no doubt that the latter is similar in nature to the force that keeps the earth and the other planets in their orbits round the sun, we arrive at the idea that bodies are not simply "heavy" but are mutually heavy or *heavy relative to one another.* The earth, being a planet, is attracted towards the sun, but it itself attracts the moon. Obviously this is only an approximate description of the true state of affairs, in which the sun, moon, and earth attract one another. Certainly, as far as the orbit of the earth round the sun is concerned, the latter may, to a high degree of approximation, be regarded as at rest, because its enormous mass hinders the coming into play of appreciable accelerations; and conversely, the moon, on account of its small mass, does not come into account. But an exact theory will have to take into consideration these influences, called "perturbations."

Before we begin to consider more closely this aspect, which signifies the chief advance of Newton's theory, we shall give Newton's law its final form. We saw that a planet situated at a distance r from the sun experiences from it an attraction of the value (21)

$$K = m\frac{4\pi^2 C}{r^2},$$

where C is a constant depending only on the properties of the sun, not on those of the planet. Now, according to the new view of mutual or relative weight the planet must likewise attract the sun. If M is the mass of the sun and c a constant dependent only on the nature of the planet, then the force exerted on the sun by the planet must be expressed by

$$K' = M\frac{4\pi^2 c}{r^2}.$$

But earlier, in introducing the concept of force (II, 1, p. 14), we made use of the principle that the reaction equals the action, which is one of the simplest and most certain laws of mechanics. If we apply it here, we must set $K = K'$, or

$$m\frac{4\pi^2 C}{r^2} = M\frac{4\pi^2 c}{r^2}.$$

From this it follows that

$$mC = Mc,$$

or

$$\frac{C}{M} = \frac{c}{m}.$$

Therefore this ratio has the same value for both bodies (sun and planets), and hence also for any body whatsoever. If we call this value $\frac{k}{4\pi^2}$, then we may write

$$4\pi^2 C = kM, \qquad 4\pi^2 c = km. \tag{25}$$

The factor of proportionality k is called the *gravitational constant.*

The Newtonian law of general gravitation then assumes the symmetrical form

$$K = k\frac{mM}{r^2}. \tag{26}$$

In words it states:

Two bodies attract each other with a force that is proportional to the mass of each body and is inversely proportional to the square of their distance apart.

4. Celestial Mechanics

It is only in this general form that the Newtonian law denotes a real advance in the description of the planetary orbits. For in the original form it was deduced from Kepler's laws by calculation and signified no more than a very short and striking résumé of these laws.

It is also possible to prove the converse law, that the motion of a body about a central body at rest and attracting it according to Newton's law, is necessarily a Kepler ellipse. (This is true in the case of a closed periodical orbit. However, some comets have hyperbolic orbits. Such orbits are not closed.) A new feature arises only when, first, we regard both bodies as moving and, second, we add further bodies to the problem. Then we get the *problem of three or*

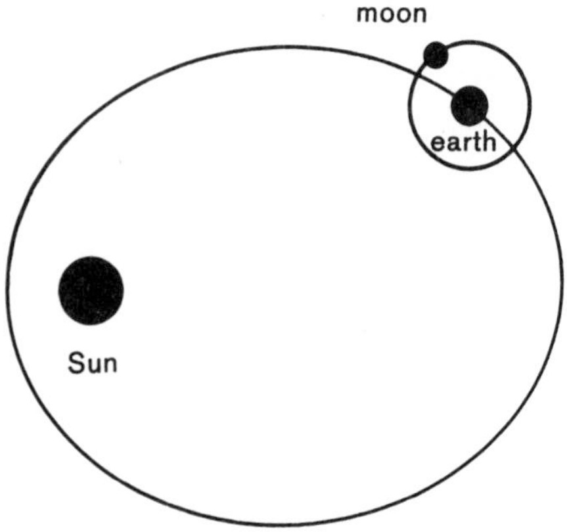

Fig. 34 *Three-body problem: sun, earth, moon.*

more bodies, which corresponds exactly to the actual conditions in the planetary system (Fig. 34). For not only are the planets attracted by the sun, and the moons by their planets, but every body, be it sun, planet, moon, or comet, attracts every other body. Accordingly, the Kepler ellipses appear to be only approximations and that only because the sun on account of its great mass overshadows by far the mutual action of all other bodies of the planetary system. But in long periods of time these mutual actions must also manifest themselves as deviations from the Kepler laws. We speak, as already remarked, of "perturbations."

In Newton's time such perturbations were already known, and in the succeeding centuries refinements in the methods of observation have accumulated an immense number of facts that have had to be accounted for by Newton's theory. That it succeeded in doing so is one of the greatest triumphs of the human mind.

It is not our aim here to pursue the development of mechanics from Newton's time to the present day, and to describe the mathematical methods that were devised to calculate the "perturbed" orbits. The most ingenious mathematicians of different countries have played a part in setting up the "theory of perturbations," and even if no rigorous solution has yet been found for the problem of three bodies, it is possible to calculate with great precision the motions for hundreds, thousands, or millions of years ahead or back. So Newton's theory has been tested in countless cases of new observations and has never failed—except in one case, of which we shall speak presently. Theoretical astronomy, as founded by Newton, was therefore long regarded as a model for the exact sciences. It achieved what had been the longing of mankind since earliest history. It lifted the veil that was spread over the future; it endowed its disciples with the gift of prophecy. Even if the subject matter of astronomic predictions was unimportant or indifferent to human life, yet it became a symbol for the liberation of the spirit from the trammels of earthly bonds. We, like the peoples of earlier times gaze upwards with awe at the stars, which reveal the law of the world.

But the world law can tolerate no exception. And there is one case, as we have already mentioned, in which Newton's theory has failed. Although the error is small, it is not to be denied. It occurs in the case of the planet Mercury, the planet nearest the sun.

The orbit of any planet may be regarded as a Keplerian elliptic motion that is perturbed by the other planets; that is, the position of the orbital plane, the position of the major axis of the ellipse, its eccentricity—in short, all "elements of the orbit"—undergo gradual changes. If we calculate these according to Newton's law they agree with the observed data for all planets except Mercury. In this case the rotation of the perihelion (Fig. 35) shows a very small but firmly

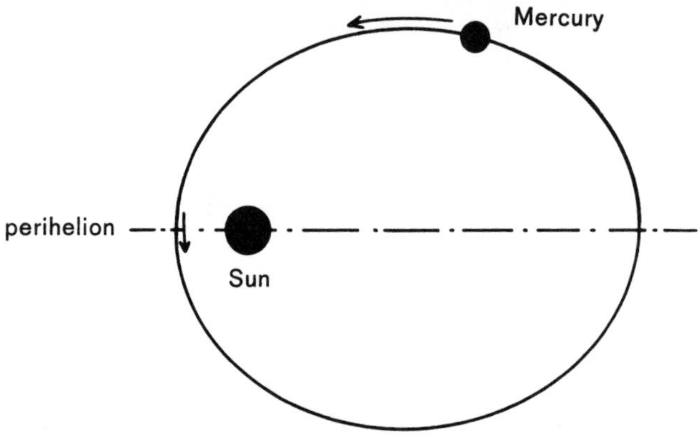

Fig. 35 *The perihelion is the point of the orbital ellipse nearest to the sun. The perihelion of Mercury shows a small rotation around the sun which cannot be explained by Newton's mechanics.*

established deviation from the value computed with Newton's law, namely 43 seconds of arc every hundred years. The astronomer Leverrier (1845)—the same who predicted the existence of the planet Neptune from calculations based on the perturbations—first calculated this deviation, and it is fully established. Yet it cannot be explained by the Newtonian attraction of the planetary bodies known to us. Hence recourse was taken to hypothetical masses whose attraction was to bring about the motion of Mercury's perihelion. Thus, for example, the zodiacal light, which is supposed to emanate from thinly distributed nebulous matter in the neighborhood of the sun, was brought into relation with this anomaly of Mercury. But this and numerous other hypotheses all suffer from the fault that they have been invented *ad hoc* and have been confirmed by no other observation.

The fact that the only definitely established deviation from Newton's law occurs in the case of Mercury, the planet nearest the sun, indicates that perhaps there is after all some fundamental defect in the law. For if one makes the plausible assumption that the deviations from Newton's law increase at the same or a higher rate as the force itself, they must be largest near the sun. Changes in the laws have been proposed, but they have been invented quite arbitrarily and can be tested by no other facts, and their correctness is not proved by accounting for the motion of Mercury's perihelion. If Newton's theory really requires a refinement we must demand that it emanate, without the introduction of arbitrary constants, from a principle that is superior to the existing doctrine in generality and intrinsic probability.

Einstein was the first to succeed in doing this. We shall turn in the last chapter to his explanation of the motion of Mercury's perihelion.

5. The Relativity Principle of Classical Mechanics

In discussing the great problems of the cosmos we have almost forgotten our point of departure—the earth. The laws of dynamics discovered by terrestrial experiments were transplanted to the astronomical space through which the earth rushes in its orbit about the sun with such great speed. How is it, then, that we notice so little of this journey through space? How is it that Galileo succeeded in finding laws on the moving earth which, according to Newton, were to be rigorously valid only in space absolutely at rest? We have called attention to this question already when mentioning Newton's views about space and time. We stated then that the apparently straight path of a sphere rolling on the table would, because of the rotation of the earth, in reality be slightly curved, for the path is straight not with respect to the moving earth but with respect to absolute space. The fact that we do not notice this curvature is due to the shortness of the path and of the time of observation, during which the earth has turned only slightly. Even if we admit this, we are still left with the motion of revolution about the sun, which proceeds with the immense speed of 30 km./sec. Why do we notice nothing of this?

This motion from the revolution is also, indeed, a rotation, and should be observable in terrestrial motions much as the rotation of the earth on its own axis, only much less so since the curvature of the earth's orbit is so very small. But in our question we do not mean the rotatory motion but the forward motion, which, in the course of a day, is practically rectilinear and uniform.

Actually, all mechanical events on the earth occur as if this tremendous forward motion did not exist, and this law holds quite generally for every system of bodies that performs a uniform and rectilinear motion through Newton's absolute space. This is called *the relativity principle of classical mechanics*, and it may be formulated in various ways. For the present, we shall state it as follows:

Relative to a coordinate system moving rectilinearly and uniformly through absolute space, the laws of mechanics have exactly the same expression as when referred to a coordinate system at rest in space.

To see the truth of this law we need only keep clear in our minds the fundamental law of mechanics, the law of impulsive forces, and the concepts that occur in it. We know that a blow produces a *change* of velocity. But such a change is quite independent of whether the velocities before and after the blow, v_1 and v_2, are referred to absolute space or to a system of reference which is itself moving with the constant velocity a. If the moving body is moving before the blow in space with the velocity $v_1=5$ cm./sec., then an observer moving with the velocity $a=2$ cm./sec. in the same direction would measure only the relative velocity $v_1'=v_1-a=5-2=3$ cm./sec. If the body now experiences a blow in the direction of motion which increases its velocity to $v_2=7$ cm./sec., then the moving observer would measure the final velocity as $v_2'=v_2-a=7-2=5$ cm./sec. Thus the change of velocity produced by the blow is $w=v_2-v_1=7-5=2$ cm./sec. in absolute space. On the other hand the moving observer notes the increase of velocity as

$$w' = v_2'-v_1' = (v_2-a)-(v_1-a)$$
$$= v_2-v_1 = w = 5-3 = 2 \text{ cm./sec.}$$

Both are of the same value.

Exactly the same holds for continuous forces and for the accelerations produced by them. For the acceleration b was defined as the ratio of the change of velocity w to the time required in changing it, and since w is independent of whatever rectilinear uniform forward motion (motion of translation) the system of reference used for the measurement has, the same holds for *b*.

The root of this law is clearly the law of inertia, according to which a motion of translation occurs when no forces act. A system of bodies all of which travel through space with the same constant velocity is, therefore, not only at rest as regards their mutual position, but also without actions of forces manifesting themselves on the bodies of the system in consequence of the motion. But if the bodies of the system exert forces on one another, the motions thereby produced will occur relatively just as if the common motion of translation were not taking place. Thus, for an observer moving with the system, it would not be distinguishable from one at rest.

The experience, repeated daily and thousands of times, that we observe nothing of the translatory motion of the earth is an impressive proof of this law. But the same fact is seen in motions on the earth. For when a motion on the earth is rectilinear and uniform with respect to the earth, it is so also with respect to space, if we disregard the rotation in the earth's motion. Everyone knows that in a ship or a railway carriage moving uniformly mechanical events occur in the same way as on the earth (considered at rest). On a moving ship, for example, a stone falls vertically: it falls along a vertical that is moving with the ship. If the ship were to move quite uniformly and without jerks of any kind the passengers would notice nothing of the motion so long as they did not observe the apparent movement of the surroundings.

6. Limited Absolute Space

The law of the relativity of mechanical events is the starting point of all our later arguments. Its importance lies in the fact that it is intimately connected with Newton's views concerning absolute space and essentially limits the physical reality of this concept from the outset.

We gave as the reason for assuming absolute space and absolute time the statement that without it the law of inertia would be meaningless. We must now enter into the question of how far these concepts deserve the term "real" in the sense in which it is used in physics. A concept refers to a physical reality only when there is something ascertainable by measurement corresponding to it in the world of phenomena. This is not the place to enter into a discussion on the philosophic concept of reality; it is at least certain that the criterion of reality just given corresponds fully with the way the word "reality" is used in the physical sciences. Every concept that does not satisfy it has gradually been eliminated from the structure of physics.

We see at once that in this sense a "fixed spot" in Newton's absolute space has no (physical) reality. This follows from the principle of relativity. Given that we had somehow arrived at the assumption that a definite system of reference is at rest in space, then a system of reference moving uniformly and rectilinearly with respect to it may with equal right be regarded as at rest. The mechanical events in both occur in the same manner, and neither system enjoys preference over the other. A definite body that seems at rest in the one system performs a uniform rectilinear motion seen from the other system, and if anyone were to assert that this body marks a spot in absolute space, another might with equal right challenge this and declare the body to be moving.

In this way the absolute space of Newton loses a considerable part of its weird existence. A space in which there is no place that can be marked by any physical means whatsoever is at any rate a very subtle and abstract idea, and not simply a box into which material things are crammed.

We must now also alter the terms used in our definition of the principle of relativity, for in it we still spoke of a coordinate system at rest in absolute space, and this is clearly without sense physically. To arrive at a definite formulation, we therefore introduce the concept of *inertial system*, which is taken to signify a coordinate system in which the law of inertia holds in its original form. There is not only the one system at rest as in Newton's absolute space but an infinite number of others all equally justified, and since we cannot very well speak of several "spaces" moving with respect to one another, we prefer to avoid the word "space" as much as possible. The principle of relativity then assumes the following form:

There are an infinite number of equivalent systems, inertial systems, executing a motion of translation (rectilinear and uniform) with respect to one another, in which the laws of mechanics hold in their simple classical form.

We here see clearly how intimately the problem of space is connected with mechanics. It is not space that is there and that impresses its form on things, but the things and their physical laws that determine space. We shall find later how this view gains more and more ground until it reaches its climax in the general theory of relativity of Einstein.

7. Galileo Transformations

Although the laws of mechanics are the same in all inertial systems, it does not of course follow that coordinates and velocities of bodies with respect to two inertial systems in relative motion are equal. If, for example, a body is at rest in a system S, then it has a constant velocity with respect to the other system S', moving relative to S. The general laws of mechanics contain only the accelerations, and these, as we saw, are the same for all inertial systems. This is not true of the coordinates and the velocities.

Hence the problem arises to find the position and the velocity of a body in an inertial system S' when they are given for another inertial system S.

The question is how to pass from one coordinate system to another one moving relative to the former. We must at this stage interpose a few remarks about equivalent (equally acceptable) coordinate systems in general and about the laws, the so-called *transformation equations*, that allow us to pass from one to the other by calculation.

In geometry coordinate systems are a means of conveniently fixing the relative positions of one body with respect to another. For this we suppose the coordinate system to be rigidly connected to one of the bodies. Then the coordinates of the points of the other body determine the relative position completely. It is, of course, immaterial whether the coordinate system chosen is rectangular, oblique, polar, or still more general. It is also immaterial how it is orientated with respect to the first body, except that either this orientation must be maintained, or, if it is changed, how the coordinate system alters its position with respect to the body must be specified. If, for example, we work with rectangular coordinates in a plane, then in place of the system S first chosen we may select a second, S', which is displaced (Fig. 36) or turned (Fig. 37) with

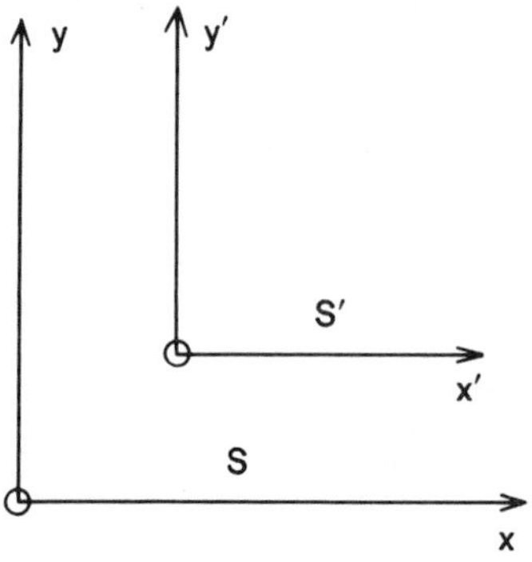

Fig. 36 *Two frames of reference* S *and* S′ *which are displaced relative to each other.*

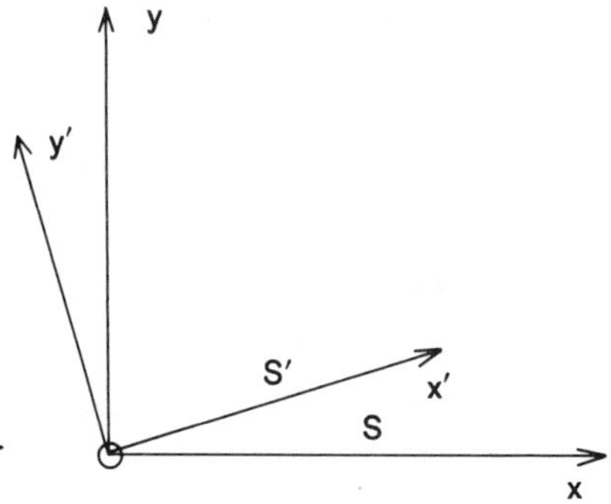

Fig. 37 *Two frames of reference* S *and* S′ *which are turned relative to each other.*

respect to S. But we must specify exactly the amount of the displacement and the turning. From these data we can then calculate what the coordinates of a point P that had the values x, y in the old system S are in the new system S'. If we call them x', y' we get formulae that allow us to calculate x', y' from x, y. We shall do this for the simplest case, namely, that in which the system S' arises from S as the result of a parallel displacement by the amount a in the x-direction (Fig. 38). Then clearly the new coordinate x' of a point P

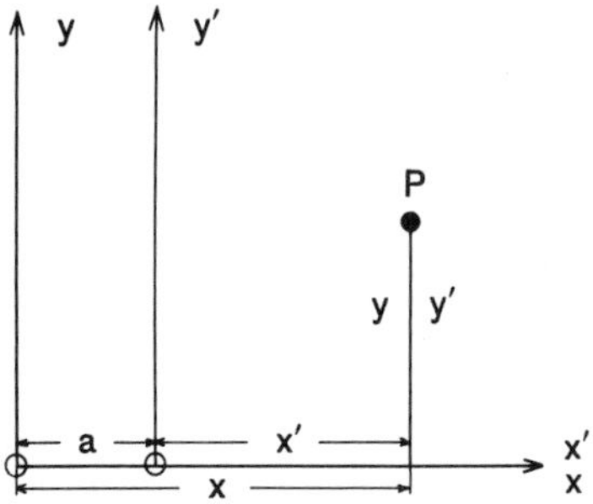

Fig. 38 S′ *is displaced by a distance* a *in the* x*-direction. The point* P *has the coordinates* x, y *in* S *and* $x'=x-a$, $y'=y$ *in* S′.

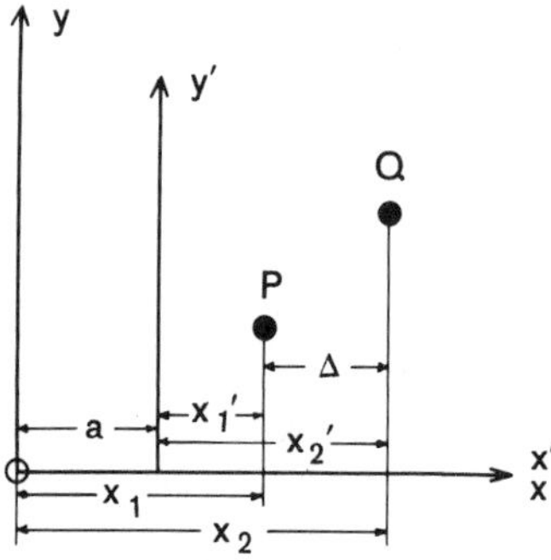

Fig. 39 *The expression for the distance* Δ *between* P *and* Q *measured along the* x*-axis is equal in both systems:* $\Delta = x_2 - x_1 = x_2' - x_1'$.

will be equal to its old x diminished by the displacement a, whereas the y-coordinate remains unaltered. Thus we have

$$x' = x-a, \qquad y' = y. \tag{27}$$

Similar, but more complicated, transformation formulae hold in other cases. We shall later have to discuss these more fully. It is important to recognize that one can find quantities for which the expressions in different coordinate systems are the same. Such a quantity is said to be *invariant* with respect to the coordinate transformation connecting the two coordinate systems. Let us consider, as an example, the transformation (27) above, that expresses a displacement along the x-axis. It is clear that the difference of the x-coordinates of two points P and Q, Δ, does not change. As a matter of fact (Fig. 39),

$$\Delta = x_2'-x_1' = (x_2-a)-(x_1-a) = x_2-x_1.$$

If the two coordinate systems S and S' are inclined to each other, then the distance s between two points P and Q is an invariant

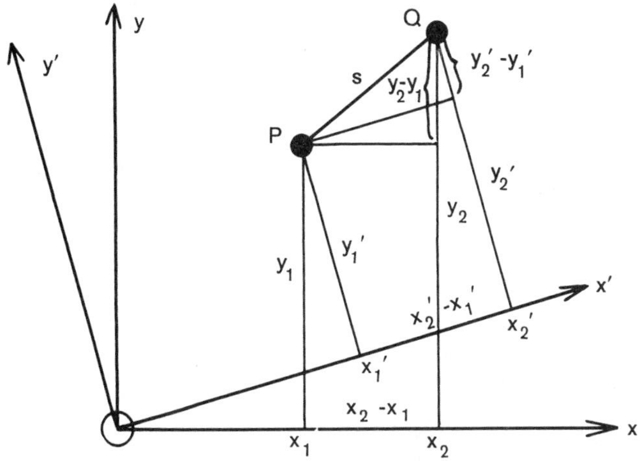

Fig. 40 *The expression for the distance* s *between* P *and* Q *is the same in both frames of reference:* $s^2 = (x_2-x_1)^2+(y_2-y_1)^2$.

(Fig. 40). It has the same expression in both systems, for, by Pythagoras' theorem, we have

$$s^2 = (x_2'-x_1')^2+(y_2'-y_1')^2 = (x_2-x_1)^2+(y_2-y_1)^2. \tag{28}$$

In the more general case, in which the coordinate system is simultaneously displaced and turned, the distance P, Q of two points again remains invariant. Invariants are particularly important because they represent geometrical relations without reference to the accidental choice of the coordinate system. They will play a considerable part later on.

If we now return after this geometrical digression to our starting point, we have to answer the question: What are the transformation laws that allow us to pass from one inertial system to another?

We defined the inertial system as a coordinate system in which the law of inertia holds. Only the state of motion is important in this connection—namely, that there is no acceleration with respect to absolute space, whereas the nature and position of the coordinate system is unessential. If we choose it to be rectangular, as is generally done, its position still remains free. We may take a displaced or a rotated system, but it must have the same state of motion. We have already used the term *system of reference* whenever we were concerned with the state of motion (and not with the nature and position of the coordinate system), and we shall use this expression systematically from now on.

If an inertial system S' is moving rectilinearly with respect to S with the velocity v, we may choose rectangular coordinates in both systems of reference such that the direction of motion becomes the x- and x'-axis, respectively. Further, we may assume that at the time $t=0$ the origins of both systems coincide. Then, in the time t the origin of the S'-system will have been displaced by the amount $a=vt$ in the x-direction: thus at this moment the two systems are exactly in the position that was treated above from a purely geometrical standpoint. Hence the equations (27) hold, in which a is now to be set equal to vt. Consequently, we get the transformation equations

$$x' = x-vt, \qquad y' = y, \qquad z' = z, \tag{29}$$

where we have added the unchanged y- and z-coordinates. This law is called a *Galileo transformation* in honor of the founder of mechanics.

We may also enunciate the *principle of relativity* as follows:

The laws of mechanics are invariant with respect to Galileo transformations.

This is due to the fact that accelerations are invariant, as we have already seen by considering the change of velocity of a moving body with respect to two inertial systems.

We showed earlier that the theory of motion, or kinematics, may be regarded as a geometry in four-dimensional $xyzt$-space, the "world" of Minkowski. In this connection it is not without interest to consider what the inertial systems and the Galileo transformations signify in this four-dimensional geometry. This is by no means difficult, for the y- and the z-coordinate do not enter into the transformation at all. It is thus sufficient to operate in the xt-plane.

We represent our inertial system S by a rectangular xt-coordinate system (Fig. 41). A second inertial system S' then corresponds to

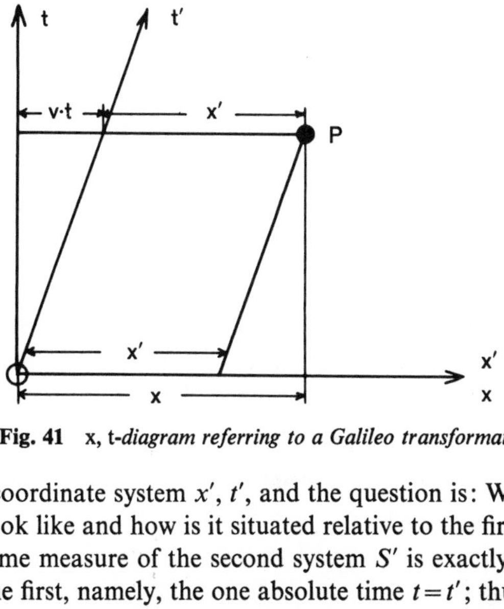

Fig. 41 x, t-*diagram referring to a Galileo transformation.*

another coordinate system x', t', and the question is: What does the second look like and how is it situated relative to the first? First of all, the time measure of the second system S' is exactly the same as that of the first, namely, the one absolute time $t = t'$; thus the x-axis, on which $t = 0$, coincides with the x'-axis, $t' = 0$. Consequently the system S' can only be an oblique coordinate system. The t'-axis is the world line of the point $x' = 0$, that is, of the origin of the system S'. The x-coordinate of this point which moves with the velocity v relative to the system S is equal to vt in this system at the time t. For any world point P the figure then gives the formula of the Galileo transformation $x' = x - vt$.

Corresponding to any other inertial system there is another oblique xt-coordinate system with the same x-axis, but a differently inclined t-axis. The rectangular system from which we started has no favored position among all these oblique systems. The unit of time is marked on all the t-axes of the various coordinate systems by the *same* parallel to the x-axis. This is in a certain sense the "calibration curve" of the xt-plane with respect to the time.

We summarize the result in the sentence:

In the xt-*plane the choice of the direction of the* t-*axis is quite arbitrary; in every* xt-*coordinate system having the same* x-*axis the fundamental laws of mechanics hold.*

From the geometric point of view this manifold of equivalent coordinate systems is extremely singular and unusual. The fixed position or the invariance of the x-axis is particularly remarkable. When we operate in geometry with oblique coordinates there is usually no reason for keeping the position of one axis fixed. But this is required by Newton's fundamental law of absolute time. All events which occur simultaneously, that is, for the same value of t, are represented by a parallel to the x-axis, since, according to Newton, time flows "absolutely and without reference to any object whatsoever."

We shall see that this unsymmetrical behavior of the world coordinates x and t, here only mentioned as a defect in mathematical perfection, is actually nonexistent. Einstein has eliminated it through his relativization of the concept of time.

8. Inertial Forces

Having recognized that the individual points in Newton's absolute space have no physical reality, we must now inquire what remains of this concept at all. There remains this: the resistance of all bodies to accelerations must be interpreted in Newton's sense as the action of absolute space. The locomotive that sets the train in motion must overcome the inertial resistance. The shell that demolishes a wall draws its destructive power from inertia. Inertial actions arise wherever accelerations occur, and these are nothing more than changes of velocity in absolute space; we may use the latter expression, for a change of velocity has the same value in all inertial systems. Systems of reference that are themselves accelerated with respect to inertial systems are thus *not* equivalent to the latter, or equivalent to each other. We can, of course, also refer the laws of mechanics to them, but they then assume a new and more complicated form. Even the path of a body left to itself is no longer uniform and rectilinear in an accelerated system (see III, 1, p. 54). This may also be expressed by saying that in an accelerated system *apparent forces*, *inertial forces*, act besides the true forces. A body on which no true forces act is yet subject to these inertial forces, and its motion is therefore in general neither uniform nor rectilinear. For example, a vehicle being set into motion or stopped is such an accelerated system. Everyone is familiar with the jerk a train makes starting or stopping, and this is nothing other than the inertial force of which we have spoken.

We shall consider the phenomena in detail for a system S moving rectilinearly with an acceleration denoted by κ. If we now measure the acceleration b of a body with respect to this moving system S, then the acceleration with respect to absolute space is obviously greater by κ. Hence the fundamental dynamical law with respect to space is

$$m(b+\kappa) = K.$$

If we write this in the form

$$mb = K-m\kappa,$$

we may say that in the accelerated system S a law of motion of Newtonian form, namely,

$$mb = K',$$

again holds, except that now we must write for the force K' the sum

$$K' = K-m\kappa,$$

where K is the true, and $-m\kappa$ the apparent, or inertial, force.

Now, if there is no true force acting, that is, if $K=0$, then the total force becomes equal to the force of inertia,

$$K' = -m\kappa. \tag{30}$$

Thus this force acts on a body left to itself. We may illustrate its action by the following considerations: We know that gravitation on the earth, the force of gravity, is determined by the formula $G=mg$, where g is the constant acceleration due to gravity. The force of inertia $K' = -m\kappa$ thus acts exactly like weight or gravity; the minus sign denotes that the force of acceleration is in a direction opposite to the motion of the system of reference S used as a basis. The value of the apparent gravitational acceleration κ is equal to the acceleration of the system of reference S. Thus the motion of a body left to itself in the system S is simply a motion such as that due to falling or being thrown.

This relationship between the inertial forces in accelerated systems and the force of gravity still appears quite fortuitous here. It actually remained unnoticed for two hundred years. But even at this stage we must state that it forms the basis of Einstein's general theory of relativity.

9. Centrifugal Forces and Absolute Space

In Newton's view the occurrence of inertial forces in accelerated systems proves the existence of absolute space or, rather, the favored position of inertial systems. Inertial forces may be seen particularly clearly in rotating systems of reference in the form of centrifugal forces. It was from them that Newton drew the main support for his doctrine of absolute space. Let us give the substance of his own words*:

The Effects which distinguish absolute from relative motion are, the forces of receding from the axe of circular motion. For there are no such forces in a circular motion purely relative, but in a true and absolute circular motion, they are greater or less, according to the quantity of the motion. If a vessel, hung by a long cord, is so often turned about that the cord is strongly twisted, then fill'd with water, and held at rest together with the water; after by the sudden action of another force, it is whirl'd about the contrary way, and while the cord is untwisting itself, the vessel continues for some time in this motion; the surface of the water will at first be plain, as before the vessel began to move: but the vessel, by gradually communicating its motion to the water, will make it begin sensibly to revolve, and recede by little and little from the middle, and ascend to the sides of the vessel, forming itself into a concave figure (as I have experienced) . . .

At first, when the relative motion of the water in the vessel was greatest it produc'd no endeavour to recede from the axe: the water shew'd no

* Andrew Motte's translation, 1729.

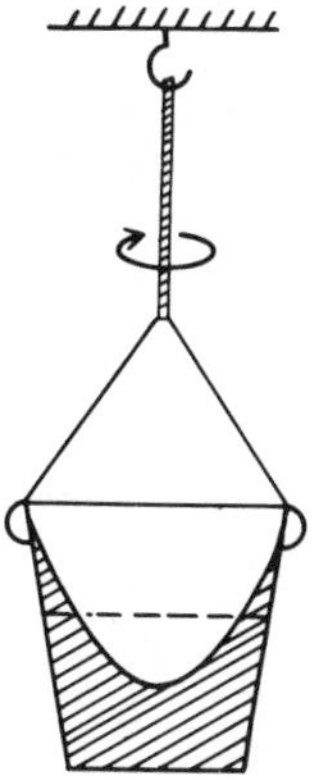

Fig. 42 *Newton's vessel experiment. Centrifugal forces drive the liquid to the wall of the vessel.*

tendency to the circumference, nor any ascent towards the sides of the vessel, but remain'd of a plain surface, and therefore its True circular motion had not yet begun. But afterwards, when the relative motion of the water had decreas'd, the ascent thereof towards the sides of the vessel prov'd its endeavour to recede from the axe; and this endeavour shew'd the real circular motion of the water perpetually increasing, till it had acquir'd its greatest quantity, when the water rested relatively in the vessel . . .

It is indeed a matter of great difficulty to discover, and effectually to distinguish, the True motions of particular bodies from the Apparent:

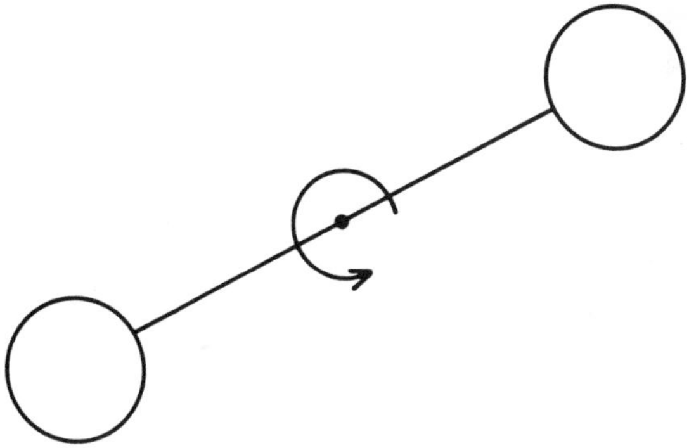

Fig. 43 *Rotation of two masses connected by a thread. The tension of the thread shows the centrifugal forces or "rotation in absolute space."*

because the parts of that immoveable space in which those motions are perform'd, do by no means come under the observations of our senses. Yet the thing is not altogether desperate; for we have some arguments to guide us, partly from the apparent motions, which are the differences of the true motions; partly from the forces, which are the causes and effeçts of the true motions. For instance, if two globes kept at a given distance one from the other, by means of a cord that connects them, were revolv'd about their common centre of gravity; we might, from the tension of the cord, discover the endeavour of the globes to recede from the axe of their motion, and from thence we might compute the quantity of their circular motions. . . . And thus we might find both the quantity and the determination of this circular motion, ev'n in an immense vacuum, where there was nothing external or sensible with which the globes could be compared.

These words express most clearly the meaning of absolute space. We have only a few words of explanation to add to them.

Concerning, first, the quantitative conditions in the case of centrifugal forces, we can at once get a survey of them if we recall the magnitude and the direction of the acceleration in the case of circular motions. It was directed towards the center and, according to formula (4), it had the value $b=\frac{v^2}{r}$, where r denotes the circular radius and v, the velocity.

Now, if we have a rotating system of reference S that rotates once in time T, then the velocity of a point at the distance r from the axis (see formula (18)) is

$$v = \frac{2\pi r}{T},$$

hence the acceleration relative to the axis, which we denoted by κ, is

$$\kappa = \frac{4\pi^2 r}{T^2}.$$

Now, if a body has the acceleration b relative to S, its absolute acceleration is $b+\kappa$. Just as in the case of rectilinear accelerated motion above there results an apparent force of absolute value

$$m\kappa = m\frac{4\pi^2 r}{T^2} \tag{31}$$

which is directed away from the axis. It is the *centrifugal force.*

It is well known that the centrifugal force also plays a part in proving that the earth rotates (Fig. 44). It drives the masses away from the axis of rotation and because of this causes the flattening of the earth at the poles and also the decrease of gravity from the poles towards the equator. We became acquainted with the latter phenomenon (without going into its cause) above, when we were dealing with the choice of the unit of force (II, 15, p. 51). According to Newton it is a proof of the earth's rotation. The centrifugal force, acting outwards, acts against gravity and reduces the weight. The decrease of the acceleration g due to gravity has the value $\frac{4\pi^2 a}{T^2}$ at the equator, where a is the earth's radius. If we here insert for a

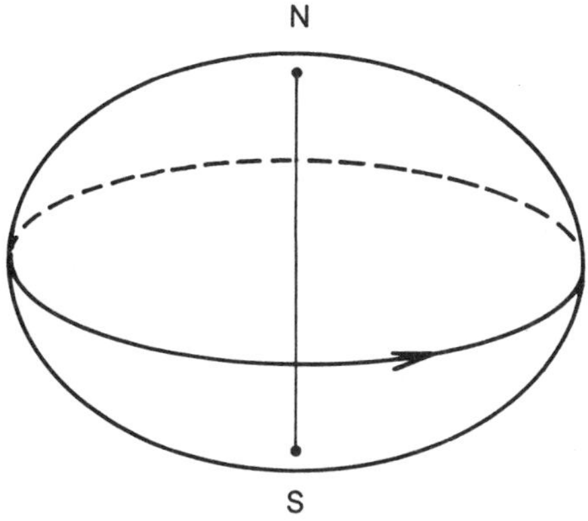

Fig. 44 *Schematic picture of flattening of the earth because of centrifugal forces due to its own rotation.*

the value given above (III, 3, (23)) $a = 6.37 \times 10^8$ cm., and for the time of rotation $T = 1$ day $= 24 \times 60 \times 60$ sec. $= 86{,}400$ sec., we get for the difference of the gravitational acceleration at the pole and at the equator the value 3.37 cm./sec.^{2}, which is relatively small compared with 981 cm./sec.^{2}; this value has to be increased slightly because of the flattening of the earth.

According to Newton's doctrine of absolute space, these phenomena are to be regarded not as resulting from motion relative to other masses, such as the fixed stars, but as resulting from absolute rotation in empty space. If the earth were at rest, and if, instead, the whole stellar system were to rotate in the opposite sense once around the earth's axis in twenty-four hours, then, according to Newton, the centrifugal forces would not occur. The earth would not be flattened and the gravitational force would be just as great at the equator as at the pole. The motion of the heavens, as viewed from the earth, would be exactly the same in both cases. And yet there would be a definite difference between them which could be observed.

The position is brought out perhaps still more clearly in Foucault's pendulum experiment (1850). According to the laws of Newtonian dynamics a pendulum swinging in a plane must permanently maintain its plane of vibration in absolute space if all deflecting forces are

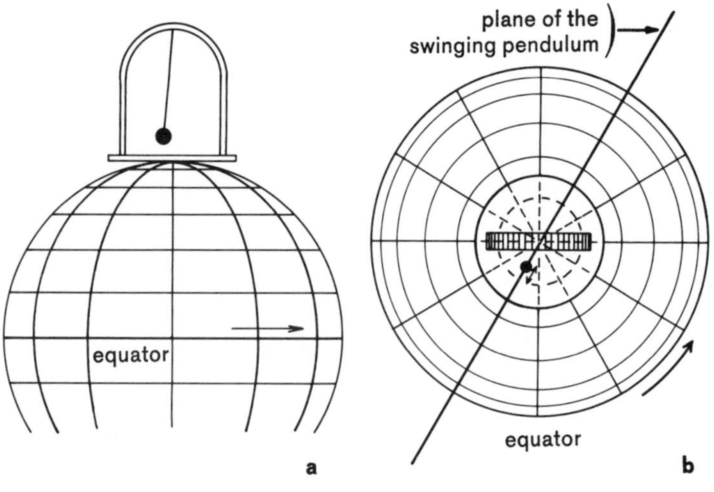

Fig. 45 *Foucault's pendulum at the North Pole. The direction of motion is unchanged during earth's rotation.*

excluded. If the pendulum is suspended at the North Pole, the earth rotates, as it were, below it (Fig. 45*a*, *b*). Thus the observer on the earth sees a rotation of the plane of oscillation in the reverse sense. If the earth were at rest but the stellar system in rotation, then, according to Newton, the position of the plane of oscillation should not alter with respect to the earth. The fact that it does so appears again to prove the *absolute* rotation of the earth.

We shall consider a further example—the motion of the moon about the earth (Fig. 46). According to Newton the moon would fall onto the earth if it had not an absolute rotation about the latter. Let us imagine a coordinate system with its origin at the center of the earth, the xy-plane containing the moon's orbit, and the x-axis always passing through the moon. If this system were absolutely

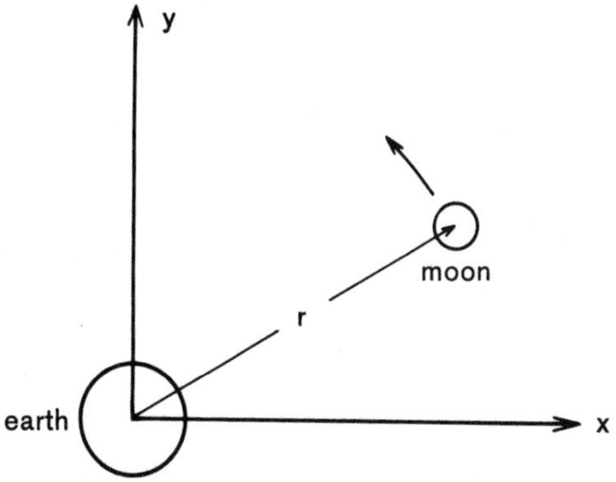

Fig. 46 *The gravitational forces from the earth on the moon are exactly compensated by centrifugal forces due to its motion around the earth.*

at rest, then the moon would be acted on only by the gravitational force directed towards the center of the earth, which, by formula (26) (p. 64), has the value

$$K = k\frac{Mm}{r^2}.$$

Thus it would fall to the earth along the x-axis. The fact that it does not do so apparently proves the absolute rotation of the coordinate system x, y. For this rotation produces a centrifugal force that keeps equilibrium with the force K, and we get

$$\frac{mv^2}{r} = k\frac{Mm}{r^2}.$$

This formula is, of course, nothing other than Kepler's third law. For if we cancel the mass m of the moon on both sides and express v by the period of revolution T, $v = \frac{2\pi r}{T}$ we get

$$\frac{4\pi^2 r}{T^2} = \frac{kM}{r^2}$$

or, by (25) (p. 63),

$$\frac{r^3}{T^2} = \frac{kM}{4\pi^2} = C.$$

A corresponding result holds, of course, for the rotation of the planets about the sun.

These and many other examples show that Newton's doctrine of absolute space rests on very concrete facts. If we run through the sequence of arguments again, we see the following:

The example of the rotating glass of water shows that the relative rotation of the water with respect to the glass is not responsible for the occurrence of centrifugal forces. It might be that greater masses in the neighborhood, say the whole earth, are the cause. The flattening of the earth, the decrease of gravity at the equator, Foucault's pendulum experiment, all show that the cause is to be sought outside the earth. But the orbits of all moons and planets likewise exist only through the centrifugal force that maintains equilibrium with gravitation. Finally, we notice the same phenomena in the case of the farthermost double stars, the light from which takes thousands of years to reach us. Thus it seems as if the occurrence of centrifugal forces is universal and cannot be due to interactions. Hence nothing remains for us but to assume absolute space as their cause.

Such modes of reasoning have been common and regarded as valid since the time of Newton, with only a few thinkers opposing them. Foremost among those who questioned the validity of these arguments was Ernst Mach. In his critical account of mechanics he has analyzed the Newtonian concepts and tested their logical bases. He starts from the view that mechanical experience can never teach us anything about absolute space. Only relative positions and relative motions may be ascertained and hence only they are physically real. Newton's proofs of the existence of absolute space, therefore, must be illusory. As a matter of fact, everything depends on whether it is admitted that if the whole stellar system were to rotate about the earth no flattening and no decrease of gravity at the equator would occur. Mach asserts rightly that such statements go far beyond possible experience. He reproaches Newton for having become untrue to his principle of allowing only verifiable facts to be considered valid. Mach himself has sought to free mechanics from this blemish. He was of the opinion that inertial forces would have to be regarded as actions of the whole mass of the universe, and sketched the outlines of an altered system of dynamics in which only relative quantities occurred. Yet his attempt could not succeed. In the first place, the importance of the relation between inertia and gravitation that expresses itself in the proportionality of weight to mass escaped him. In the second place, he was unacquainted with the relativity theory of optical and electromagnetic phenomena which eliminated the prejudice in favor of absolute time. A knowledge of both of these facts was necessary to build up the new mechanics, and the discovery of both was the achievement of Einstein.

CHAPTER IV

THE FUNDAMENTAL LAWS OF OPTICS

1. The Ether

Mechanics is both historically and logically the foundation of physics, but it is nevertheless only a part of it, and indeed a small part. Hitherto, to solve the problem of space and time, we have made use only of mechanical observations and theories. We must now inquire what the other branches of physical research teach us about the problem.

It is, above all, the realms of optics, of electricity, and of magnetism that are connected with the problem of space; this is because light and the electric and magnetic forces traverse empty space. Vessels from which the air has been pumped are completely transparent for light, no matter how high the vacuum. Electric and magnetic forces, too, act across such a vacuum. The light of the sun and the stars reaches us after its passage through empty space.

The fact that certain physical events propagate themselves through astronomic space led long ago to the hypothesis that space is not empty but is filled with an extremely fine imponderable substance, the ether, which is the carrier or medium of these phenomena. In so far as this concept of ether is still used nowadays, it is taken to mean nothing more than certain physical states or "fields" in empty space. If we were to adopt this abstract concept from the very outset, most of the problems that are historically connected with the ether would remain unintelligible. The earlier ether was indeed regarded as a real substance, not only endowed with physical states but also capable of motion.

We shall now describe the development, first, of the principles of optics, and, second, of those of electrodynamics. This will cause us to digress a little from the problem of space and time, but will allow us to take it up again later fortified with new facts and laws.

2. The Corpuscular and the Undulatory Theory

I say therefore that semblances of things
And tenuous shapes are thrown off from their surface . . .
Therefore it must needs be that in like manner
Idols can course through inexpressible space
In a moment of time . . .
. . . but because we can perceive them
With our eyes only, therefore it comes to pass
That to whatever side we turn our sight,
There all things strike it with their shape and colour.*

That is what we read in Lucretius' poem *On the Nature of Things* (Book IV), that poetic guide to Epicurean philosophy written in the first century B.C. The lines quoted contain a sort of corpuscular theory of light which is elaborated by the powerful imagination of the poet but at the same time developed in a true scientific spirit. Yet we can no more call these verses a scientific doctrine than we can other ancient speculations about light. There is no sign of an attempt to determine the phenomena quantitatively, the first characteristic of objective effort. In fact, it is particularly difficult to dissociate the subjective sensation of light from the physical phenomenon and to render it measurable.

The science of optics may be dated from the time of Descartes. His *Dioptrics* (1638) contains the fundamental laws of the propagation of light: the laws of reflection and refraction. The former was already known to the ancients, and the latter had been found experimentally shortly before by Snell (about 1618). Descartes evolved the idea of the ether as the carrier of light, an idea that was the precursor of the *undulatory theory*. It was hinted at by Robert Hooke (1667), and clearly formulated by Christiaan Huygens (1678). Their great contemporary, Newton, who was somewhat younger, is regarded as the author of the opposing doctrine, the *corpuscular theory*. Before discussing the struggle between these theories we shall explain the essence of each in rough outline.

The *corpuscular theory* asserts that luminescent bodies send out fine particles that move in accordance with the laws of mechanics and produce the sensation of light when they strike the eye. The *undulatory*, or *wave theory*, on the other hand, sets up an analogy

* From the translation by R. C. Trevelyan, Cambridge University Press, 1937.

between the propagation of light and the motion of waves on the surface of water or sound waves in air. For this purpose it has to assume the existence of an elastic medium that permeates all transparent bodies; this is the *luminiferous ether*. The individual particles of this substance merely oscillate about their positions of equilibrium. That which moves on as the light wave is the *state of motion* of the particles and not the particles themselves. Fig. 47 illustrates the process for a series of points that can vibrate up and down. Each horizontal line in the diagram corresponds to a moment of time, say, $t=0, 1, 2, 3 \ldots$ sec. Each individual point executes a vibration vertically. The points all taken together present the aspect of a wave that advances towards the right, from moment to moment.

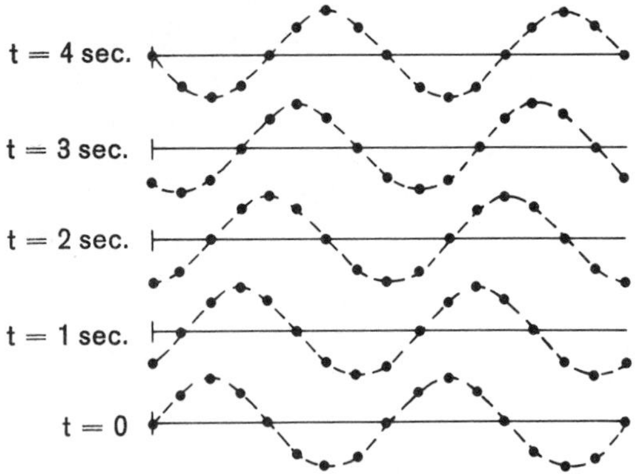

Fig. 47 *A wave moving to the right.*

Now there is a significant objection to the wave theory. It is known that waves run around obstacles. It is easy to see this happening with waves on the surface of water, or again when sound waves "go around corners." But a ray of light travels in a straight line. If we interpose a sharp-edged opaque body in its path we get a shadow with a definite boundary.

This fact moved Newton to discard the wave theory. He did not himself decide in favor of a definite hypothesis, but merely established that light is something that moves away from a luminescent body "like ejected particles." His successors, however, interpreted his opinion as being in favor of the corpuscular theory, and the authority of his name gained acceptance for this theory for a whole century.

Yet, at that time Grimaldi had already discovered (the result was published posthumously in 1665) that light can also "bend round corners." At the edges of sharp shadows a weak illumination in the form of alternating bright and dark stripes or fringes is seen; this phenomenon is called the *diffraction* of light. It was this discovery in particular that made Huygens a zealous pioneer of the wave theory. He regarded as the first and most important argument in favor of it the fact that two rays of light cross each other without interfering with each other just as two trains of water waves do, whereas between bundles of emitted particles, collisions, or at least disturbances of some kind would necessarily occur. Huygens succeeded in explaining the reflection and the refraction of light on the basis of the wave

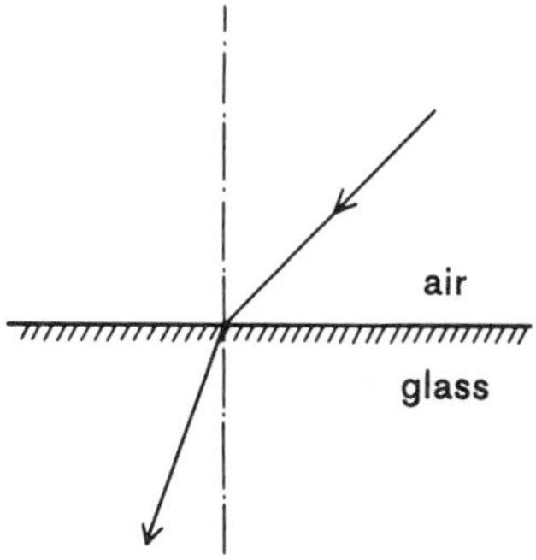

Fig. 48 *A ray of light changes its direction when passing from air to glass.*

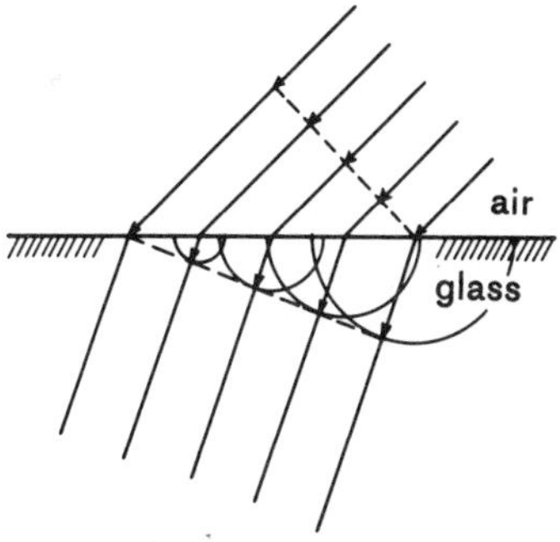

Fig. 49 *The refraction of a ray of light passing from air into glass explained by wave theory.*

theory. He made use of the principle, now called after his name, that every point on which the light impinges is to be regarded as the source of a new spherical wave of light. This resulted in a fundamental difference between the corpuscular and the wave theory, a difference that later led to the final experimental decision in favor of the latter.

It is known that a ray of light which passes through the air and strikes the plane bounding surface of a denser body such as glass or water is bent or refracted so that it is more steeply inclined to the bounding surface (Fig. 48). The corpuscular theory accounts for this by assuming that the corpuscles of light experience an attraction from the denser medium at the moment they enter into it. Thus they are accelerated by an impulse perpendicular to the bounding surface and hence deflected towards the normal. It follows from this that they must move more rapidly in the denser than in the less dense medium. Huygens's construction on the wave theory depends on just the opposite assumption (Fig. 49). When the light wave strikes the bounding surface it excites elementary waves at every point. If these are transmitted more slowly in the second, denser medium, then the plane that touches all these spherical waves, and represents the refracted wave according to Huygens, is deflected in the right sense.

Fig. 50 *Chain of mass points in equilibrium with distance* l.

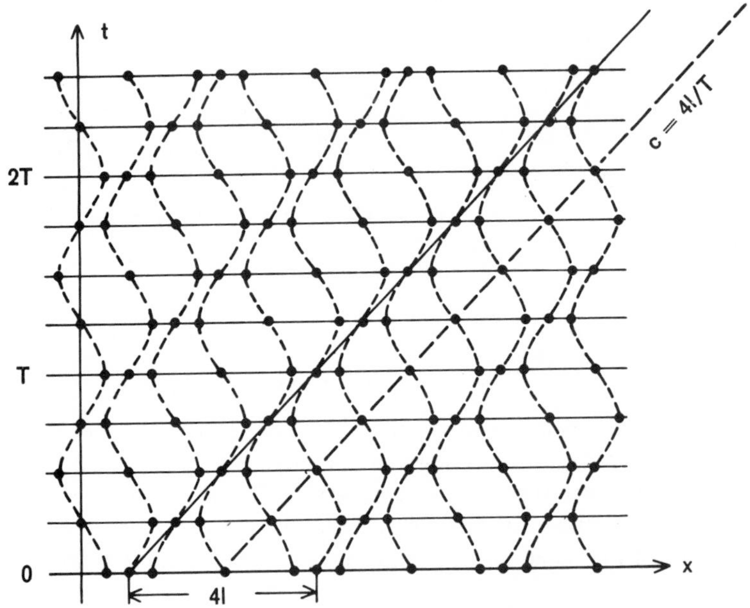

Fig. 51 *Longitudinal wavelike motion of the chain in Fig. 50. Each point moves periodically around its equilibrium position with period* T. *There is a shift in time between the oscillations of different points. The state of the chain, e.g., maximum (—) and minimum (- - -) of density, is moving to the right with velocity* $c = \frac{4\,l}{T}$.

Huygens also interpreted the *double refraction* of Iceland spar, discovered by Erasmus Bartholinus in 1669, on the basis of the wave theory, by assuming that light can propagate itself in the crystal with two different velocities in such a way that the one elementary wave is a sphere, the other a spheroid. Double refraction means that a ray of light entering, for example, a plate of fluorspar splits up into two rays. Huygens discovered that these two rays are different from each other and also from natural light. We can demonstrate this using another plate of spar. If one ray coming from the first plate strikes the second plate perpendicularly, then two rays emerge. The intensity of these two rays varies when the second crystal is turned about the direction of the incoming ray. In a certain position the intensity of one ray may even be zero (no double refraction). So the rays split up by double refraction show marked directional properties not observed with natural light. Newton remarked (1717) that not all directions about a ray of light are equivalent. He interpreted this as evidence against the wave theory, for at that time only waves of compression and rarefaction (like sound waves) were thought of, in which the particles swing "longitudinally" in the direction of propagation of the wave (Figs. 50 and 51). In this case it is clear that no direction perpendicular to the propagation is preferred.

3. The Velocity of Light

The determination of the most important property of light and the one which will form the nucleus of our following reflections, namely, the *velocity of light*, was made independently of the controversy between the two hypotheses about the nature of light. The fact that the velocity was very great was clear from all observations about the propagation of light. Galileo had endeavored (1607) to measure it with the aid of lantern signals, but without success, for light traverses earthly distances in extremely short fractions of time. Hence the measurement succeeded only when the enormous distances between heavenly bodies in astronomic space were used.

Olaf Römer (1676) was the first to calculate the velocity of light c from astronomical observations, from the eclipses of the satellites of Jupiter. Fig. 52 shows the situation for one eclipse. An eclipse

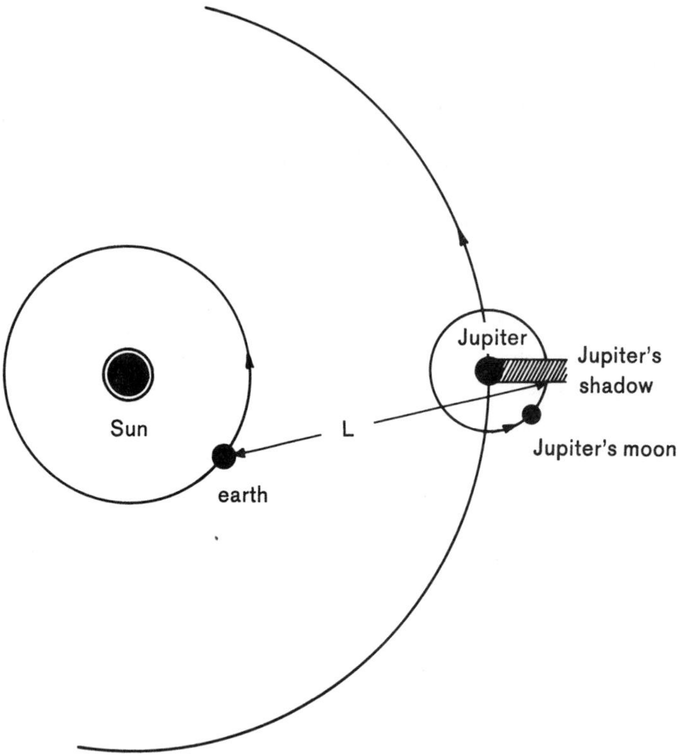

Fig. 52 L *is the distance between earth and the point where Jupiter's moon enters the shadow. Since light travels with velocity* c, *the eclipse is observed on earth with a time delay* $\frac{L}{c}$ *as compared with an observer on Jupiter's moon.*

occurs each time the moon enters the shadow of Jupiter; seen from Jupiter this happens at intervals equal to the time of revolution τ of the satellite. If L is the distance from Jupiter to the earth, these signals arrive on the earth the time L/c later, and if l is the change of L during the time τ of the moon's revolution, the terrestrial observer sees the eclipses in slowly changing time intervals $\tau + l/c$.

The times of revolution as observed on the earth are therefore longer or shorter than the real times (as observed on Jupiter) according to whether the distance L increases or decreases.

The time needed for n revolutions observed on earth is

$$t_n = n\tau + l_n/c,$$

where l_n is the total change of L during this time.

Now the two unknown quantities τ and c can be determined by two properly chosen observations. First, one counts the number N of eclipses during the time interval t_N in which the distance L between Jupiter and earth has become the same again; since Jupiter moves comparatively slowly, this will be about a year, the time of revolution of the earth in its orbit. Then $l_N = 0$ and $\tau = t_N/N$ (where t_N is approximately a year). Thus τ can be found.

Now one counts the number N' of eclipses during half a year, beginning with a position of nearest approach of earth and Jupiter. Then $l_{N'}$ is the diameter of the earth's orbit ($\sim 3 \times 10^8$ km.), and one has $t_{N'} = N'\tau + l_{N'}/c$ or

$$c = \frac{l_{N'}}{t_{N'} - N'\tau}.$$

The retardation time $t_{N'} - N'\tau$ has been found to be 17 min. $\sim$ 1000 sec. Therefore one obtains $c = 3 \times 10^8$ km./10^3 sec. = 300,000 km./sec. The exact value, which Römer approximated very closely, is

$$c = 299{,}793 \text{ km./sec.} \tag{32}$$

James Bradley discovered (1727) another effect of the finite velocity of light, namely, that all fixed stars appear to execute a common annual motion that is evidently a counterpart to the rotation of the earth around the sun. It is very easy to understand from the point of view of the emission theory how this effect comes about. We shall give this interpretation here, but we must remark that it is just this phenomenon that raises certain difficulties for the wave theory, about which we shall have much to say. We know (III, 7) that a motion which is rectilinear and uniform in our system of reference S is so also in another system S', if the motion of the latter is a translation with respect to S. But the magnitude and direction of the velocity is different in the two systems. It follows from this that a stream of light corpuscles coming from a fixed star and striking the earth, appears to come from another direction. We shall consider this deflection or *aberration* for the particular case when the light impinges perpendicular to the motion of the earth (Fig. 53). When the earth is at rest, the telescope must be directed towards the star (Fig. 53*a*). But if the earth has, for instance, a velocity v to the right, then in the position of Fig. 53*a* the star cannot be seen in the telescope, for the light corpuscle which reaches the objective strikes the wall of the telescope and not the eyepiece. To observe the star the telescope must be turned (Fig. 53*b*). Let the telescope on the objective of which the light particle impinges be in position 1. Now, while the light traverses the length l of the telescope in the time l/c, the earth and with it the telescope moves by a distance $v \times l/c$ into position 2. The light particle strikes the eyepiece only when $v \times l/c$

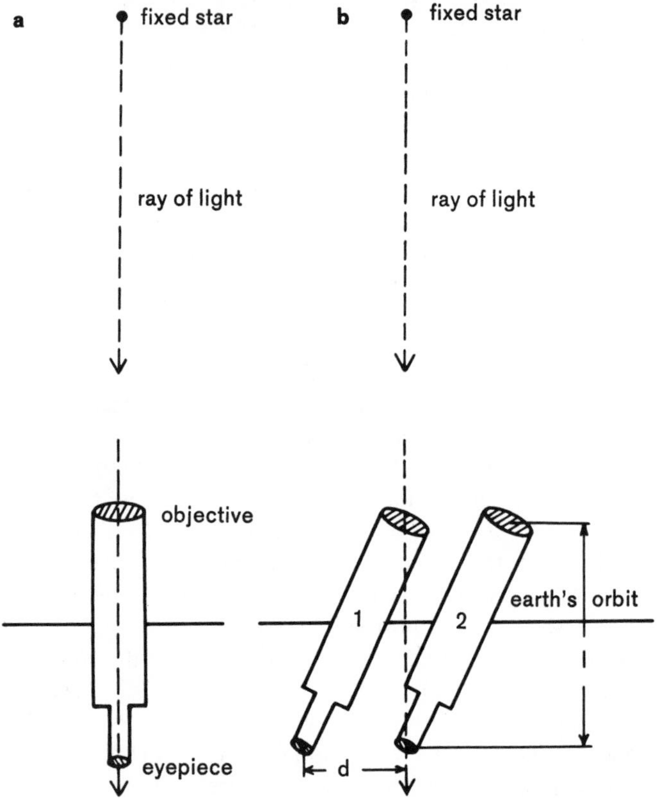

Fig. 53*a* *Observation of a fixed star on the earth at rest.*
b *Observation of a fixed star on the moving earth. The telescope has to be inclined so that the path of the light from the fixed star passes objective* and *eyepiece.*

equals the shift d of the telescope. Therefore we have

$$\frac{d}{l} = \frac{v}{c}.$$

The angle of the telescope's deflection is determined by the ratio v/c, and the direction of the telescope axis does not point to the true position of the star but to a point of the sky that is displaced in the direction of v.

The ratio v/c, also called the *aberration constant*, will be denoted by β:

$$\beta = \frac{v}{c}. \tag{33}$$

It has a very small numerical value, for the velocity of the earth in its orbit about the sun amounts to about $v = 30$ km./sec., whereas the velocity of light, as already mentioned, amounts to 300,000 km./sec. Hence β is of the order 1:10,000.

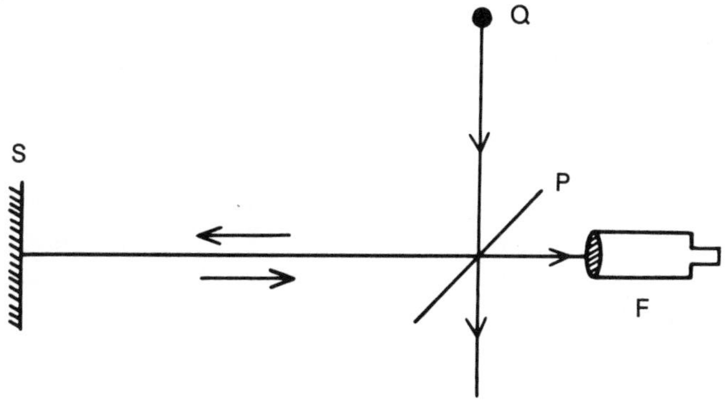

Fig. 54 *Apparatus to measure the velocity of light. A light ray emitted from* Q *meets a semitransparent mirror* P. *Part of the ray passes through, while another part is reflected to* S. *The ray to* S *is reflected a second time in* S, *passes through* P, *and is observed in* F.

Therefore, the apparent positions of all the fixed stars are always a little displaced in the direction of the earth's motion at that moment, and hence describe a small elliptical figure during the annual revolution of the earth around the sun. By measuring this ellipse the ratio β may be found, and since the velocity v of the earth in its orbit is known from astronomic data, the velocity of light c may be determined from it. The result agrees closely with Römer's measurement.

We shall next anticipate the historical course of events and give a short account of the terrestrial measurements of the velocity of light. To accomplish these it was necessary to find a technical device that allowed precise measurements of the extremely short times required for light to traverse earthly distances of a few kilometers or even only a few meters. Fizeau (1849) and Foucault (1865) used two different methods to carry out these measurements, and confirmed the numerical value of c found by the astronomical methods. We will not discuss the process in detail here. We shall merely call attention to one point: In both processes the ray of light is travelling from the source Q to a distant mirror S, where it is reflected and returns to its starting point (Fig. 54). It traverses the same path twice, and hence it is only the mean velocity during the motion to and fro that is measured. The following remark, which is important for later considerations, arises from this circumstance: If we suppose that the velocity of light is not the same in both directions, because the earth itself is in motion (we shall discuss this point later, in IV, 9, p. 128), then this influence will be wholly or partially cancelled in the motion to and fro. In view of the smallness of the velocity of the earth in comparison with that of light, we need take no account of the earth's motion in the measurements of the velocity of light.

These measurements were later repeated with improved apparatus, and a considerable degree of accuracy was obtained. Nowadays they can be carried out in a room of moderate length. The result is the value (32) given above. Foucault's method also made possible measurements of the velocity of light in water. It was found to be *smaller* than that obtained for air. Thus a definite decision on the most important point in the dispute between the corpuscular and the wave theory was reached in favor of the latter. This occurred, in fact, at a time when the triumph of the wave theory had already long been assured on other grounds.

4. Elements of the Wave Theory—Interference

Newton's greatest achievement in optics was the resolution, by means of a prism, of white light into its colored constituents, and the exact examination of the spectrum, which led him to the conviction that the individual spectral colors were the indivisible constituents of light. He is the founder of the theory of color, the physical content of which is still completely valid today—in spite of Goethe's attacks. The power of Newton's discoveries paralyzed the independence of thought of the succeeding generations. His refusal to accept the wave theory blocked the road to its acceptance for nearly a century. Nevertheless it found isolated supporters, such as, for example, the great mathematician Leonhard Euler in the eighteenth century.

The revival of the wave theory is due to the works of Thomas Young (1802), who adduced the principle of *interference* to explain the colored rings and fringes which Newton had already observed in thin layers of transparent substances. We shall at this stage deal in some detail with the phenomenon of interference because it plays a decisive part in all finer optical measurements, particularly in researches that constitute the foundations of the theory of relativity.

We explained the nature of waves before by saying that they are the result of individual particles of a body executing periodic oscillations about their positions of equilibrium; the instantaneous position of the phase of the motion is different for neighboring particles and moves forward with constant velocity. The time that a definite particle requires for one vibration to and fro is called the *time of vibration* or the *period* and is denoted by T. The number per second of vibrations, or the *frequency*, is designated by ν. Since the time of a vibration multiplied by their number per second must give exactly 1, we have $\nu T = 1$, thus

$$\nu = \frac{1}{T} \quad \text{or} \quad T = \frac{1}{\nu}. \tag{34}$$

Instead of vibration number or frequency we often say "color," because a light wave of a definite frequency produces a definite sensation of color in the eye. We shall not enter into the complicated question as to how "physical colors," as we may call the great manifold of psychological impressions of color, come about through the conjoined action of simple periodic vibrations. The waves that start from a small source of light have the form of spheres. Physically defined, this means that all particles on a sphere with the source as center, or on a "wave surface," are in the same state of vibration (e.g., all on a crest, or all in a trough): they have the same phase (Fig. 55). By means of refraction or other influences a part of such a spherical wave may be deformed so that the wave surfaces have some other shape. The simplest wave surface is evidently the plane, and it is clear that a sufficiently small piece of any arbitrary wave surface, particularly of a spherical surface, may always be regarded as approximately a plane. Hence we consider in particular the propagation of plane waves (Fig. 56). The direction perpendicular to the planes of the waves, that is, the normal to the waves, is at the same time the direction of propagation. It is clearly sufficient to consider the state of vibration along a straight line parallel to this direction.

Whether the vibration of the individual particle occurs parallel or perpendicular to the direction of propagation, whether it is longitudinal or transverse, will be left quite open at this stage. In the figures we shall simply draw wave lines and call the greatest displacements upwards and downwards crests and hollows.

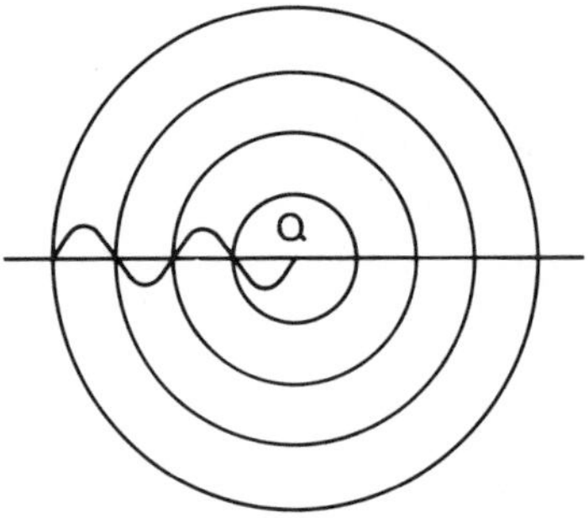

Fig. 55 *The phases of a spherical wave emitted from* Q. *Maxima, minima, and more generally points of equal phase are situated on spheres around* Q. *This phenomenon may be seen on a water surface when a stone is thrown into it.*

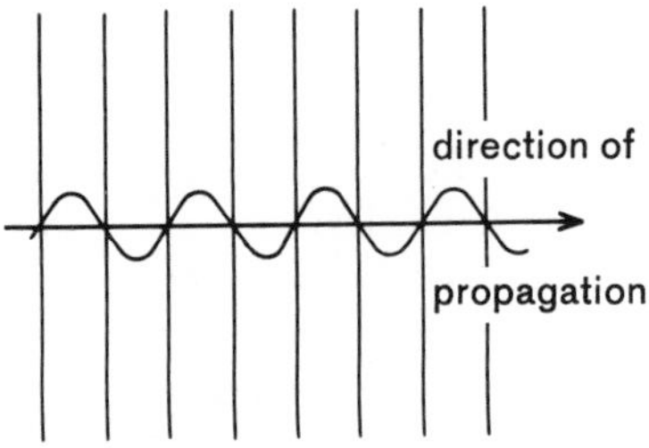

Fig. 56 *The phases of a plane wave. The phases are equal on planes perpendicular to the direction of propagation.*

The distance from one crest to the next is called *wave length* and is designated by λ. The distance between any two consecutive planes of equal phase obviously has the same value λ.

During a vibration of a definite particle to and fro, the duration of which is T, the whole wave moves forward exactly one wave length λ (Fig. 47 shows this for the first half of a period). Since the velocity in any motion is equal to the ratio of the path traversed to the time required to do so, the wave velocity c is equal to the ratio of the wave length to the time of vibration (see Fig. 51, where $\lambda = 4l$ and $c = 4l/T$).

$$c = \frac{\lambda}{T} \quad \text{or} \quad c = \lambda\nu. \tag{35}$$

If a wave passes from one medium into another, say, from air into glass, the time rhythm of the vibrations is carried over the bounding surface, that is, T (or ν) remains the same. On the other hand, the velocity c and hence, on account of formula (35), also the wave length λ change. Thus any method of measuring λ may serve to compare the velocity of light in various substances or under various circumstances. We shall make use of this fact later.

We are now in a position to understand the nature of interference phenomena, the discovery of which helped the wave theory to prevail. Interference may be described by the paradoxical statement: Light added to light does not necessarily give intensified light, but may lead to a weaker illumination, even to darkness.

The reason for this is that, according to the wave theory, light is not a stream of material particles but a state of motion. Two vibration impulses that occur together may, however, destroy the motion just as two people who wish to do contrary things impede

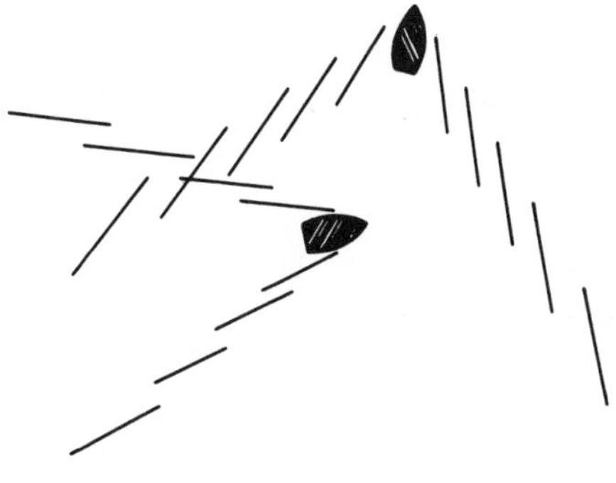

Fig. 57 *Penetration of two trains of waves produced by moving ships on a lake.*

each other and produce nothing. Let us imagine two trains of waves intersecting. This phenomenon can be conveniently observed if we look from a hillock down into a lake in which the waves caused by two ships meet (Fig. 57). These two wave systems interpenetrate without disturbing each other. In the region where both exist simultaneously a complicated motion arises, but as soon as the one wave has passed through the other, it moves on as if nothing had happened to it. If we fix our attention on one particular vibrating particle, we see that it experiences independent impulses from both waves. Hence its displacement at any point is simply the sum of the displacements that it would have under the influence of the individual waves. These two wave motions are said to superpose without disturbing each other. It follows that at points where crest and crest and also at points where hollow and hollow meet, where two equal phases encounter each other, the elevations and the depressions are twice as great (Fig. 58). But at points where crest and hollow meet, the displacements cancel each other (Fig. 59).

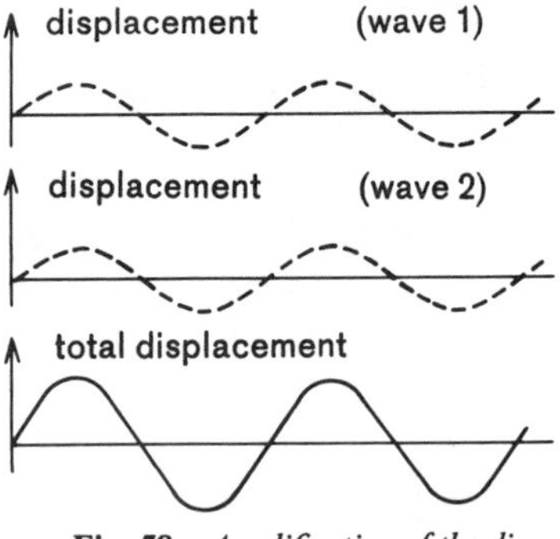

Fig. 58 *Amplification of the displacement when two waves with equal phases interfere.*

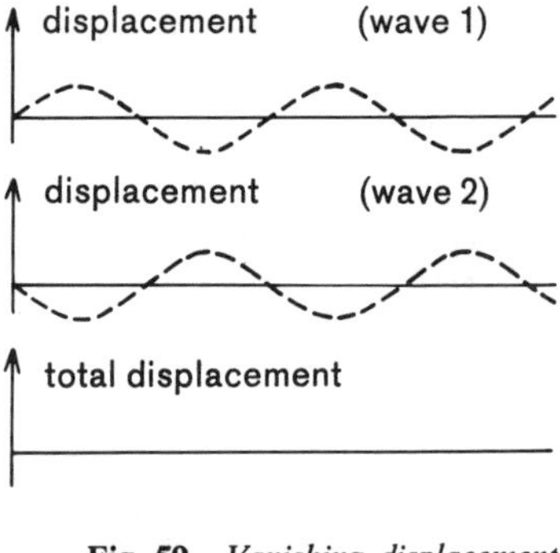

Fig. 59 *Vanishing displacement by interference of two waves of the same displacement which have opposite phases.*

If we wish to observe interference of light it will not do simply to take two sources of light and to allow the trains of waves emerging from them to interpenetrate. No observable interference phenomenon would occur, because the actual light waves are not absolutely regular. Rather, the state of vibration suddenly changes after a series of regular vibrations, corresponding to the accidental phenomena that occur at the source during the emission of light. These irregular changes effect a corresponding fluctuation of the interference phenomena much too quick for the eye to follow; hence we see only uniform illumination.

To obtain observable interferences we must resolve a ray of light by artificial means, by reflection or refraction, into two rays and afterwards make them come together again. Then we see that the irregularities of the vibrations in both rays occur in exactly the same time rhythm, and thus it follows that the interference phenomena do not fluctuate in space but remain fixed. Wherever the waves strengthen or cancel each other at a certain moment, they do so at every moment. If we direct the eye, armed with a magnifying glass or a telescope, at such a point, we see fringes or rings, provided we use light of one color (monochromatic light) such as is approximately emitted by a Bunsen flame colored yellow by common salt. In ordinary light, which is composed of many colors, the interference fringes corresponding to the various wave lengths do not exactly coincide. At one point red is intensified, say, and blue is extinguished; at other points other colors occur, and hence colored fringes arise. It would, however, take us away from our path of inquiry to pursue these interesting phenomena further.

The simplest arrangements for producing interferences were invented by Fresnel (1822), whose researches have furnished the foundation for the theory of light that has remained unassailed up to the present day. His time, the first decades of the nineteenth century, must in many respects have resembled our own. Just as nowadays, through the development of quantum theory and nuclear physics, our knowledge of physical nature is undergoing a process of deepening and enlargement which appears to be a complete revolution in the realm of physical law, so a hundred years ago thousands of individual observations, theoretical experiments, physical or metaphysical speculations coalesced for the first time into complete and uniform ideas and theories, the application of which suggested an undreamed-of abundance of new observations and experiments. At that time there appeared Lagrange's *Analytical Mechanics* and Laplace's *Celestial Mechanics*, the two works that brought Newton's ideas to their consummation. From them there was developed on the one hand, by Navier, Poisson, Cauchy, and Green, the mechanics of deformable bodies and the theory of fluids and elastic substances; on

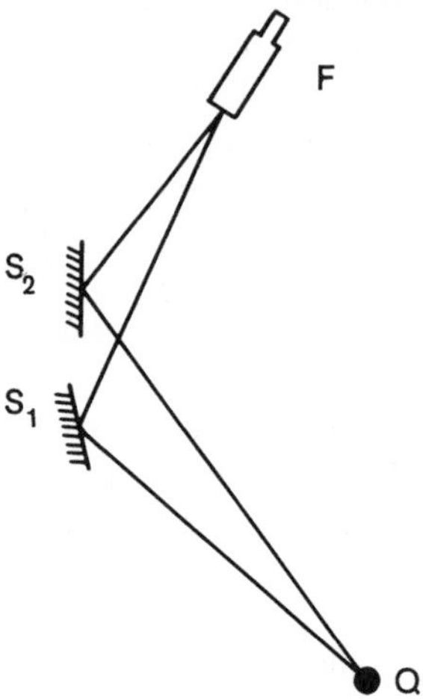

Fig. 60 *Fresnel's mirror experiment.*

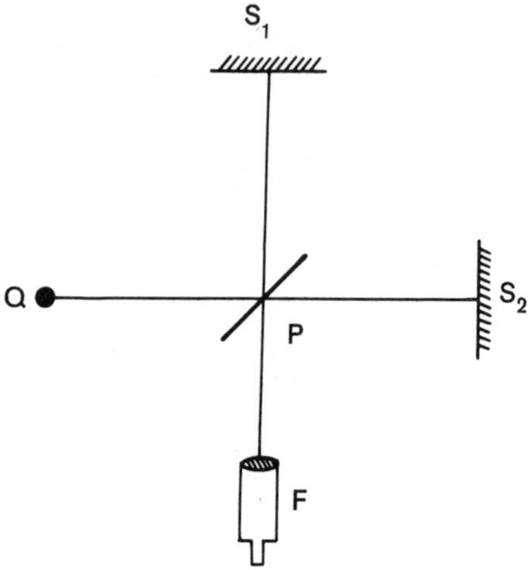

Fig. 61 *Michelson's interferometer.*

the other hand, by the works of Young, Fresnel, Arago, Malus, and Brewster, the theory of light. At the same time began the era of electromagnetic discoveries which we shall speak about later.

Fresnel allowed a ray of light to be reflected at two mirrors, S_1 and S_2 (Fig. 60), slightly inclined to each other. At the points where they meet, these two reflected rays give interference fringes that can be seen with a magnifying glass. Many similar apparatus have been constructed. We shall here enter only into the discussion of a problem which is important for our purpose, that of the experimental methods of measuring minute changes in the velocity of light. The apparatus used is called an *interferometer*. It depends on the fact that the wave length alters proportionately with the velocity of the light, and this can be observed by a displacement of interferences. An example of an apparatus of this kind is the interferometer of Michelson. It consists in the main (Fig. 61) of a glass plate P that is slightly silvered so as to allow one-half of the light from the source Q to pass through while the other half is reflected.

These two rays travel on to two mirrors S_1 and S_2, where they are reflected and again encounter the semitransparent glass plate P, which again splits them in two, sending one-half of each ray into the observing telescope F. If the two paths PS_1 and PS_2 are exactly equal, the two component rays arrive at the telescope in the same phase of vibration and recombine to form the original light again. But if the path of the first ray be lengthened by displacing the mirror S_1, then the crests and hollows of the two trains of waves no longer coincide when the rays are recombined at F, but are displaced with respect to each other and strengthen or weaken each other more or less. If the mirror S_1 is moved slowly, we see alternate patches of light and darkness in the telescope F. The distance of the positions of S_1 for two successive dark fields is exactly equal to the wave length of the light. In this way Michelson has made measurements of wave length that exceed almost all other physical measurements in accuracy. This is done by counting the changes of light and darkness during a considerable shift comprising many thousands of wave lengths of the mirror S_1. The error of observation of an individual wave length then becomes just as many thousand times smaller.

Actually what is seen in the telescope of the interferometer is not simply a light or a dark field, but a system of bright and dark fringes. This is due to the fact that the two rays are not exactly parallel and the waves are not exactly plane. Each of the separate parts of the two rays have thus to traverse paths of different length. We shall not, however, enter into the geometric details, but mention this circumstance only because it is customary to speak of interference fringes.

We have here to give several numerical data. By the above method it is found that the wave length of the yellow light that is sent out by a Bunsen flame colored with common salt (NaCl), and the source of which are sodium atoms, is about $\frac{6}{10{,}000}$ mm. $= 6\times10^{-5}$ cm. *in vacuo*. All visible light lies within the small region of wave lengths stretching from about 4×10^{-5} (violet) to 8×10^{-5} cm. (red). Thus in the language of acoustics this comprises one octave; that is, it is the region between one wave and another that is twice as long. From formula (35) there then follows for the vibration number of yellow sodium light the immense number $\nu=\frac{c}{\lambda}=\frac{3\times10^{10}}{6\times10^{-5}}=5\times10^{14}$, or 500 trillion vibrations per second. The most rapid acoustic vibrations that are still audible vibrate only about 20,000 times per second.

The astonishing accuracy of optical methods of measurement rests on the fact that large numbers of wave lengths are contained in the path of light of an interferometric apparatus. For example, it allows us to ascertain that the velocity of light in a gas alters if there is a very small change of pressure or temperature (due, say, to the apparatus being touched by the hand). To show this the gas is passed into a cylinder between the glass plate P and the mirror S_1. It is then seen that for even the slightest increase of pressure the interference changes, bright fringes being converted into dark ones and vice versa.

We shall meet with Michelson's interferometer again when we have to decide the question as to whether the earth's motion influences the velocity of light.

5. Polarization and Transversality of Light Waves

Although interference phenomena hardly allow any interpretation other than that of the wave theory, its general recognition was impeded by two difficulties which, as we have already seen, were regarded by Newton as being decisive contradictions to it: first, the general rectilinear propagation of light; second, the explanation of *polarization phenomena.* The first difficulty disappeared when the wave theory itself was worked out in more detail; for it was found that waves do, indeed, "bend round corners," but only in regions that are of the order of magnitude of the wave length. As this is very small in the case of light, our ordinary unaided vision receives the impression of sharp shadows and rectilinearly bounded rays. Only minute observation is able to detect the interference fringes of diffracted light along the edges of the shadow. The merit of elaborating the theory of diffraction is due to Fresnel, later Kirchhoff (1882), and, more recently, Sommerfeld (1895). They have deduced the finer phenomena mathematically and have defined the limits within which the concept *ray of light* may be applied.

The second difficulty concerned the phenomena due to the polarization of light. When we spoke earlier of waves, we always had in mind longitudinal waves such as are known in the case of sound. For a sound wave consists of rhythmical condensations and rarefactions, during which the individual particles of air move to and fro in the direction of propagation of the wave. Transverse waves were, indeed, also known—for example, the waves on a surface of water, or the vibrations of a stretched string, in which the particles vibrate at right angles to the direction of propagation of the wave. But in this case we are dealing not with waves in the interior of a substance but in part with phenomena on a surface (water waves) and in part with motions of whole configurations (vibration of strings). Observations or theories about the propagation of waves in elastic solid bodies were not yet known. This accounts for the circumstance, which appears strange to us, that it was so long before optical waves were recognized as transverse vibrations. In fact, it is remarkable that the motivation for the development of the mechanics of ordinary solid elastic bodies came from experiments and concepts concerning the dynamics of the imponderable and intangible ether.

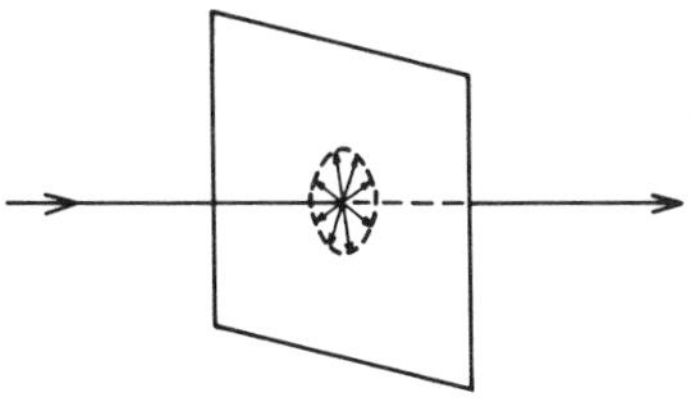

Fig. 62 *In natural light perpendicular to the ray no direction is preferred.*

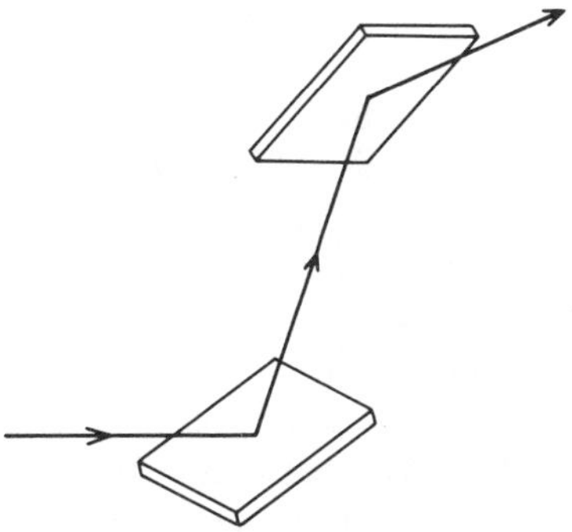

Fig. 63 *Polarization experiment: If one turns the first or the second plate about the corresponding incident ray, the intensity of the reflected ray changes.*

We explained earlier (p. 91) what constitutes the nature of polarization. The two rays that emerge out of a doubly refracting crystal of calcite do not behave like ordinary light when they pass through a second such crystal, that is, they do not again resolve into the equally intense rays but into two of unequal intensity, one of which may under certain circumstances vanish entirely.

In ordinary "natural" light the various directions within the plane of a wave, i.e., the plane perpendicular to the ray, are of equal value or equivalent (Fig. 62). In a beam of polarized light, this means that for one of the two beams emerging from double refraction by calcite, for example, this is no longer the case. Malus discovered (1808) that polarization is not a peculiarity of the light that has passed through a doubly refracting crystal, but may also be produced by simple reflection. He looked through a plate of calcite at the reflection in a window of the setting sun and, on turning the calcite, noticed the varying intensity of the two images of the sun. This does not occur if one looks directly at the sun through calcite. Brewster (1815) showed that light which has been reflected from a glass plate at a certain angle is reflected from a second plate by a varying amount, if the latter is turned about the incident ray (Fig. 63). The plane perpendicular to the surface of the mirror and containing the incident and the reflected ray is called the *plane of incidence.* The reflected ray is then said to be polarized in the plane of incidence; this implies no more than that it behaves differently towards a second mirror according to the position of the second plane of incidence to the first. This behavior cannot be explained by the corpuscular theory, for a light particle striking the surface of the glass plate must either enter the plate or be reflected.

The two rays that emerge from a crystal of calcite are polarized perpendicular to each other. If we let them both fall onto a mirror at an appropriate angle, the one is not reflected at all just when the other is reflected to its full amount.

Fresnel and Arago made the decisive experiment (1816) when they attempted to obtain interferences from two such rays polarized perpendicular to each other. They did not succeed. Fresnel and also Young then drew the inference (1817) that light vibrations must be transverse.

As a matter of fact this deduction makes the peculiar behavior of polarized light intelligible at once. The vibrations of the ether particles do not occur in the direction of propagation but perpendicular to it, that is, in the plane of the wave (Fig. 62). But every motion of a point in a plane may be regarded as composed of two motions in two directions perpendicular to each other. In dealing with the kinematics of a point (II, 3) we saw that its motion is determined uniquely when its rectangular coordinates, which vary with time, are given. Now a doubly refracting crystal clearly has the property of transmitting the vibrations of light in it at different velocities in two mutually perpendicular directions. Hence, by Huygens'

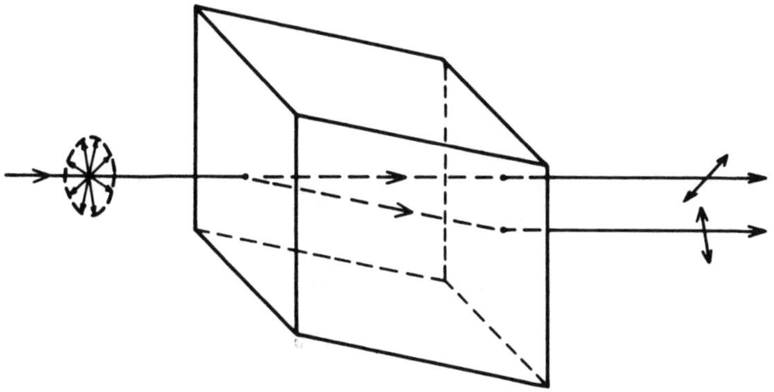

Fig. 64 *The two rays formed by double refraction are polarized perpendicular to each other.*

principle, when these vibrations enter the crystal they will suffer different deflections, or be refracted differently—that is, they will be separated in space. Each of the emergent rays then consists only of vibrations that take place in a certain plane passing through the direction of the ray, and the planes belonging to these two rays are mutually perpendicular (Fig. 64). Two such vibrations clearly cannot influence each other: they cannot interfere. Now, if a polarized ray enters into a second crystal it is transmitted without

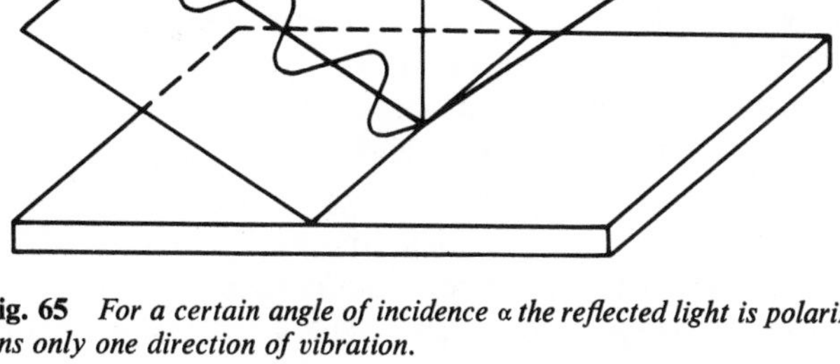

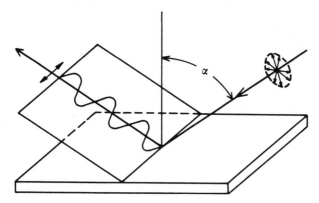

Fig. 65 *For a certain angle of incidence* α *the reflected light is polarized. It contains only one direction of vibration.*

being weakened only if its direction of vibration is in just the right position with respect to the crystal, that in which this vibration can propagate itself. In all other positions the ray is split in two, the intensity of the two rays varying with the position of the second plate.

Similar conditions obtain in reflection. If this occurs at the appropriate angle, then of the two vibrations, one of which is parallel, the other perpendicular with respect to the incident plane, only one is reflected; the other penetrates the mirror and is absorbed, in the case of a metallic mirror, or is transmitted, in the case of a glass plate (Fig. 65). Whether the reflected vibration is that which takes place in the incident plane or perpendicular to it cannot, of course, be ascertained. (In Fig. 65 the latter is assumed to be the case.) But this question of the position of vibration with respect to the plane of incidence or the direction of polarization has given rise to elaborate researches, theories, and discussions, as we shall presently see.

6. The Ether as an Elastic Solid

After the transversality of light waves had been recognized and proved by numerous experiments, there arose in Fresnel's mind the vision of a future *dynamical theory of light* which was to derive in conformity with the method of mechanics the properties of optical phenomena from the properties of the ether and the forces acting in it. The ether was necessarily a kind of elastic solid, for it is only in such a substance that mechanical transverse waves can occur. But in Fresnel's time the mathematical *theory of elasticity of solid bodies* had not yet been developed. Possibly he also thought from the outset that the analogy of the ether with material substances was not to be carried too far. At any rate he preferred to investigate the laws of the propagation of light empirically and to interpret them by means of the idea of transverse waves. Above all, it was to be expected that the optical phenomena in crystals would shed light on the behavior of the ether. Fresnel's work in this field is to be ranked among the most remarkable achievements of systematic physical research both in experimental as well as in theoretical respects. Yet we must not digress too far in pursuing details, but must keep in view our problem: How is the ether constituted?

Fresnel's results appeared to confirm the analogy of the nature of light waves with that of elastic waves. This gave a powerful stimulus to the working out of the theory of elasticity, which had already been started by Navier (1821) and Cauchy (1822), and to which Poisson (1828) devoted his attention. Cauchy then at once applied the laws derived from elastic waves to optics (1829). We shall try to give an idea of the content of this ether theory.

The difficulty of doing so here is that the proper and adequate means of describing changes in continuous deformable bodies is the method of differential equations. Since we do not want to assume this to be known, all that can be done is to illustrate it by a simple example, and then affirm that in the general case a similar result holds, though in a rather more complicated way. The nonmathematical reader may perhaps then get a rough idea of what is involved. It will not, however, give him a real impression of the efficiency and power of the physical models and the mathematical methods. Although it is impossible to satisfy the nonmathematician, we cannot refrain from attempting to illustrate the mechanics of continua, because all subsequent theories, not only of the elastic ether but also electrodynamics in all its ramifications and, above all, Einstein's theory of gravitation, are built on these concepts.

A very thin stretched string is a one-dimensional elastic body. We shall use it to develop the theory of elasticity. To link up with ordinary mechanics, which deals only with individual rigid solids, we suppose the string to be not continuous but of an atomistic structure, as it were. Let it consist of a series of equal small bodies that are arranged in a line at equal distances from each other (Fig. 66). The particles are to possess inertial mass and each is to exert

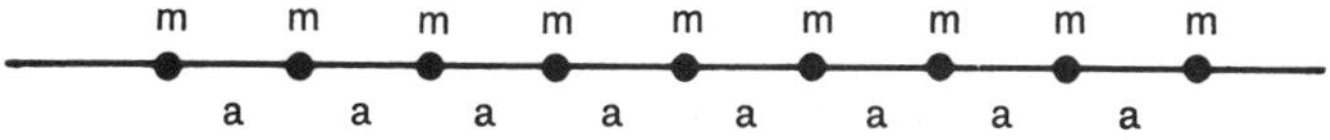

Fig. 66 *A chain of particles with mass* m *in equal distances* a.

forces on its two neighbors: these forces are to be such that they resist both an increase and a decrease of the distance between these particles. If we wish to have a concrete picture of such forces, we need only think of small spiral springs that are fixed between the particles. These resist compression as well as extension. But such a representation must not be taken literally. Forces of this kind, indeed, constitute the essential phenomena of elasticity.

Now if the first particle is displaced a little in the longitudinal or in the transverse direction, it immediately acts on the second particle; the latter in its turn passes the action onto the next, and so forth. The disturbance of the equilibrium of the first particle thus passes along the whole series like a short wave and finally also reaches the last particle. This process does not, however, occur instantaneously. At each particle a small fraction of time is lost because the particle, owing to its inertia, does not instantly respond to the impulse. For the force does not produce an instantaneous displacement but an acceleration, that is, a change of velocity during a small interval of time, and the change of velocity again requires time to produce a displacement. Only when this displacement has reached its full

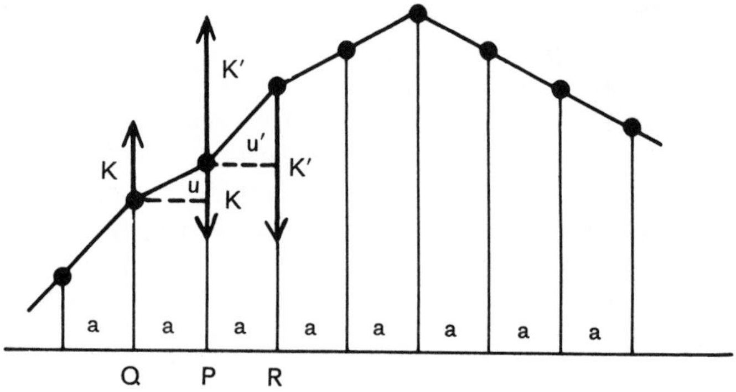

Fig. 67 *The particles* Q *and* R *act on the particle* P *by forces* K *and* K′. *The action of these two forces accelerates* P.

value does the force act to its full extent on the next particle, and from then onwards the process repeats itself with a loss of time that is dependent on the mass of the particles. If the force that arises through the displacement of the first particle were to influence the last particle of the series directly, the action would occur instantaneously. According to Newton's theory of gravitation this is actually supposed to be the case in the mutual attraction of the heavenly bodies. The force with which one acts on the other is always directed at the point momentarily occupied by the other and is determined by the distance separating these points at that moment. Newtonian gravitation is said to be an *action at a distance*, for it acts between points at a distance although there is no intervening medium to convey this action.

In contrast with this our series of equidistant points is the simplest model of *contiguous action* or *action by contact.* For the action exerted by the first point on the last is transferred by the intervening masses, and hence does not occur instantaneously but with a loss of time. The force exerted by a particle on its neighbors is certainly still imagined as an action at a distance, although only at a small distance. We may, however, suppose these distances between the particles to grow smaller and smaller, their number becoming correspondingly greater and greater, but in such a way that their total mass remains the same. Thus the chain of particles goes over into what is called a *continuum.* The forces act between infinitely near particles, and the laws of motion assume the form of differential equations. They express mathematically the physical concept of contiguous action.

We shall pursue this limiting process of the laws of motion in more detail for the case of our chain of mass particles. Let us consider purely transverse displacements (Fig. 67). In the theory of elasticity it is assumed that a particle P is pulled back by its neighbor Q more strongly in proportion to the amount that Q is displaced transversely beyond P. If u is the excess of the transverse displacement of P beyond that of Q, and if a is the original distance between the particles along the straight line, then the restoring force is to be proportional to the ratio $\frac{u}{a}=d$, which is called the *deformation.* We set

$$K = p\frac{u}{a} = pd,$$

where p is a constant which is clearly equal to the force if the deformation d be chosen equal to 1. The quantity denoted by the symbol p is called the elastic constant.

Now the same particle likewise experiences a force $K'=p\frac{u'}{a}=pd'$ from its other neighbor R. With the exception of the particular case in which the deflection of P is exactly a maximum, the particle R will be more strongly displaced than P, and hence will not pull back the latter, but will tend to increase its displacement. Thus K' will work against K.

The resultant force on the particle P is the difference of these forces

$$K - K' = p(d-d').$$

And this determines the motion of P according to the fundamental dynamical formula: mass times acceleration equals force,

$$mb = K - K' = p(d-d').$$

Now let us suppose the number of particles to be increased more and more, but their mass to be decreased in the same ratio so that the mass per unit of length always retains the same value. Let there be n particles per unit of length, so that $na = 1$, that is, $n = \frac{1}{a}$. The mass per unit of length is $mn = \frac{m}{a}\cdot$ This linear quantity is called the *density of mass*, and is designated by ρ. By dividing the above equation by a, we get

$$\frac{m}{a}b = \rho b = \frac{K-K'}{a} = p\frac{d-d'}{a},$$

and here we have expressions quite similar to those which occurred in the definitions of the concepts velocity and acceleration. For just as the velocity was the ratio of the path u to the time $\tau, v = \frac{u}{\tau}$, where the time τ is to be considered quite short, so we have here the deformation $d = \frac{u}{a}$, the ratio of the relative displacement to the original distance, where the latter is to be regarded as extremely small. Just as the acceleration was before defined as the ratio of the change of velocity to the time, $b = \frac{w}{\tau} = \frac{v-v'}{\tau}$, so we have here the quantity $f = \frac{d-d'}{a}$, which measures in a fully analogous manner, the change of the deformation from point to point.

Exactly as the velocity v and the acceleration b retain their meaning and their finite values for time intervals that are arbitrarily small, so the quantities d and f retain their meaning and finite values no matter how small the distance a becomes. All these are so-called *differential coefficients* ($v=\frac{u}{\tau}$ and $d=\frac{u}{a}$ are differential coefficients of the first order, $b=\frac{v-v'}{\tau}$ and $f=\frac{d-d'}{a}$ are differential coefficients of the second order).

Thus the equation of motion becomes a differential equation of the second order,

$$\rho b = pf, \tag{36}$$

both with respect to the change of time as well as to the change of position of the event. *All* laws of contiguous action in theoretical physics are of this type. If, for example, we are dealing with elastic bodies that are extended in all directions, we get two analogously formed terms for the other two space dimensions. Moreover, precisely similar laws hold in the theory of electric and magnetic events. Finally, the gravitational theory of Einstein has also been brought into such a form.

We should also remark that laws of action at a distance may be written in a form similar to that of formulae for contiguous action. For instance, if we drop ρb in our equation (36), that is, if we assume that the density of mass is extremely small, then a displacement of the first particle will at the same moment produce a force acting on the last particle, because the inertia of the intervening particles has dropped out. Thus we really have the transmission of a force with infinite velocity, a true action at a distance. Nevertheless, the law $pf=0$ appears in the form of a differential equation as a contiguous action. Such *laws of pseudocontiguous action* will be met with in the theory of electricity and magnetism, where they have really prepared the way for the true laws of contiguous action. The essential factor in the latter is the inertial term that is responsible for the finite velocity of transmission of disturbances of equilibrium, that is, the generation of waves.

In the law (36) two quantities occur that determine the physical character of the substance: the mass per unit of volume, or density ρ, and the elastic constant p. If we write $b=\frac{p}{\rho}f$, we see that for a given deformation, i.e., for a given f, the acceleration becomes greater in proportion as p becomes greater and ρ becomes smaller. Now p is a measure of the elastic rigidity of the substance, and ρ is a measure of the inertial resistance, and it is clear that an increase of the rigidity accelerates the motion, whereas an increase of the inertia retards it. Accordingly the velocity c of a wave will depend only on the ratio $\frac{p}{\rho}$. For the more quickly the wave travels, the greater are the accelerations of the individual particles of the substance. The exact law for this relationship is found by the following considerations.

Each individual point mass executes a simple periodic motion of the kind that we investigated earlier (II, 11, p. 37). We showed there that in it the acceleration is connected with the deflection x according to formula (11)

$$b = -(2\pi\nu)^2 x,$$

where ν is the number of vibrations per second. If we insert in place of ν the time of vibration according to formula (34) (p. 97), $T=\frac{1}{\nu}$, we get

$$b = -\left(\frac{2\pi}{T}\right)^2 x.$$

The same argument that has here been used for time may also be applied for space, and must lead to corresponding relations. We have simply to replace the acceleration b (the second time-coefficient) by the quantity f (the second space-coefficient) and the time of vibration T (the "time period") by the wave length λ (the "space period"). We thus get the formula

$$f = -\left(\frac{2\pi}{\lambda}\right)^2 x.$$

If we form the quotient of the two expressions for b and f, the factor $(2\pi)^2 x$ and the negative sign cancel out, and there remains

$$\frac{b}{f} = \frac{\lambda^2}{T^2}.$$

Now, on the one hand we have by formula (35), that $\frac{\lambda}{T}=c$; on the other hand by (36) (p. 113), that $\frac{b}{f}=\frac{p}{\rho}$. Hence it follows that

$$c^2 = \frac{p}{\rho} \quad \text{or} \quad c = \sqrt{\frac{p}{\rho}}. \tag{37}$$

This relation holds for all bodies, no matter whether they be gaseous, liquid, or solid. But there is the following difference: In *liquids and gases* there is no elastic resistance to the lateral displacement of the particles, but only to the change of volume, i.e., compressions and rarefactions. Hence *only longitudinal waves* can propagate themselves in such substances, their velocities being determined according to formula (37) by the elastic constant p, which is decisive in such changes of volume.

On the other hand, in *solid bodies*, on account of the elastic rigidity which opposes lateral displacements, *three waves, one longitudinal and two transverse*, with different velocities, can be transmitted in each direction. This is due to the fact that the compressions and rarefactions of the longitudinal waves involve an elastic constant p different from that which comes from the lateral distortions due to the transverse vibrations.

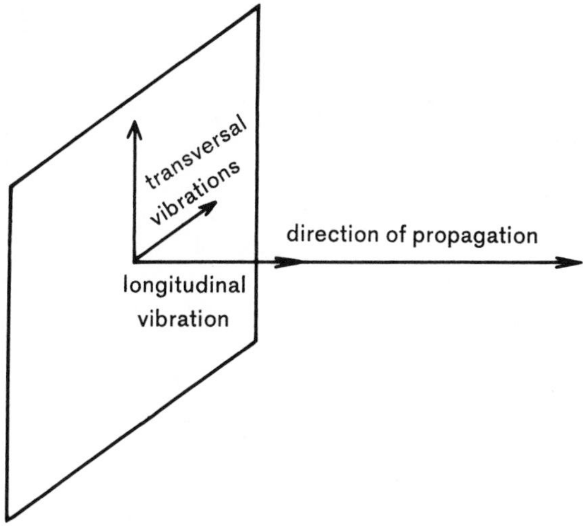

Fig. 68 *In solids the atoms can vibrate in the direction of propagation and perpendicular to it.*

Moreover, in noncrystalline bodies one can choose the direction of the vibration for transverse waves arbitrarily; all these waves have the same velocity c_t; the longitudinal wave has a different velocity c_l (Fig. 68).

All these statements are confirmed by experiments on acoustic waves in solid bodies.

We now return to the starting point of our reflections, the elastic theory of light, a theory which consists of treating the ether, the carrier of light vibrations, as a solid elastic body. The light waves are then regarded as sound waves in this hypothetical medium.

Now, what properties are to be ascribed to this elastic ether? In the first place the enormous velocity of propagation c requires either that the elastic rigidity p be very great or that the density of mass ρ be very small, or that both conditions hold simultaneously. But since the velocity of light is different in different substances, either the ether within a material body must be condensed or its elasticity must be changed; or again both may be true simultaneously. We see that different courses are here open to us. The number of possibilities is still further increased by the fact that, as we saw earlier (IV, 5, p. 104), experiment cannot decide whether the vibrations of polarized light are parallel or perpendicular to the plane of polarization (the incident plane of the polarizing mirror).

Corresponding to the indefinite nature of the problem we find a great number of different theories of the elastic ether in history. We have already mentioned the pioneers Navier, Cauchy, Poisson, and add to them the names Green and Neumann.

Nowadays we feel surprise at the amount of ingenuity and labor that was expended on the problem of comprehending optical phenomena in terms of an elastic ether having the same properties as those possessed by material elastic solids. For we have learned since then that the nature of elastic solids is by no means simple; they are not continuous but have actually an atomistic structure. The physics of the ether has proved to be simpler and more easily intelligible than the physics of matter.

One obvious objection to the hypothesis of an elastic ether arises from the necessity of ascribing to it the great rigidity it must have to account for the high velocity of waves. Such a substance would necessarily offer resistance to the motion of heavenly bodies, particularly to that of the planets. Astronomy has never detected departures from Newton's laws of motion that would point to such a resistance. Stokes (1845) tried to dispose of this objection by remarking that the concept of solidity of a body is in some way relative. A piece of pitch (sealing wax and glass behave similarly) when struck with a hammer splits cleanly. But if it is loaded with a weight, the weight sinks gradually, although very slowly, into the pitch which behaves like a viscous fluid. Now, the forces that occur in light vibrations change extremely quickly (600 billion times per sec.) compared with the relatively slow processes that occur in planetary motions in the course of time; the ratio of this rate of change of forces is much more extreme than that of the hammer blow to the static pressure of weight. Therefore the ether may function for light as an elastic solid and yet give way completely to the motion of the planets.

Now even if we could bring ourselves to accept the idea of astronomic space filled with a kind of pitch, serious difficulties arise from the laws of propagation of light themselves. Above all, we have to take into account that in elastic solids a longitudinal wave always occurs conjointly with the two transverse waves. If we now consider the refraction of a wave at the boundary of two media, and if we assume that the wave vibrates purely transversally in the first medium, then in the second medium a longitudinal wave must arise together with the transverse wave. All attempts to escape this consequence of the theory by making more or less arbitrary alterations have been doomed to failure. Extraordinary hypotheses were suggested, such as that the ether opposes to compression an infinitely small or an infinitely great resistance compared with its rigidity towards transversal distortions. In the former case the longitudinal waves would travel infinitely slowly, in the latter infinitely fast, and would thus not manifest themselves as light. A physicist, MacCullagh (1839), went so far as to construct an ether that departed altogether from the model of elastic bodies. For whereas in these the particles oppose a resistance to every change of their distance from their neighbors, MacCullagh's ether possessed the property of opposing rotational motions of the neighbors around the central particle. We cannot here enter into the theory. However strange it may appear, it is nevertheless of importance as the forerunner of the electromagnetic theory of light. It leads to almost the same formulae as the latter, and is actually able to give an account of optical phenomena that is to a considerable degree correct. But its weakness is that it disclosed no relationship between optical and other physical phenomena. It is clear that by means of arbitrary constructions ether models can be found that allow a certain domain of phenomena to be represented. However, such inventions are of value only when they lead to a connection between formerly unconnected physical phenomena. This is the great advance achieved by Maxwell (of which we have to speak later) when he showed light to be an electromagnetic phenomenon.

7. The Optics of Moving Bodies

We have now to discuss the theory of the elastic ether in regard to the space-time problem and relativity. In our optical investigations so far we have taken no account of the position or motion of the bodies that emit, receive, or propagate light; now we shall turn our attention to these questions.

The space of mechanics is regarded as empty wherever there are no material bodies present. The space of optics is filled with ether. The ether is considered to be a kind of matter that has a certain mass, density, and elasticity. Accordingly we can immediately apply Newton's doctrine of space and time to the universe filled with ether. This universe then no longer consists of isolated masses that are separated by empty spaces but is completely filled with the thin but almost rigid mass of the ether in which the coarse masses of matter are floating. The ether and matter act on each other with mechanical forces and move according to Newtonian laws. Thus Newton's standpoint is logically applicable to optics. The question is only whether observation is in agreement with it.

This question, however, cannot be answered simply by unambiguous experiments, for the state of motion of the ether outside and inside matter is not known, and we are free to think out hypotheses about it. Thus we must put the question in the form: Is it possible to make assumptions about the mutual actions of the motions of the ether and of matter such that all optical phenomena are thereby explained?

We now recall the principle of relativity of classical mechanics. According to it absolute space exists only in a restricted sense, for all inertial systems that move rectilinearly and uniformly with respect to one another may be regarded with equal right as being at rest in space. The first hypothesis that suggests itself concerning the luminiferous ether is the following:

The ether in astronomic space far removed from material bodies is at rest in an inertial system.

If this were not the case, parts of the ether would be accelerated. Centrifugal forces would arise in it and would bring about changes of density and elasticity, and we should expect that the light from stars would have given us indications of this.

Formally this hypothesis satisfies the classical principle of relativity. If the ether is counted among material bodies, then motions of translation of bodies with respect to the ether are just as much relative motions as those of two bodies with respect to each other, and a common motion of translation of the ether and all matter should not be detectable either mechanically or optically.

But the physics of material bodies *alone*, without the ether, need no longer satisfy the principle of relativity. A common translation of all matter in which the ether does not participate (i.e., a relative motion with respect to the latter) could very well be ascertained by optical experiments. Then the ether would practically define a system of reference that is absolutely at rest. The question that is important above all else for our discussion is whether the observable optical phenomena depend only on the relative motions of material bodies or whether the motion in the sea of ether can be detected.

A light wave has three characteristics:

1. The vibration number or frequency ν.
2. The velocity c.
3. The direction of propagation.

The wave length λ, another characterizing quantity, is defined as the quotient of velocity c and frequency ν (equation (35)).

We shall now investigate systematically how these three characteristics change relative to one another and to the transmitting medium (the ether in space or a transparent medium) due to motions of light-emitting and -receiving bodies.

We shall apply a method which looks somewhat complicated but which will be useful later on. We consider a train of waves propagated in x-direction with the velocity c and consisting of exactly n waves of wave length λ. In Fig. 69 we have chosen $n=4$. This

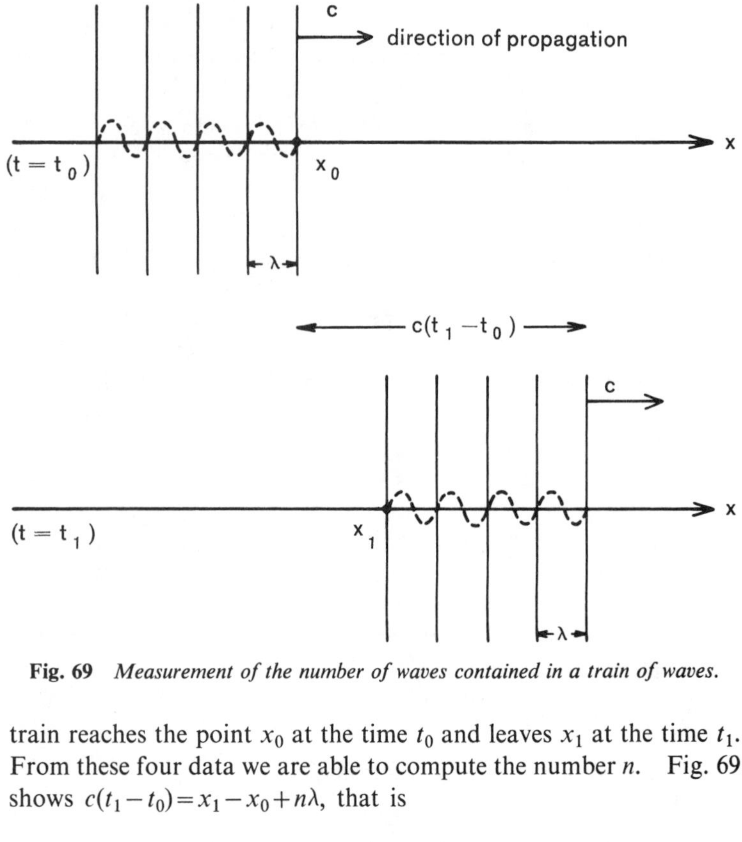

Fig. 69 *Measurement of the number of waves contained in a train of waves.*

train reaches the point x_0 at the time t_0 and leaves x_1 at the time t_1. From these four data we are able to compute the number n. Fig. 69 shows $c(t_1-t_0)=x_1-x_0+n\lambda$, that is

$$n = \frac{c}{\lambda}\left(t_1-t_0-\frac{x_1-x_0}{c}\right) = \nu\left(t_1-t_0-\frac{x_1-x_0}{c}\right). \tag{38}$$

Here we have used $c=\nu\lambda$ according to (35). Our equation combines two simple methods to obtain n. If we observe the train of waves at a fixed point ($x_1=x_0$), we get n as the time t_1-t_0 required for the whole train to pass this point divided by the corresponding time $T=\frac{1}{\nu}$ for one wave length. If we make the observation at a fixed time $(t_1=t_0)$, x_1-x_0 is the length of the train, and n equals x_1-x_0 divided by λ.

Now the number of waves in a train is a quantity which is quite independent of the coordinate system. In a moving system we can compute n exactly in the same way and we must get the same value. The number of waves cannot be four for a resting observer while it is five for a moving one. Thus the expression (38) is an invariant in the sense that we have previously attached to this word.

This comes out most clearly if we use Minkowski's mode of expression. According to this the departure of the first wave from the point x_0 at the time t_0 is an event, a world point; the arrival of the last wave at the time t_1 at the point x_1 is another event, a second world point. But world points exist without relation to definite coordinate systems. And since the wave number is determined by the two world points x_0, t_0 and x_1, t_1, it is independent of the system of reference, i.e., it is invariant.

From this one can deduce, either by intuition or by applying Galileo transformations, all theorems about the behavior of the three characteristics of the wave—the frequency, direction, and velocity—when the system of reference is changed.

8. The Doppler Effect

The fact that the observed frequency of a wave depends on the motion both of the source of light and of the observer, each with respect to the intervening medium, was discovered by Christian Doppler (1842). The phenomenon may easily be observed in the case of sound waves. The whistle of a locomotive seems higher when it is approaching the observer and becomes deeper at the moment of passing. The rapidly approaching source of sound carries the individual phases of the waves forward so that the crests and hollows succeed each other more rapidly. The motion of an observer towards the source has a similar effect; he then receives the waves in more rapid succession.

Now the same phenomenon must hold in the case of light. The frequency of the light determines its color; the rapid vibrations correspond to the violet end of the spectrum, the slower vibrations to the red. Hence when a light source is approaching an observer, or vice versa, the color of the light is displaced a little towards violet; when either is receding, it is displaced a little towards red. This phenomenon has actually been observed.

Now the light which comes from luminescent gases does not consist of all possible vibrations but of a number of separate frequencies. The spectrum that a prism or a spectral apparatus depending on interference exhibits is not a continuous band of color like the rainbow but consists of separate, sharp, colored lines. The frequency of these spectral lines is characteristic of the chemical elements that are emitting light in the flame (spectral analysis by Bunsen and Kirchhoff, 1859). The stars, in particular, have such line spectra, whose lines coincide with those of elements known on the earth. From this it is to be inferred that the matter in the furthermost depths of astronomic space is composed of the same primary constituents. The lines of the stars, however, do not coincide exactly with the corresponding lines on the earth but show small displacements towards the one side for one half of the year and towards the other during the other half. These changes of frequency are the results of the Doppler effect of the earth's motion about the sun. During the one half of the year the earth moves towards a definite star, and hence the frequencies of all the light waves coming from this star are increased and the spectral lines of the star appear shifted towards the side of higher frequencies (the violet end), whereas during the second half of the year the earth moves away from the star, and hence the spectral lines are then displaced towards the other side (the red end).

This wonderful image of the earth's motion in the spectrum of the stars does not, indeed, present itself in an unadulterated form. For it is clear that there will be superposed on it the Doppler effect due to the emission of the light by a moving source. Now, if the stars are not all at rest in the ether, their motion must again manifest itself in a displacement of the spectral lines. This is added to that due to the earth's motion, but does not show the annual change, and hence may easily be distinguished and separated from the former. Astronomically this phenomenon is more important still, for it gives us information about the velocities of even the most distant stars so far as the motion entails an approach towards or a recession from the earth. It is not our object, however, to enter more closely into these investigations.

We are interested above all in the question of what happens when the observer and the source of light move in the same direction with the same velocity. Does the Doppler effect then vanish; does it depend only on the relative motion of the material bodies; or does it fail to vanish and thereby betray the motion of bodies through the ether? In the former case the principle of relativity would remain valid for the optical phenomena that occur between material bodies.

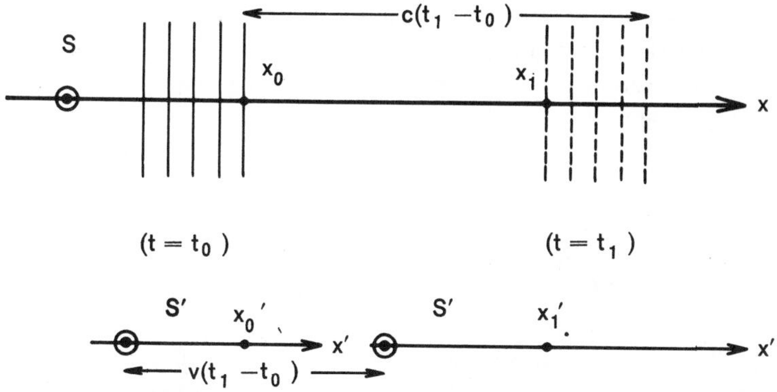

Fig. 70 *Observation of a train of waves from two systems.* S *rests and* S′ *moves with velocity* v *in the direction of propagation.*

The ether theory gives the following answer to this question: The Doppler effect depends not only on the relative motion of the source of light and of the observer but also to a slight extent on the motions of both with respect to the ether. This influence, however, is so small that it escapes observation; moreover, in the case of a common translation of the source of light and of the observer it is precisely equal to zero.

The latter point is so self-evident that it need hardly be emphasized. It is necessary only to reflect that the waves pass by any two points at rest relative to each other in the same rhythm, irrespective of whether the two points are at rest in the ether or move with a common motion. Nevertheless the principle of relativity does not hold rigorously, but only approximately, for the bodies emitting and absorbing the light. We shall prove this.

For this purpose we make use of the theorem derived above concerning the invariance of the wave number.

An observer in a coordinate system at rest in the ether observes a limited train of waves, which reaches the point x_0 at the time t_0 and leaves x_1 at t_1 (Fig. 70). Then we know from (38) that the number of waves n is given by

$$n = \nu\left(t_1 - t_0 - \frac{x_1 - x_0}{c}\right).$$

Another observer, moving in the x-direction with the velocity v, measures the same number n in the same manner. But he perceives another frequency ν' and velocity c'. At the time t_0 the waves reach x_0', and at t_1 they leave x_1'. Therefore

$$n = \nu'\left(t_1 - t_0 - \frac{x_1' - x_0'}{c'}\right)$$

and

$$\nu\left(t_1 - t_0 - \frac{x_1 - x_0}{c}\right) = \nu'\left(t_1 - t_0 - \frac{x_1' - x_0'}{c'}\right). \tag{39}$$

The Galileo transformation (29) connects x_1, x_0 with x_1', x_0'. Assuming the origins $x=0$ and $x'=0$ of the two systems coincide at the time $t=0$, we have $x_1' = x_1 - vt_1$ and $x_0' = x_0 - vt_0$.

From (39) we can calculate the relation of the characteristics of the waves in the two systems. First the two observers may choose to observe at the same time $t_1 = t_0$, then $x_1 - x_0 = x_1' - x_0'$ and (39) shows

$$\frac{\nu}{c} = \frac{\nu'}{c'}. \tag{40}$$

Secondly the observation may be made at a fixed space point in the moving system $x_1' = x_0'$. The Galileo transformation gives $x_1 - x_0 = x_1' - x_0' + v(t_1 - t_0)$. Therefore with $x_1' - x_0' = 0$ one has $x_1 - x_0 = v(t_1 - t_0)$. Introducing this in (39) the result is

$$\nu\left(1 - \frac{v}{c}\right) = \nu'. \tag{41}$$

This relation between the frequencies ν and ν' shows how the frequency is diminished when the observer has a velocity v in the direction of the light.

From (40) and (41) one obtains the obvious result

$$c' = c - v. \tag{42}$$

This and another self-evident fact, the equality of the wave lengths in the two systems $\lambda' = \lambda$, would be sufficient to derive (41). But we have chosen a method using the invariance of the number of waves because they can be used later in the theory of relativity. There we will see that the relations $c' = c - v$ and $\lambda' = \lambda$ are not at all self-evident but are actually replaced by others.

Conversely we now consider a source of light that vibrates with the frequency ν_0, and moves in the direction of the x-axis with the velocity v_0. Let an observer at rest in the ether measure the frequency ν. This case is immediately reducible to the preceding one. For it is quite immaterial for our argument whether it is the light source or the observer that is moving; it depends only on the rhythm with which the waves impinge on the point of observation. The moving point is now the source of light. We thus get the formula for this case from the preceding case if we replace v by v_0 and ν' by ν_0:

$$\nu_0 = \nu\left(1 - \frac{v_0}{c}\right).$$

But here ν_0 is given as the frequency of the source of light, and ν, the observed frequency, is being sought. Thus we must solve for ν and we get

$$\nu = \frac{\nu_0}{1 - \frac{v_0}{c}}. \tag{43}$$

The observed frequency, therefore, appears magnified, since the denominator is less than 1.

We see now that it is not immaterial whether the observer moves in one direction or the source in the opposite direction with the same velocity. For if the resting source radiates with frequency ν_0, an observer moving to the right with velocity v sees a frequency ν_B (putting $\nu' = \nu_B$, $\nu = \nu_0$ into (41))

$$\nu_B = \nu_0\left(1 - \frac{v}{c}\right).$$

If, however, the source is moving away to the left with velocity v from the resting observer, we must insert $v_0 = -v$ into (43) and get

$$\nu_B = \nu_0 \cdot \frac{1}{1+\frac{v}{c}}.$$

These two frequencies are not equal. In all practical cases the difference is certainly very small. We saw earlier (IV, 3, p. 95) that the ratio of the velocity of the earth in its orbit around the sun compared with that of light is $\beta = \frac{v}{c} = 1:10{,}000$, and similar small values of β hold for all cosmic motions. We may then write as a very close approximation

$$\frac{1}{1+\beta} = 1-\beta;$$

for if we neglect $\beta^2 = \frac{1}{100{,}000{,}000} = 10^{-8}$ compared with 1, we have $(1+\beta)(1-\beta) = 1-\beta^2 = 1$.

This neglecting the square of $\beta = \frac{v}{c}$ will play an important part later on. It is almost always permissible because such exceedingly small quantities as $\beta^2 = 10^{-8}$ are open to observation in only a few cases. Indeed, the phenomena of optics (and electrodynamics) of moving bodies are nowadays classified according to whether they are of the order β or β^2. The former quantities are said to be of the *first order*, and the latter of the *second order* in β. In this sense we may assert the following:

The Doppler effect depends only on the relative motion of the source of light and of the observer if quantities of the second order are neglected.

If we assume a simultaneous motion of the source of light (velocity v_0) and the observer (velocity v), the observed frequency ν' is obtained by inserting (43) in (41):

$$\nu' = \nu\left(1-\frac{v}{c}\right) = \nu_0 \frac{1-\frac{v}{c}}{1-\frac{v_0}{c}} \sim \nu_0\left(1-\frac{v}{c}\right)\left(1+\frac{v_0}{c}\right) \sim \nu_0\left(1+\frac{v_0-v}{c}\right).$$

If the source of light and the observer have the same velocity $v_0 = v$, this becomes precisely equal to 1 and we get $\nu' = \nu_0$. Thus the observer notices nothing of a common motion with the source relative to the ether. But as soon as v differs from v_0, a Doppler effect appears. In the first order it depends only on the difference $v - v_0$, but this does not hold if one includes terms of second order. Hence the motion relative to the ether could be observed, if the difference were not of the second order and hence were not much too small to be measured.

We see that the Doppler effect gives no practical method of establishing motions with respect to the ether in astronomic space.

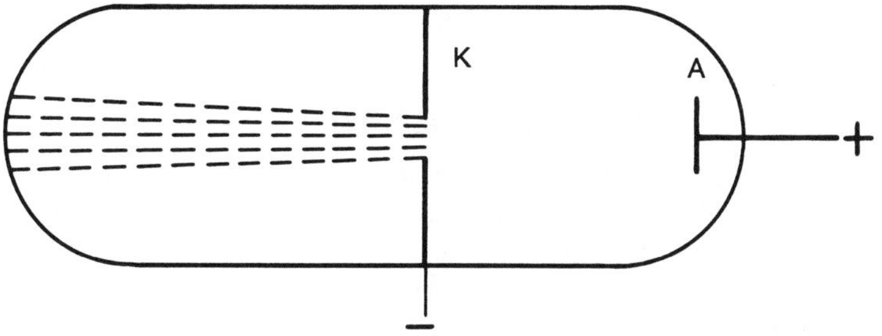

Fig. 71 *Vacuum tube with cathode* K *and anode* A. *Charged atoms and molecules pass with large velocities through the hole in the cathode to the left.*

The optical Doppler effect has been detected with terrestrial sources of light. To do this required sources of light moving with extremely great speed in order that the ratio $\beta = \frac{v}{c}$ might attain a perceptible value. For this purpose J. Stark (1906) used the so-called *canal rays.* If two electrodes are fixed in an evacuated tube containing hydrogen of very small density, and if one of the electrodes is perforated and made the negative terminal (cathode) of an electric discharge (Fig. 71), we get, first, the so-called cathode rays which start at the cathode, and second, as Goldstein discovered in 1886, a reddish luminescence penetrating through the hole or holes of the cathode, due to positively charged hydrogen atoms or molecules moving at a great speed. The velocity of these canal rays is of the order $v=10^8$ cm./sec., thus β has the value

$$\beta = \frac{10^8}{3\times 10^{10}} = \frac{1}{300},$$

which is fairly high compared with the astronomic values.

Stark investigated the spectrum of canal rays and found that the bright lines of hydrogen exhibited the displacement that was to be expected from the Doppler effect. This discovery became of great importance for atomic physics, but it does not belong to our proper theme.

Finally we have to mention that Beloposki (1895) and Galitzin (1907) proved the existence of a sort of Doppler effect with the help of sources of light on the earth and moving mirrors.

9. The Convection of Light by Matter

We have next to investigate the second characteristic of a train of light waves, its *velocity*. According to the ether theory, the velocity of light is a quantity that is determined by the mass density and the elasticity of the ether. Thus it has a fixed value in the ether of astronomic space, but may have a different value in every material body, depending on how the body influences the ether in its interior and carries it along.

From (42) we know the velocity of light in astronomical space. It is c if the observer is resting and is $c'=c-v$ if the observer moves with velocity v in the direction of light.

This may also be interpreted by regarding an observer moving through the ether as being in an ether wind that blows away the light waves just as air brushes past a quickly moving automobile and carries the sound with it.

Now, this furnishes us with a means of establishing the motion of, say, the earth or the solar system relative to the ether. We have two essentially different methods of measuring the velocity of light, an astronomical and a terrestrial one. The former, the old method of Römer, makes use of the eclipses of Jupiter's satellites; it measures the velocity of the light that traverses the space between Jupiter and the earth. In the latter method the source of light and the observer participate in the motion of the earth. Do these two methods give exactly the same result or are there deviations that betray motion relative to the ether?

Maxwell (1879) called attention to the fact that by observing the eclipses of Jupiter's moons it should be possible to ascertain a motion of the whole solar system with respect to the ether. Let us suppose the planet Jupiter to be at the point *A* of its orbit (Fig. 72),

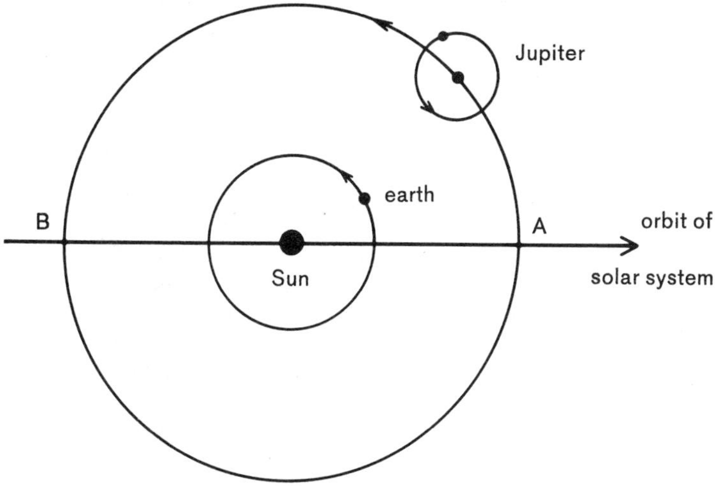

Fig. 72 *Motion of solar system in the ether.*

the point nearest the orbit of the sun in the motion of the solar system in the direction shown. (It has been assumed in the diagram that the orbit of Jupiter intersects the orbit of the solar system at *A*.) In the course of a year Jupiter moves only a short distance away from *A*, since its time of revolution in its own orbit is about twelve years. In one year the earth traverses its orbit once, and by observing eclipses (compare Fig. 52) it is possible to find the time required by the light to travel across the diameter of the earth's orbit. Now, since the whole solar system moves with the sun towards *A*, the light from Jupiter to the earth runs contrary to this motion and its velocity appears increased. Let us now wait for six years until Jupiter is situated at the opposite point *B* of its orbit. The light now runs in the same direction as the solar system, and thus requires a longer time to cross the earth's orbit, so its velocity appears smaller.

When Jupiter is at A the eclipses of one of its satellites during half a year (of the earth) must be delayed by $t_1 = \frac{l}{c+v}$, where l denotes the diameter of the earth's orbit. When Jupiter is at B the delay is $t_2 = \frac{l}{c-v}$. If the solar system were at rest in the ether, both delays would be equal to $t_0 = \frac{l}{c}$. Their actual difference, namely,

$$t_2 - t_1 = l\left(\frac{1}{c-v} - \frac{1}{c+v}\right) = \frac{2lv}{c^2 - v^2} = \frac{2lv}{c^2(1-\beta^2)}$$

for which, by neglecting β^2 in comparison with 1, we may write

$$t_2 - t_1 = \frac{2lv}{c^2} = 2t_0\beta.$$

This allows us to determine β and hence also the velocity $v = \beta c$ of the solar system relative to the ether. Now light takes about 8 min. to travel from the sun to the earth, thus $t_0 = 16$ min. or $= 1000$ sec. (in round numbers). Thus from a time difference $t_2 - t_1 = 1$ sec. we should get

$$\beta = \frac{1}{2000} \quad \text{or} \quad v = \beta c = \frac{300{,}000}{2000} = 150 \text{ km./sec.}$$

The velocities of the stars relative to the solar system, which may be deduced from the Doppler effect, are mostly of the order 20 km./sec., but velocities up to 300 km./sec. occur in certain clusters of stars and spiral nebulae. The accuracy of the astronomical determinations of time has thus far not been great enough to establish a delay in the eclipses of a satellite of Jupiter to the extent of 1 sec. or less in the course of half a year. Yet it is not out of the question that refinement of the methods of observation will yet disclose such a delay.

An observer situated on the sun who happened to know the value of the velocity of light in the ether at rest would also be able to ascertain the motion of the solar system through the ether by means of the eclipses of Jupiter's satellites. To do this he would have to measure the delay in the eclipses during half a revolution of Jupiter in his orbit. The same formula $t_2 - t_1 = 2t_0\beta$ is valid for this, but now t_0 denotes the time that the light requires to traverse the diameter of

Jupiter's orbit. This value of t_0 is (about $2\frac{1}{2}$ times) greater than the value used above for the earth's orbit, 16 min., and the delay t_2-t_1 becomes greater in the same proportion. But for the same reason the time of revolution of Jupiter, during which the eclipses must be observed consecutively, is much greater than (about 12 times as great as) an earth year, so that this method, which could also be applied by an observer on the earth, seems to have no advantage.

At any rate the fact that the accuracy that is nowadays attainable has revealed not even a delay of several seconds proves that the velocity of the solar system with respect to the ether is not much greater than the greatest known velocities of the stars relative to each other.

We next turn our attention to the terrestrial methods of measuring the velocity of light. It is easy to see why they do not allow us to draw conclusions about the motion of the earth through the ether. We have already indicated the grounds for this when these methods were first mentioned (IV, 3, p. 96), for the light traverses one and the same path in its journey to and fro. It is only a mean velocity during the path to and fro that is actually measured. The deviation of this from the velocity of light c in the ether is, however, a quantity of the second order with respect to β and is not open to observation. For if l is the length of the path, then the time that the light requires for the first journey, in the direction of the earth's motion, is equal to $\frac{l}{c-v}$ and the time for the return journey is $\frac{l}{c+v}$, thus the whole time is

$$l\left(\frac{1}{c+v}+\frac{1}{c-v}\right) = \frac{2lc}{(c+v)(c-v)} = \frac{2lc}{c^2-v^2}.$$

The mean velocity is $2l$ divided by this time, thus it is

$$\frac{c^2-v^2}{c} = c\left(1-\frac{v^2}{c^2}\right),$$

and hence it differs from c by a quantity of the second order.

Besides the direct measurement of the velocity of light, there are numerous other experiments in which the velocity of light comes into play. All interference and diffraction phenomena are brought about by making light waves travelling along different paths meet at the same place and causing them to be superposed on each other.

Refraction at the boundary of two bodies arises because light has different velocities in them; thus this velocity enters into the action of all optical apparatus that contain lenses, prisms, and similar parts. Is it not possible to think of a scheme in which the motion of the earth and the "ether wind" produced by it make themselves noticeable?

Many experiments have been designed and carried out to discover this motion. The general result of experiments with terrestrial sources of light teaches us that not the slightest influence of the ether wind is ever observable. It is true that in most cases we are dealing with experimental arrangements that allow only quantities of the first order in β to be measured. That this must always lead to a negative result easily follows from the fact that the true duration of the motion of the light from one place to another is never measured, but that only the sum and difference of the trip there and the trip back over the same light path are found. For the reason given above we thus see that the quantities of the first order always cancel out.

But we might expect a positive result if we took a source in the heavens instead of on the earth. If we point a telescope at a star toward which the velocity v of the earth is directed at that moment (Fig. 73), the velocity of light in the lenses of the telescope relative

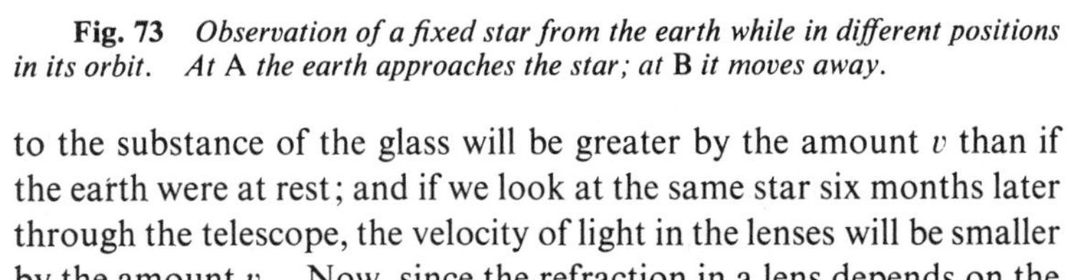

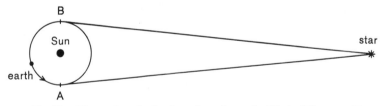

Fig. 73 *Observation of a fixed star from the earth while in different positions in its orbit. At* A *the earth approaches the star; at* B *it moves away.*

to the substance of the glass will be greater by the amount v than if the earth were at rest; and if we look at the same star six months later through the telescope, the velocity of light in the lenses will be smaller by the amount v. Now, since the refraction in a lens depends on the velocity of light, we might expect the focus of the lens to have a different position in these two cases. This would be an effect of the first order. For the difference of the velocity of light in the two cases would be $2v$, and its ratio to the velocity in the ether at rest would be $\frac{2v}{c}=2\beta$.

Arago actually carried out this experiment, but found no change in the position of the focus. How is this to be explained?

We made the assumption above that the velocity of light in a body moving with velocity v in the opposite direction to the light ray is greater by v than if the body were at rest in the ether. In other words, we have assumed that material bodies penetrate the ether without carrying it along in the slightest degree, just like a net carried through water by a boat.

The results of experiment teach us that this is manifestly not the case. Rather, the ether must participate in the motion of matter. It is only a question of how much.

Fresnel established that to explain Arago's observation and all other effects of the first order it was sufficient to assume that the ether is only partly carried along by matter. We shall forthwith discuss in detail this theory, which has been brilliantly confirmed by experiment.

It was Stokes (1845) above all others who later adopted the more radical position that the ether in the interior of matter shares completely in its motion. He assumed that the earth carries along with itself the ether which is in its interior, and that this ether motion gradually decreases outwards until the state of rest of the ether in the universe is reached. It is clear that then all optical phenomena on the earth occur exactly as if the earth were at rest. But in order that the light that comes from the stars may not experience deflections and changes of velocity in the transitional stratum between the ether of space and the ether convected by the earth, special hypotheses concerning the motions of the ether must be made. Stokes found a hypothesis which satisfies all optical conditions. But it was shown later not to be in agreement with the laws of mechanics. Numerous attempts at rescuing Stokes's theory have led to no result, and it would have succumbed to internal difficulties even if Fresnel's theory had not been confirmed by Fizeau's experiment (see p. 139).

Fresnel's idea of partial convection cannot easily be deduced from Arago's experiment because refraction in lenses is a complicated process which involves not only the velocity but also the direction of the waves. But there is an equivalent experiment which was carried out later by Hoek (1868) and which is much easier to follow.

The principle underlying the arrangement of the apparatus is that of the interferometer (Fig. 74). The light falls from the source Q

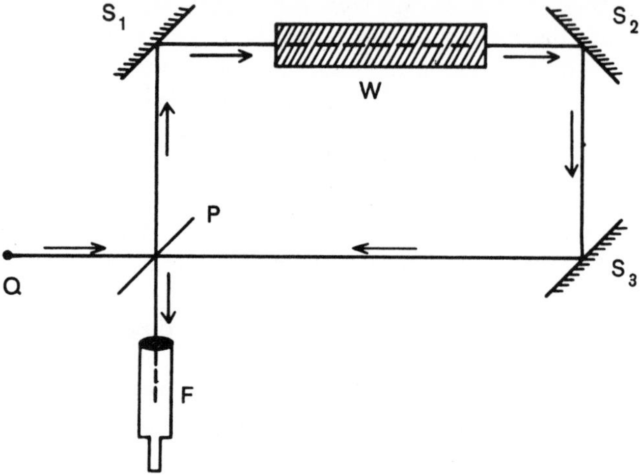

Fig. 74 *Interferometer experiment by Hoek.*

onto a half-silvered glass plate P inclined at 45° to the direction of the ray. This glass plate divides the ray into two parts. The reflected ray (ray 1) strikes consecutively the mirrors S_1, S_2, S_3, that form the corners of a square with P, and on its return to P is partly reflected into the telescope F. The transmitted ray (ray 2) traverses the same path in the opposite sense and interferes with ray 1 in the field of vision. A transparent body—say a tube W filled with water—is next interposed between S_1 and S_2, and the whole apparatus is mounted so that the straight line connecting S_1 with S_2 can be placed alternately in the same direction as, and opposite to, the earth's motion about the sun. Let the velocity of light in water that is at rest be c_1. This value is a little less than the velocity *in vacuo*, and the ratio of the one to the other $\frac{c}{c_1}=n$ is called the *refractive index* of water. The velocity of light in air differs inappreciably from c, and thus the refractive index of air is almost exactly equal to 1. Now the water is carried along by the earth in its orbit. If the ether in the water were not to participate in this motion at all, then the velocity of light in the water relative to the absolute ether (in outside space) would be unaltered; that is, it would be equal to c_1, and, for a ray travelling in the direction of the earth's motion, it would be $c_1 - v$ relative to the earth. If the ether is carried completely with the water, the velocity of light relative to the ether would be $c_1 + v$, and c_1 relative to the earth. We shall assume neither of these cases to begin with, but shall leave the amount of convection undetermined. Let the velocity of light in the moving water relative to the absolute ether be a little greater than c_1, say $c_1 + \phi$, and hence $c_1 + \phi - v$ relative to the earth. We wish to determine the unknown *convection coefficient* ϕ from experiment. If it is zero, no convection occurs; if it is v, complete convection occurs. Its true value must lie between these limits. We shall, however, make one assumption, namely, that the convection in air may be neglected in comparison with that in water.

Now, let l be the length of the tube of water. Then the ray 1 requires the time $\dfrac{l}{c_1 + \phi - v}$ to traverse the tube if the earth is moving in the direction from S_1 to S_2. To traverse the corresponding air distance between S_3 and P the same ray requires the time $\dfrac{l}{c+v}$. Thus the total time that the ray 1 requires to traverse the two equal paths in water and in air is

$$\frac{l}{c_1 + \phi - v} + \frac{l}{c+v}.$$

The ray 2 travels in the reverse direction. It first traverses the air-distance in the time $\dfrac{l}{c-v}$, then the water distance in the time $\dfrac{l}{c_1 - \phi + v}$, and hence altogether it requires for the same distances in the air and water the time

$$\frac{l}{c_1 - \phi + v} + \frac{l}{c-v}.$$

Now, experiment shows that the interferences do not shift in the slightest when the apparatus is turned into the direction opposite to that of the earth's velocity or, indeed, into any other position whatsoever. From this it follows that the rays 1 and 2 take equal times, independent of the position of the apparatus with respect to the earth's orbit, that is,

$$\frac{l}{c_1+\phi-v}+\frac{l}{c+v} = \frac{l}{c_1-\phi+v}+\frac{l}{c-v}.$$

We can calculate ϕ from this equation. If we neglect terms of the second and higher order, we obtain*

$$\phi = \left(1-\frac{1}{n^2}\right)v. \tag{44}$$

This is the famous convection formula of Fresnel, who actually found it by a different, more speculative, process. Before we discuss his assumption, let us see what the formula asserts. According to it the convection is the greater the more the refractive index exceeds the value 1 which it has *in vacuo*. For air, c_1 is almost equal to c, and n almost equal to 1, thus ϕ is almost zero, as we presumed above. The greater the refractive power, the more complete is the convection of the light. Now the velocity of light in a moving body, measured relative to the absolute ether is

$$c_1+\phi = c_1+\left(1-\frac{1}{n^2}\right)v,$$

and relative to the moving body it is

$$c_1+\phi-v = c_1+\left(1-\frac{1}{n^2}\right)v-v = c_1-\frac{v}{n^2}.$$

This last formula will serve us as a link to Fresnel's interpretation. He assumed that the density of the ether in a material body is different from the density in free ether; let the former be ρ_1 and the latter ρ.

* Remembering the approximation from page 126: $\frac{1}{1+\beta}=1-\beta$ for small β, one can write approximately $\frac{1}{c}\left(1-\frac{v}{c}\right)$ for $\frac{1}{c+v}$, $\frac{1}{c_1}\left(1-\frac{\phi-v}{c_1}\right)$ for $\frac{1}{c_1+\phi-v}$ and so forth. Then one gets immediately

$$2\frac{\phi-v}{c_1{}^2}+2\frac{v}{c^2} = 0$$

or

$$\phi = \left(1-\frac{c_1{}^2}{c^2}\right)v = \left(1-\frac{1}{n^2}\right)v.$$

We next imagine the moving body, say, in the form of a beam, whose length is parallel to the direction of motion; let its base face be of area F. In the motion of the beam through the ether the front face advances by the distance $v\tau$ in the time τ (Fig. 75), and this sweeps out a volume $Fv\tau$ (area of the face multiplied by the height).

a

b

c

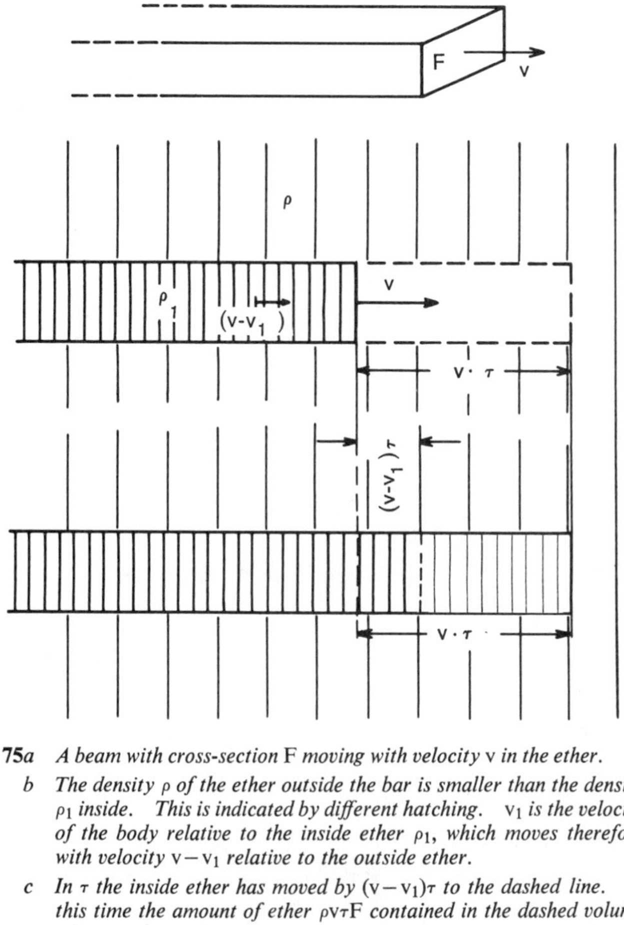

Fig. 75*a* *A beam with cross-section* F *moving with velocity* v *in the ether.*

b *The density* ρ *of the ether outside the bar is smaller than the density* ρ_1 *inside. This is indicated by different hatching.* v_1 *is the velocity of the body relative to the inside ether* ρ_1*, which moves therefore with velocity* $v-v_1$ *relative to the outside ether.*

c *In* τ *the inside ether has moved by* $(v-v_1)\tau$ *to the dashed line. In this time the amount of ether* $\rho v\tau F$ *contained in the dashed volume of Fig. 75b has flown into the bar, where it gets the density* ρ_1 *and the velocity* v_1. *Thus the flux of ether from outside* $\rho v\tau F$ *must just cover the added amount of ether inside the thin-hatched volume,* $\rho_1 v_1 \tau F$.

This volume contains an amount of ether $\rho F v \tau$. This enters into the beam through the front face. Here it assumes a new density and will thus move on with a different velocity v_1 with respect to the body, since, for the same reasons as above, its mass must also equal $\rho_1 F v_1 \tau$, and we get

$$\rho_1 F v_1 \tau = \rho F v \tau$$

or

$$v_1 = \frac{\rho}{\rho_1} v.$$

This is in a certain sense the strength of the ether wind in the beam moving with the velocity v. Light which moves with the velocity c_1 relative to the condensed ether moves relative to the body with the velocity

$$c_1 - v_1 = c_1 - \frac{\rho}{\rho_1} v.$$

Now we have seen that according to the result of Hoek's experiment the velocity of light relative to the moving body is

$$c_1 - \frac{1}{n^2} v.$$

Consequently we must have

$$\frac{\rho}{\rho_1} = \frac{1}{n^2} = \frac{c_1{}^2}{c^2}.$$

Thus the condensation $\frac{\rho_1}{\rho}$ is equal to the square of the coefficient of refraction.

Furthermore, we can conclude from this that the elasticity of the ether must be the same in all bodies. For formula (37) (p. 115) tells us that in every elastic medium $c^2 = \frac{p}{\rho}$. Thus, in ether $p = c^2 \rho$, in matter $p_1 = c_1{}^2 \rho_1$. But according to the above result concerning the condensation of ether in matter these two expressions are the same.

This mechanical interpretation of the convection coefficient by Fresnel has exerted a great influence on the elaboration of the elastic theory of light. But we must be aware that it is open to strong objections. As is well known, rays of light of different color (frequency) have different refractive indices n, that is, different velocities. Hence it follows that the convection coefficient has a different value for each color. But this is incompatible with Fresnel's interpretation, for then the ether would have to flow with a different velocity in the body according to the color. Thus there would be just as many ethers as there are colors, and that is surely impossible.

The convection formula (44), however, is founded on the results of experiment without regard to the mechanical interpretations. We shall see that it can be derived in the electromagnetic theory of light from ideas concerning the atomic structure of matter and electricity.

It is very difficult to test Fresnel's formula by means of experiments on the earth because it requires that transparent substances be moved with extreme rapidity. Fizeau succeeded in carrying out the experiment (1851) by means of a sensitive interferometer arrangement.

The apparatus used by him is quite similar to that of Hoek, except that both light paths S_1S_2 and S_3P are furnished with tubes in which the water can circulate; they are arranged so that ray 1 runs with the water, and ray 2 against it. Fizeau tested whether the water carries the light along with it by observing whether the interference fringes were displaced when the water was set into rapid motion.

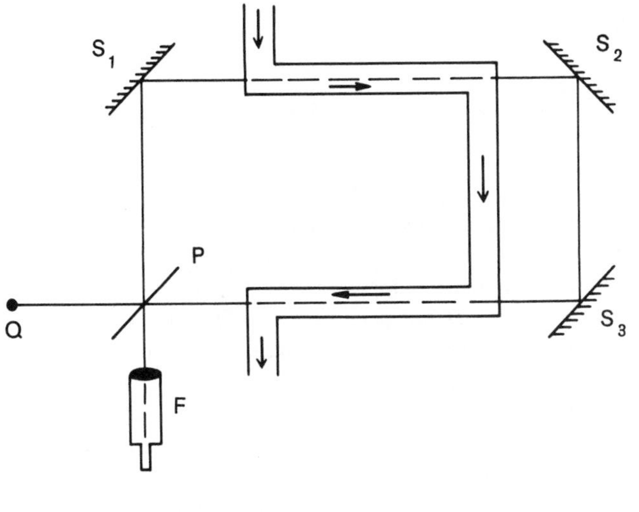

Fig. 76 *Fizeau's experiment to determine the convection coefficient.*

Such a displacement actually occurred, but very much less than would correspond to complete convection. Exact measurement disclosed perfect agreement with Fresnel's convection formula (44).

10. Aberration

We shall now discuss the influence of the motion of bodies on the direction of light rays, in particular the question of whether the motion of the earth through the ether can be ascertained by observing any phenomena accompanying changes of direction. Here again we must distinguish whether we are dealing with an astronomical or a terrestrial source of light.

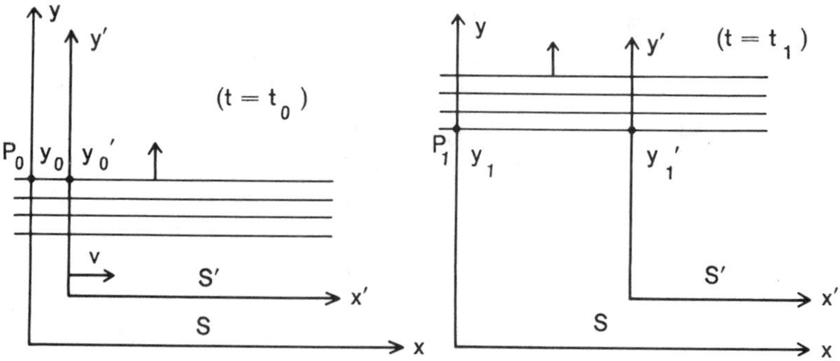

Fig. 77 *Observation of a train of waves moving in the* y*-direction in two frames of reference* S *and* S′ *moving relative to each other in the* x*-direction.*

The apparent deflection of the light that reaches the earth from the stars is the aberration which we have already discussed from the point of view of the corpuscular theory (IV, 3, p. 93). The explanation there given is very simple; it is more complicated from the point of view of the wave theory, for it is easy to see that a deflection of the wave planes does not occur at all. This is seen most readily in the case where the rays arrive perpendicular to the motion of the observer. For then the wave planes are parallel to the motion and are so perceived by the moving observer (Fig. 77). Calculation tells us the same. Let us consider a stationary coordinate system S and a system S' moving in such a way that the x-axis and x'-axis coincide with the direction of motion of S', and let us observe waves, which move in the direction of the y-axis, at two points P_0 and P_1; then we know from (38) that the number of waves n is invariant. We write this

$$n = \nu\left(t_1 - t_0 - \frac{y_1 - y_0}{c}\right) = \nu'\left(t_1 - t_0 - \frac{y_1' - y_0'}{c'}\right),$$

where we have written y_1, y_0, y_1', y_0' instead of x_1, x_0, x_1', x_0' because the waves move parallel to the y-axis. The Galileo transformation connecting S and S' is $x' = x - vt$, $y' = y$ according to (29). Now we calculate the characteristics of the waves as we have done in section 8 of this chapter. First we measure n at a fixed point, that is, we take $P_0 = P_1$ and $y_1 - y_0 = y_1' - y_0' = 0$. Thus we obtain $\nu = \nu'$. Then we make the observation at the same time $t_1 = t_0$: we get $\nu \dfrac{y_1 - y_0}{c} = \nu' \dfrac{y_1' - y_0'}{c}$. The Galileo transformation says $y_1 - y_0 = y_1' - y_0'$, therefore $\dfrac{\nu}{c} = \dfrac{\nu'}{c'}$. With $\nu = \nu'$ this gives the result $c = c'$.

Hence the moving observer sees a wave of exactly the same frequency, velocity, and direction. If the latter were altered, then the wave number in S', besides depending on y', would also have to depend on x'.

Thus it seems as if the wave theory is unable to account for the simple phenomenon of aberration, which has been known for almost 200 years.

But the position is not quite so bad as this would indicate. The reason for the failure of the argument just given is that the optical instruments with which the observations are made, and which include the naked eye, do not establish the position of the wave that arrives but accomplish something totally different.

The function of the eye or of the telescope is called optical imaging; it consists of combining the rays emitted by a luminescent object into an image. In this process part of the vibrational energy of the particles of the object is transported by the light waves to the corresponding particles of the image. The paths along which this transport of energy takes place are actually the physical rays. But energy is a quantity which, according to the law of conservation, can move about and be transformed just like a substance, but cannot be created or destroyed. Hence it seems reasonable to apply the laws of the corpuscular theory to the motion of energy. As a matter of fact, the simple derivation of the aberration formula given earlier is quite correct if we define the light rays as the energy paths of the light waves and apply the laws of relative motion to them as if they were streams of projected particles.

But we may also obtain this aberration formula without applying this concept of rays as energy paths, by following the refraction of the individual waves in the lenses or prisms of the optical instrument. For this we require a definite convection theory. Stokes's theory of complete convection can account for aberration only by making assumptions about the motion of the ether which are hardly admissible. We have already called attention to these difficulties. Fresnel's theory gives a law of refraction of waves of light at the surface of moving bodies from which the aberration formula follows exactly. The substance of the body through which the light passes does not affect the result, although the value of the convection coefficient is different in every substance. To test this directly, Airy (1871) filled a telescope with water and ascertained that the aberration retained its normal value. The aberration is, of course, no longer an effect of the first order if the lightwave and the observer have no motion relative to each other. From this it also follows that in all optical experiments with sources of light on the earth no deflection of the rays through the ether wind occurs. Fresnel's theory accounts for these facts in a way that agrees with experiment. It is unnecessary to enter into details.

11. Retrospect and Further Development

We have treated the luminiferous ether as a substance that obeys the laws of mechanics. Thus it satisfies the law of inertia, and hence where there is no matter, as in astronomic space, it will be at rest with respect to an appropriate inertial system. Now if we refer all phenomena to a different inertial system, exactly the same laws hold for the motions of bodies and of the ether, and also for the propagation of light, but of course only in so far as they concern accelerations and effects of mutual forces. We know that the velocity and direction of a motion are quite different with respect to different inertial systems, for we may regard every body that is moving in a straight line to be at rest merely by choosing a suitable system of reference, namely, one that moves with it. Thus in this almost trivial sense the classical principle of relativity must hold for the ether regarded as a mechanical substance.

From this it follows, however, that the velocity and direction of light rays must appear different in every inertial system. Thus it was to be expected that it would be possible to ascertain the velocity of the earth or of the solar system by observing optical phenomena at the surface of the earth which are determined by the velocity and direction of light. But all experiments performed with this end in view led to a negative result. Hence it appears that the velocity and direction of light rays are quite independent of the motion of the celestial body on which the observations are carried out. Or, in other words, optical phenomena depend only on the relative motions of material bodies.

This is a principle of relativity which seems quite similar to the classical principle of mechanics, and yet it has a different meaning. For it refers to velocities and directions of motion, and in mechanics these are not independent of the motion of the system of reference.

Now there are two possible points of view. One of these starts from the assumption that optical observations actually introduce something that is fundamentally new—namely, that light behaves differently from material bodies as regards direction and velocity. If the optical observations are accepted as convincing evidence, this point of view must be adopted, providing all speculations about the *nature* of light are left out of consideration. We shall see that Einstein finally pursued this path. However, it requires freedom from the conventions of traditional theory which is attained only when the Gordian knot of constructions and hypotheses has become so intricate that the only solution left is to cut it.

In our discussion above, however, we were still thinking in terms of the most flourishing period of the theory of the mechanical ether. This theory was compelled to regard the optical principle of relativity as a secondary, in a certain sense half-accidental, phenomenon brought about by the compensating effect of causes that were acting in opposite directions. The fact that it was possible to dismiss the anomalies of optical phenomena in this way was due somewhat to the circumstance that one could still make hypotheses about how the ether moves and is influenced in its motion by moving bodies. Now it is the achievement of Fresnel's convection hypothesis that it accounts for the optical principle of relativity, as far as quantities of the first order are concerned. Until the accuracy of optical measurements did attain the great improvement necessary to measure quantities of the second order, this theory satisfied all demands of experiment, with one possible exception, to which, curiously enough, very little attention was paid. If, however, improved accuracy in astronomical measurement should arrive at the result that by observing the eclipses of Jupiter's satellites according to the old method of Römer (see p. 91) no influence of the motion of the solar system on the velocity of light could be revealed, then certainly the ether theory would be confronted with a problem that would appear insoluble. For it is clear that this effect of the first order could be argued away by no hypothesis about the convection of the ether.

So we recognize the importance of the experimental task of measuring the dependence of optical events on the earth's motion up to quantities of the second order. Only the solution of this problem can decide whether the optical principle of relativity holds rigorously or only approximately. In the former case Fresnel's ether theory would fail; we should then be confronted with a new situation.

Historically, this occurred about 100 years after Fresnel's time. In the meanwhile the ether theory was developed in other directions. For at the outset there was not one ether but a whole series, an optical, a thermal, an electrical, a magnetic ether, and perhaps a few more. A special ether was invented as a carrier for every phenomenon that occurred in space. At first all these ethers had nothing to do with one another, but existed independently side by side in the same space. This state of affairs could not, of course, last. Relationships were soon found between the phenomena of different branches of physics that were at first separate, and so there emerged, finally, one ether as the carrier of all physical phenomena that occur in space free of matter. In particular, light was found to be an electromagnetic process of vibration and its carrier identical with the medium that transmits electric and magnetic forces. These discoveries gave the ether theory strong support. The ether came to be identified with Newtonian space. It was conceived to be at absolute rest and to transmit not only electromagnetic effects but also indirectly to generate the Newtonian inertial and centrifugal forces.

We shall now describe the development of this theory. The process has features resembling the trial of a case in court. The ether is alleged to be the universal culprit; the pieces of evidence accumulate overwhelmingly until at the end the undeniable proof of an alibi—namely, Michelson and Morley's experiment about the quantities of second order, and its interpretation by Einstein—puts an end to the whole business.

CHAPTER V

THE FUNDAMENTAL LAWS OF ELECTRODYNAMICS

1. Electro- and Magnetostatics

The fact that a certain kind of ore, magnetite, attracts iron, and that rubbed amber (*elektron* in Greek) attracts and holds light bodies was already known to the ancients. But the sciences of magnetism and electricity are products of more recent times which had been trained by Galileo and Newton to ask rational questions of nature with the help of experiment.

The fundamental facts of electric phenomena, which we shall now recapitulate briefly, were established after the year 1600. At that time friction was the exclusive means of producing electrical effects. Gray discovered (1729) that metals, when brought into contact with bodies that have been electrified by friction, themselves acquire similar properties. He showed that electricity can be conducted in metals. This led to the classification of substances as *conductors* and *nonconductors* (*insulators*). It was discovered by du Fay (1730) that electrical action is not always *attraction* but may also be *repulsion.* To account for this fact he assumed the existence of two fluids (nowadays we call them *positive* and *negative electricity*), and he established that similarly charged bodies repel each other, while oppositely charged bodies attract each other.

We shall define the concept of *electric charge* quantitatively. In doing so we will not follow the oftentimes very circuitous steps of argument that led historically to the enunciation of the concepts and laws, but rather we shall select a series of definitions and experiments in which the logical sequence emerges most clearly.

Let us imagine a body M that has somehow been electrified by friction. This now acts attractively or repulsively on other electrified bodies. To study this action we shall take small test bodies, say spheres, whose diameters are very small compared with the distance of their closest approach to the body M. If we bring a test body P near the body M, P experiences a statical force of definite magnitude and direction which may be measured by the methods of mechanics, say, by balancing it against a weight with the help of levers or threads. It is found at once qualitatively that the *force decreases* with *increasing distance PM*.

We next take two such test bodies P_1 and P_2, bring them in turn to the same point in the vicinity of M, and measure in each case the forces K_1 and K_2 as regards size and direction. We shall henceforth adopt the *convention* that *opposite forces* are to be regarded as being in the same direction and having *opposite signs*. Experiment shows that the two forces have the same direction but that their values may be different and they may have different signs.

Now let us bring the two test bodies to a different point near M and let us again measure the forces K_1' and K_2' as regards value and direction. Again they have the same direction, but in general they have different values and a different sign.

If we next form the ratio $K_1:K_2$ of the forces at the first point, and then the ratio $K_1':K_2'$ at the second, it is found that both have the same value, which may be positive or negative:

$$\frac{K_1}{K_2} = \frac{K_1'}{K_2'}.$$

From this result we may conclude:

1. The direction of the force exerted by an electrified body M on a small test body P does not depend at all on the nature and the amount of electrification of the test body, but only on the properties of the body M.

2. The ratio of the forces exerted on two test bodies brought to the same point in turn is quite independent of the choice of the point, that is, of the position, nature, and electrification of the body M. It depends only on the properties of the test bodies.

We now choose a definite test body, electrified in a definite way, and let its charge be the unit of charge or amount of electricity q. With the aid of this test body we measure the force that the body M exerts at many places. Let this force be denoted by K_q. Then this also determines the direction of the force K exerted on any other test body P. The ratio $K:K_q$, however, depends only on the test body P and defines the ratio e of the electric charge of P and the unit of charge q. This may be positive or negative depending upon whether K and K_q are in the same or in opposite directions. Thus we have in any position: $\frac{K}{e}=\frac{K_q}{q}$.

From this one concludes that $\frac{K}{e}$ depends only on the electrical nature of the body M. Therefore we call the quotient $\frac{K}{e}=\frac{K_q}{q}$ the *electrical field strength E*. This quantity E determines the electrical action of M in the surrounding space, or as we usually say, its *electric field*. From $\frac{K}{e}=E$ follows

$$K = eE. \tag{45}$$

As for the choice of the unit charge, it would be almost impossible to fix this in a practical way by a decree concerning the electrification of a definite test body; a mechanical definition would be preferable. This can be arrived at as follows:

We first give two test bodies equal charges. The criterion of equal charges is that they are subject to the same force from the third body M when placed at the same point near M. The two bodies will then repel each other with the same force. We now say that their charge equals the unit of charge q if this repulsion is equal to the unit of force when the distance between the two test bodies is equal to unit length. No assumption is made here about the dependence of the force on the distance.

Through these definitions the amount of electricity or the electric charge becomes a measurable quantity just as length, mass, or force may be measured.

The most important law about amounts of electricity, which was enunciated independently in 1747 by Watson and Franklin, is that in every electrical process equal amounts of positive and negative electricity are always formed. For example, if we rub a glass rod with a piece of silk, the glass rod becomes charged with positive electricity; an exactly equal negative charge is then found on the silk.

This empirical fact may be interpreted by saying that the two kinds of electrification are *not generated* by friction but are *only separated.* They may be thought of as two *fluids* that are present in all bodies in equal quantities. In nonelectrified "neutral" bodies they are everywhere present to the same amount so that their outward effects are counterbalanced. In electrified bodies they are separated. One part of the positive electricity, say, has flowed from one body to another; just as much negative has flowed in the reverse direction.

But it is clearly sufficient to assume *one* fluid that can flow independently of matter. Then we must ascribe to matter that is free of this fluid a definite charge, say positive, and to the fluid the opposite charge, that is, negative. Electrification consists of the flowing of negative fluid from one body to the other. The first body will then become positive because the positive charge of the matter is no longer wholly compensated; the other becomes negative because it has an excess of negative fluid.

The struggle between the supporters of these two hypotheses, the one-fluid theory and the two-fluid theory, lasted a long time, and of course remained futile and purposeless until it was decided by the discovery of new facts. We shall not enter further into these discussions, but shall only state briefly that characteristic differences were finally found in the behavior of the two kinds of electricity; these differences indicated that positive electrification is actually firmly attached to matter but that negative electrification can move more or less freely. This doctrine still holds today. We shall revert to this point later in dealing with the theory of electrons.

Another controversy arose around the question of how the electrical forces of attraction and repulsion are transmitted through space. The first decades of electrical research came before the Newtonian theory of attraction. Action at a distance seemed unthinkable. Metaphysical theorems were held to be valid (for example, that matter can act only at points where it is present) and diverse hypotheses were evolved to explain electrical forces—for example, that emanations flowed from the charged bodies and exerted a pressure when they impinged on bodies, and similar assumptions. But after Newton's theory of gravitation had been established, the idea of a force acting directly at a distance gradually became a habit of thought. For it is, indeed, nothing more than a thought habit when an idea impresses itself so strongly on minds that it is used as the ultimate principle of explanation. It does not then take long for metaphysical speculation, often in the garb of philosophic criticism, to maintain that the correct or accepted principle of explanation is a logical necessity and that its opposite cannot be imagined. But fortunately progressive empirical science does not as a rule trouble about this, and when new facts demand it, it often has recourse to ideas that have been condemned. The development of the doctrine of electric and magnetic forces is an example of such a cycle of theories. First came a theory of contiguous action based on metaphysical grounds, later a theory of action at a distance on Newton's model. Finally this became transformed, owing to the discovery of new facts, into a general theory of contiguous action again. This fluctuation is no sign of weakness. For it is not the pictures that are connected with the theories which are the essential features, but the empirical facts and their conceptual relationships. Yet if we follow these we see no fluctuation but only a continuous development full of inner logical consistency. We may justifiably pass by the first theoretical attempts of pre-Newtonian times because the facts were known too incompletely to furnish really convincing starting points. But the rise of the theory of action at a distance in Newtonian mechanics is founded quite solidly on facts of observation. Research which had at its disposal only the experimental means of the eighteenth century was bound to come to the decision that the electric and magnetic forces act at a distance in the same way as gravitation. Even nowadays it is still permissible, from the point of view of the highly developed theories of contiguous action of Faraday and Maxwell, to represent electro- and magnetostatic forces by means of actions at a distance, and when properly used, they lead to correct results.

The idea that electric forces act like gravitation at a distance was first conceived by Aepinus (1759). He did not succeed in setting up the correct law for the dependence of electric actions on the distance, but he was able to explain the phenomenon of electrostatic induction qualitatively. This consists of a charged body acting attractively not only on other charged bodies but also on uncharged bodies, particularly on conducting bodies: a charge of the opposite sign is induced on the side of the influenced body nearest the acting body, whereas a charge of the same sign is driven to the farther side

(Fig. 78); hence, since the forces decrease with increasing distance, the attraction outweighs the repulsion.

The exact law of this decrease was presumably first found by Priestley, the discoverer of oxygen (1767). He discovered the law in an ingenious indirect way which was more convincing than a direct measurement would have been. Independently Cavendish (1771) derived the law by similar reasoning. But it received its name from the physicist who first proved it by measuring the forces directly, Coulomb (1785).

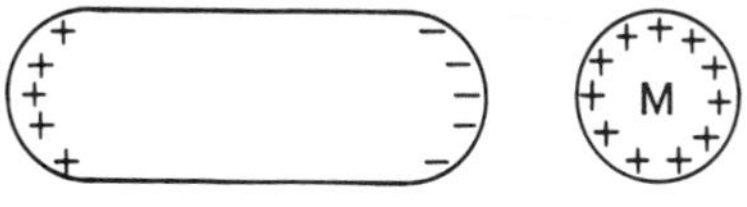

Fig. 78 *A charged body* M *influences charges on an originally uncharged body.*

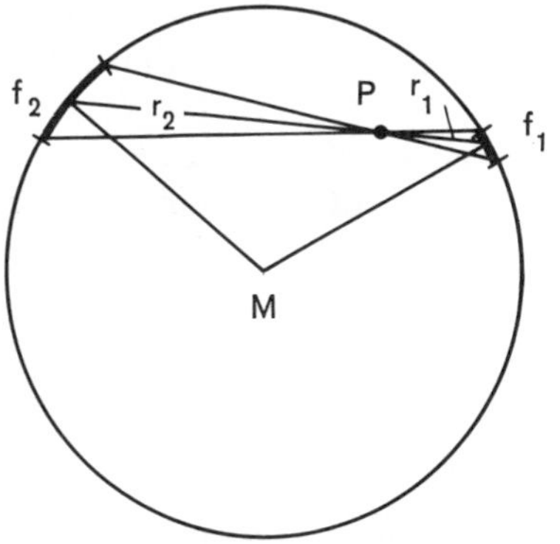

Fig. 79 *Derivation of Coulomb's law.*

The argument of Priestley and Cavendish ran somewhat as follows: If an electric charge is given to a conductor, then it cannot remain in equilibrium in the interior of the conducting substance, since particles of the same charge repel each other. Rather, they must tend to the outer surface where they distribute themselves in a certain way so as to be in equilibrium. Now, experiment teaches very definitely that no electric field exists within a space that is enclosed on all sides by metallic walls, no matter how strongly the envelope is charged. The charges on the outer surface of the empty space must thus distribute themselves so that the force exerted at each point in the interior vanishes. Now, if the empty space has the particular form of a sphere, the charge, for reasons of symmetry, can only be distributed uniformly over the surface. If ρ is the charge per unit area of surface (density of charge), then the amounts of electricity on two portions f_1, f_2 of the surface are ρf_1, and ρf_2. The force that the portion of surface f_1 exerts on a test body P situated in the interior of the sphere and carrying the charge e is then $K_1 = \frac{e\rho f_1}{q^2} R_1$, where R_1 denotes the force which acts between two units of charge q placed at P and f_1, and which somehow depends on the distance r_1 between P and f_1. Now, corresponding to each portion of surface f_1 there is an opposite portion f_2 which is obtained by connecting the points of the boundary of f_1 with P and producing these lines through P until they intersect the sphere again. The two areas f_1 and f_2 are thus cut out of the surface of the sphere by the same double cone with its apex at P (Fig. 79), and the angles between them and the axis of the double cone are equal. The areas of f_1 and f_2 are thus in the ratio of the squares of the distances from P:

$$\frac{f_2}{f_1} = \frac{r_2^2}{r_1^2}.$$

The charge ρf_2 on f_2 exerts the force $K_2 = \frac{e\rho f_2}{q^2} R_2$ on P, where R_2 depends on r_2 in some way; K_2 is of course oppositely directed to K_1. The idea readily suggests itself that all forces acting on P can neutralize each other only if the forces due to two opposite portions of area exactly counterbalance, that is, when $K_1 = K_2$. It is possible to prove this assumption, but that would take us too far. If we take it for granted, then it follows that $f_1 R_1 = f_2 R_2$, or

$$\frac{R_1}{R_2} = \frac{f_2}{f_1} = \frac{r_2^2}{r_1^2}.$$

Hence

$$R_1 r_1^2 = R_2 r_2^2 = C,$$

where C is a quantity independent of the distance r. This determines R_1 and R_2, namely,

$$R_1 = \frac{C}{r_1^2}, \qquad R_2 = \frac{C}{r_2^2}.$$

Hence, in general, the force R between two unit charges at a distance

r apart must have the value $R=\frac{C}{r^2}$, and the force K between two charges e_1 and e_2 at the same distance r is therefore $K=\frac{C}{r^2}\frac{e_1e_2}{q^2}$.

In conformity with our convention about the unit of electric charge we must set $C=1\times$ unit of force $\times$ (unit of length)^2; then we define the dimensions of charge by putting $C=q^2$. Now the force between two unit charges a unit distance apart is to be equal to one unit of force. With this convention the force that two bodies carrying charges e_1 and e_2 and at a distance r apart exert on each other is

$$K = \frac{e_1e_2}{r^2}. \tag{46}$$

This is *Coulomb's law*. In its formulation we assume, of course, that the greatest diameter of the charged bodies is small compared with their distances apart. This restriction means that we have to do, just as in the case of gravitation, with an idealized elementary law. To deduce from it the action of bodies of finite extent we must consider the electricity distributed over them to be divided into small parts, then calculate the effects of all the particles of the one body on all those of the others in pairs and sum them.

Formula (46) fixes the dimensions of quantity of electricity, since we have for the repulsion of two equal charges $\frac{e^2}{r^2}=K$, that is, $e=r\sqrt{K}$, hence

$$[e] = [l\sqrt{K}] = \left[l\sqrt{\frac{ml}{t^2}}\right] = \left[\frac{1}{t}\sqrt{ml}\right].$$

This, at the same time, fixes the unit of charge in the C.G.S. system; it must be written $\frac{\text{cm.}\sqrt{\text{gm. cm.}}}{\text{sec.}}$.

The electric intensity of field E, defined by $K=eE$, has the dimensions

$$[E] = \left[\frac{K}{e}\right] = \left[\frac{K}{l\sqrt{K}}\right] = \left[\frac{\sqrt{K}}{l}\right] = \left[\frac{\sqrt{ml}}{lt}\right] = \left[\frac{1}{t}\sqrt{\frac{m}{l}}\right],$$

and its unit is $\frac{1}{\text{sec.}}\sqrt{\frac{\text{gm.}}{\text{cm.}}}$.

After Coulomb's law had been established, electrostatics became a mathematical science. Its most important problem is this: Given the total quantity of electricity on conducting bodies, to calculate the distribution of charges on them under the action of their mutual influence, and also the forces due to these charges. The development of this mathematical problem is interesting in that it very soon became changed from the original formulation based on the theory of action at a distance to a theory of pseudocontiguous action, that is, in place of the summations of Coulomb forces there were obtained differential equations in which the field E or a related quantity called potential occurred as the unknown. However, we cannot discuss these purely mathematical questions any further here but only mention the names of Laplace (1782), Poisson (1813), Green (1828), and Gauss (1840), who have played a prominent role in their solution. We shall emphasize only one point. In this treatment of electrostatics, which is usually called the theory of potential, we are not dealing with a true theory of contiguous action in the sense which we attached to this expression before (IV, 6, p. 108), for the differential equations refer only to the change in the intensity of field from place to place and contain no term that expresses a change in time. Hence they entail no transmission of electric force with finite velocity but, in spite of their differential form, they represent an instantaneous action at a distance.

The theory of magnetism was developed in the same way as that of electrostatics. We may, therefore, express ourselves briefly.

A lozenge-shaped magnetized body, a *magnet needle*, has two *poles*, that is, points from which the magnetic force seems to start out, and the law holds that like poles repel, unlike poles attract one another. If we break a magnet in half, the two parts do not carry opposite magnetic charges, but each part shows a new pole near the new surface and again represents a complete magnet with two equal but opposite poles. This holds, no matter into how many parts the magnet be broken.

From this it has been concluded that there are indeed two kinds of magnetism as in the case of electricity except that they cannot move freely, and that they are present in the smallest particles of matter, molecules, in equal quantities, but separated by a small distance. Thus each molecule is itself a small magnet with a north and a south pole (Fig. 80). In a body that is not magnetized all the elementary magnets are in complete disorder. Magnetization consists of bringing them into the same direction. Then the effects of the alternate north (+) and south (−) poles counterbalance, except at the two ends which therefore seem to be the sources of the magnetic effects.

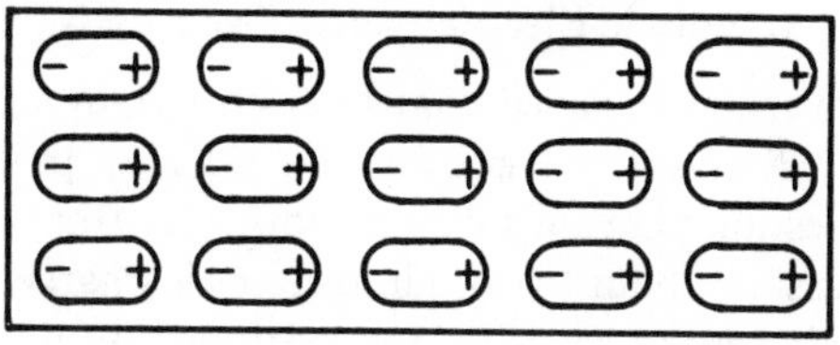

Fig. 80 *A magnetized body consisting of elementary magnets.*

By using a very long, thin magnetized needle one can be sure that in the vicinity of the one pole the force of the other becomes negligible. Hence in magnetism, too, we may operate with test bodies, namely, with the poles of very long, thin magnetic rods. These allow us to carry out all the measurements that we have already discussed in the case of electricity. We thus succeed in defining the amount of magnetism or the *pole strength* p and the *magnetic intensity of field* H. The magnetic force that a pole p experiences in the field H is

$$K = pH.$$

The unit of pole is chosen so that two unit poles at a unit distance apart exert a repulsive force of 1 on each other. The law according to which the force between two poles p_1 and p_2 changes with the distance was also found by Coulomb from direct measurement. As with Newton's law of attraction, it has the form

$$K = \frac{p_1 p_2}{r^2}. \tag{46a}$$

Clearly the dimensions of magnetic quantities are the same as those of the corresponding electric quantities, and their units have the same notation in the C.G.S. system.

The mathematical theory of magnetism runs almost parallel with that of electricity. The most essential difference is that magnetism is attached to the molecules, and that the measurable accumulations that condition the occurrence of poles in the case of finite magnets arise only owing to the summation of molecules that point in the same direction. One cannot separate the two kinds of magnetism and make a body, for example, a north pole.

2. Voltaic Electricity and Electrolysis

The discovery of so-called contact electricity by Galvani (1780) and Volta (1792) is so well known that we may pass by it here. However interesting Galvani's experiments with frog's legs and the resulting discussion about the origin of electric charges may be, we are here more concerned with formulating concepts and laws. Hence we shall recount only the facts.

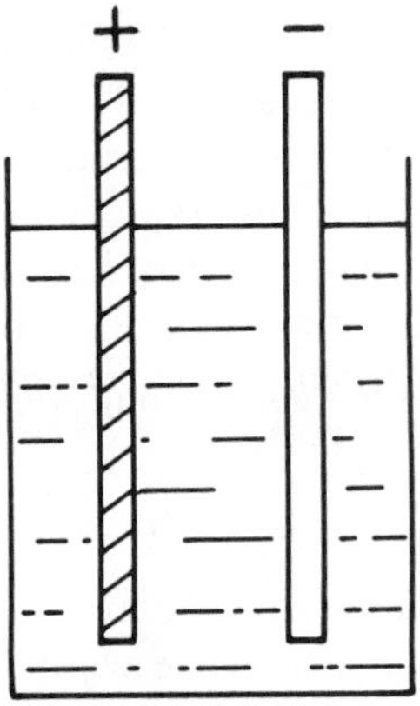

Fig. 81 *Voltaic cell.*

If two different metals are dipped into a solution (Fig. 81), say, copper and zinc into dilute sulphuric acid, the metals manifest electric charges that have exactly the same properties as frictional electricity. According to the fundamental law of electricity, charges of both signs occur on the metals (poles) to the same amount. The system composed of the solution and the metals, which is called a *voltaic element* or *cell*, thus has the power of separating the two kinds of electricity. Now, it is remarkable that this power is apparently inexhaustible, for if the poles are connected by a wire so that their charges flow around and neutralize each other, as soon as the wire is again removed, the poles are still charged. Thus the element continues to keep up the supply of electricity as long as the wire connection is maintained. Hence a continuous flow of electricity must be taking place. How this is to be imagined in detail depends on whether the one-fluid or the two-fluid theory is accepted. In the former case only one current is present; in the latter, two opposite currents, one of each fluid, flow.

Now, the *electric current* manifests its existence by showing very definite effects. Above all it heats the connecting wire. Everyone knows this fact from the metallic filaments in our electric bulbs. Thus the current continually produces heat energy. From what does the voltaic element derive the power of producing electricity continually and thereby indirectly generating heat? According to the law of conservation of energy, wherever one kind of energy appears during a process, another kind of energy must disappear to the same extent.

The source of energy is the chemical process in the cell. One metal dissolves as long as the current flows; at the same time a constituent of the solution separates out on the other. Complicated chemical processes may take place in the solution itself. We have nothing to do with these but content ourselves with the fact that the voltaic element is a means of generating electricity in unlimited quantities and of producing considerable electric currents.

We shall now have to consider, however, the reverse process, in which the electric current produces a chemical decomposition. For example, if we allow the current between two indecomposable wire leads (*electrodes*), say of platinum, to flow through slightly acidified water, the latter resolves into its components, hydrogen and oxygen, the hydrogen coming off at the negative electrode (*cathode*), the oxygen at the positive electrode (*anode*). The quantitative laws of this process of "*electrolysis*," discovered by Nicholson and Carlisle (1800), were found by Faraday (1832). The far-reaching consequences of Faraday's researches for the knowledge of the structure of matter are well known; it is not the consequences themselves that lead us to discuss these researches but the fact that Faraday's laws furnished the means of measuring electric currents accurately, and hence allowed the structure of electromagnetic theory to be completed.

This experiment of electrolytic dissociation can be carried out not only with a voltaic current, but just as well with a discharge current, which occurs when oppositely charged metallic bodies are connected by a wire. Care must be taken that the quantities of electricity that are discharged are sufficiently great. We have apparatus for storing electricity, so-called *condensers*, whose action depends on the induction principle, and which give such powerful discharges that measurable amounts are decomposed in the electrolytic cell. The amount of the charge that flows through the cell may be measured by the methods of electrostatics discussed above. Now Faraday discovered the law that twice the charge produces twice the dissociation, three times the charge three times the dissociation—in short, that the amount m of dissociated substance (or of one of the products of dissociation) is proportional to the quantity e of electricity that has passed through the cell:

$$Cm = e.$$

The constant C depends on the nature of the substances and of the chemical process.

A second law of Faraday regulates this dependence. It is known that chemical elements combine in perfectly definite proportions to form compounds. The quantity of an element that combines with 1 gm. of the lightest element, hydrogen, is called its *equivalent weight*. For example, in water (H_2O) 8 gm. of oxygen (O) are combined with 1 gm. of hydrogen (H), hence oxygen has the equivalent weight 8 gm. Now Faraday's law states that the same quantity of electricity that separates out 1 gm. of hydrogen is able to separate out an equivalent weight of every other element, for example, 8 gm. of oxygen.

Hence the constant C need only be known for hydrogen, and then we get it for every other substance by dividing this value by the equivalent weight of the substance. We have for hydrogen with its equivalent weight $\mu_0 = 1$ gm.

$$C_0\mu_0 = e$$

and for any other substance with the equivalent weight μ

$$C\mu = e.$$

By dividing these equations we get

$$\frac{C\mu}{C_0\mu_0} = 1, \qquad \text{i.e.,} \quad C = C_0\frac{\mu_0}{\mu}.$$

Thus $C_0\mu_0 = C_0 \times 1$ gm. is the exact quantity of electricity that separates out 1 gm. of hydrogen. The numerical value of C_0 has been determined by exact measurements and amounts in the C.G.S. system to

$$C_0 = 2.90 \times 10^{14} \text{ units of charge per gram.} \tag{47}$$

Now we may combine Faraday's two laws into the one formula:

$$e = C_0\frac{\mu_0}{\mu}m. \tag{48}$$

Thus electrolytic dissociation furnishes us with a very convenient measurement of the quantity of electricity e that has passed through the cell during a discharge. We need only determine the mass m of a product of decomposition that has the equivalent weight μ and then we get the desired quantity of electricity from equation (48). It is of course a matter of indifference whether this electricity is obtained from the discharge of charged conductors (condensers) or whether it comes from a voltaic cell. In the latter case the electricity flows continuously with constant strength; this means that the charge $e = J \times t$ passes through any cross-section of the conducting circuit and hence also through the decomposing cell in the time t. Here the quantity

$$J = \frac{e}{t} = C_0\frac{\mu_0}{\mu}\frac{m}{t} \tag{49}$$

is called the *intensity of current* or *current strength*, for it measures how much electrical charge flows through the cross-section of the conductor per unit time.

Its dimensions are

$$[J] = \left[\frac{e}{t}\right] = \left[\frac{l}{t}\sqrt{K}\right] = \left[\frac{l}{t^2}\sqrt{ml}\right],$$

and its unit is

$$\frac{\text{cm.}\sqrt{\text{gm. cm.}}}{\text{sec.}^2}$$

3. Resistance and Heat of Current

We must next consider the process of conduction or current itself. It has been customary to compare the electric current with the flowing of water in a pipe and to apply the concepts there valid to the electrical process. If water is to flow in a tube there must be some driving force. If it flows from a higher vessel through an inclined tube to a lower vessel, gravitation is the driving force (Fig. 82). This is greater, the higher the upper surface of the water

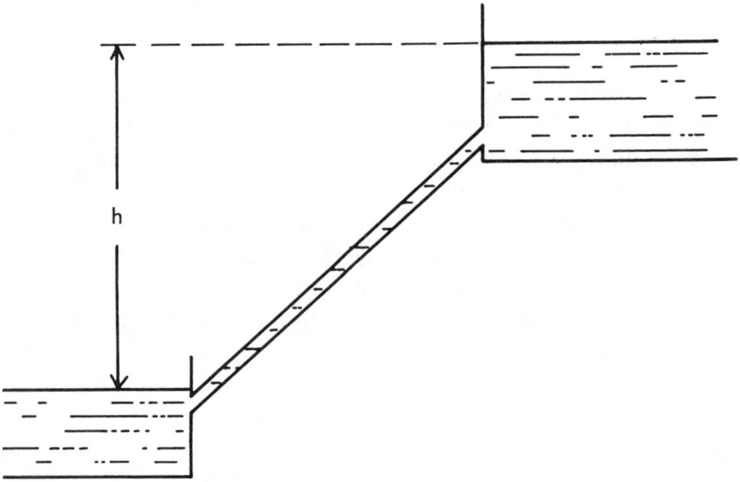

Fig. 82 *The current strength of the water is proportional to the potential difference* V *and therefore to the difference* h *in height of the two levels.*

is above the lower. But the velocity of the current of water, or its current strength, depends not only on the forces exerted by gravitation but also on the resistance that the water experiences in the conducting tube. If this is long and narrow, the amount of water passing through per unit of time is less than in the case of a short wide tube. The current strength J is thus proportional to the difference V of potential energy that drives the water (which is proportional to the difference in height h of the two levels; see p. 48) and inversely proportional to the *resistance* W. We set

$$J = \frac{V}{W} \quad \text{or} \quad JW = V, \tag{50}$$

in which the unit of resistance chosen is that which allows one unit of current to flow when the difference of level is one unit of height.

G. S. Ohm (1826) applied precisely the same ideas to the electric current. The difference of level that effects the flow corresponds to the electric force. We define the sign of the current as positive in the direction from the positive to the negative pole. For a definite piece of wire of length l we must set $V = El$, where E is the field strength, which is regarded constant along the wire. For if the same electric field acts over a greater length of wire, it furnishes a stronger impulse to the flowing electricity. The force V is also called the *electromotive force* (*difference of potential* or level). It is, moreover, identical with the concept of electric potential which we mentioned above (p. 154).

Since the current strength J and the electric intensity of field E, hence also the potential difference or electromotive force $V = El$, are measurable quantities, the proportionality between J and V expressed in Ohm's law may be tested experimentally.

The resistance W depends on the material and the form of the conducting wire; the longer and thinner it is, the greater is W. If l is the length of the wire and f the size of the cross-section, then W is directly proportional to l, and inversely proportional to f. We set

$$\sigma W = \frac{l}{f} \quad \text{or} \quad W = \frac{1}{\sigma}\frac{l}{f}, \tag{51}$$

where the factor of proportionality σ depends only on the material of the wire and is called the *conductivity*.

If we substitute W from (51) and $V = El$ in (50), we get

$$JW = J\frac{l}{f\sigma} = V = El.$$

By cancelling l we obtain

$$\frac{J}{f\sigma} = E \quad \text{or} \quad \frac{J}{f} = \sigma E.$$

But $\frac{J}{f}$ denotes the current strength per unit cross section. This is called the *current density*, and is denoted by j. We thus have

$$j = \sigma E. \tag{52}$$

In this form Ohm's law is left with only one constant whose value depends on the conducting material, namely, the conductivity, but in no other way depending on the form and size of the conducting body (wire).

In the case of insulators $\sigma = 0$. But ideal insulators do not exist. Very small traces of conductivity are always present except in a complete vacuum. There is an unbroken sequence leading from bad conductors (such as porcelain or amber) to the metals, which have enormously high conductivity.

We have already pointed out that the current heats the conducting wire. The quantitative law of this phenomenon was found by Joule (1841). It is clearly a special case of the law of conservation of energy, in which electric energy becomes transformed into heat. Joule's law states that the heat developed per unit of time by the current J in traversing the potential difference V is

$$Q = JV, \tag{53}$$

where Q is to be measured not in calories but in mechanical units of work. We shall make no further use of this formula, and state it here merely for the sake of completeness.

4. Electromagnetism

Up to the early nineteenth century, electricity and magnetism were regarded as two regions of phenomena which were similar in some respects but quite separate and independent. A bridge was eagerly sought between the two regions, but for a long time without success. At last Oersted (1820) discovered that the magnetic needle is deflected by voltaic currents. In the same year Biot and Savart discovered the quantitative law of this phenomenon, which Laplace formulated in terms of action at a distance. This law is very important for us, for the reason that in it there occurs a constant, peculiar to electromagnetism and of the nature of a velocity, which showed itself later to be identical with the velocity of light.

Biot and Savart established that the current flowing in a straight wire neither attracts nor repels a magnetic pole, but strives to drive it around in a circle about the wire (Fig. 83), so that the positive pole moves in the sense of a right-handed screw turned from below (contrary to the hands of a watch) about the (positive) direction of the current. The quantitative law can be brought into the simplest form by supposing the conducting wire to be divided into a number of short pieces of length l and writing down the effect of these current elements, from which the effect of the whole current is obtained by summation. We shall state the law of a current element only for the special case in which the magnetic pole lies in the plane that passes through the middle part of the element and is perpendicular

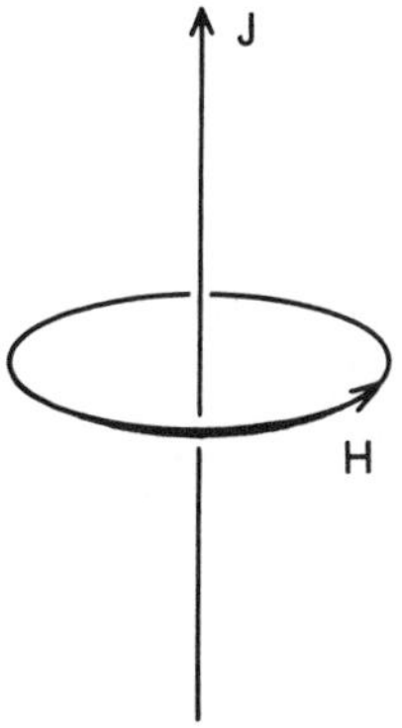

Fig. 83 *The magnetic field* H *surrounding a current* J.

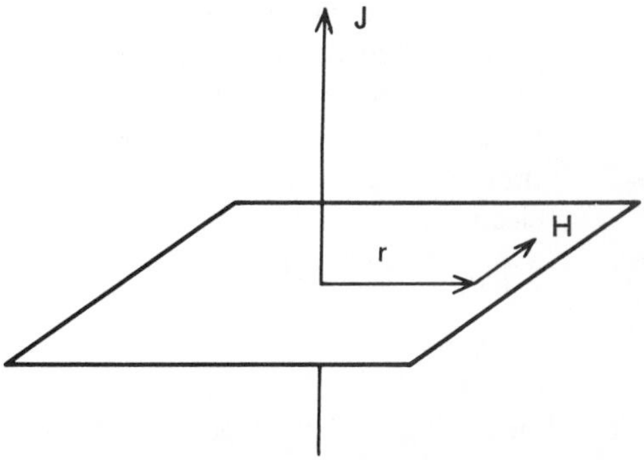

Fig. 84 *The direction of* H *is perpendicular to the directions of* J *and the radius vector* r.

to its direction (Fig. 84). Then the force that acts on the magnet pole of unit strength, i.e., the magnetic intensity of field H in this plane, is perpendicular to the line connecting the pole with the mid-point of the current element, and is directly proportional to the current intensity J and to its length l, and inversely proportional to the square of the distance r:

$$cH = \frac{Jl}{r^2}. \tag{54}$$

Outwardly this formula has again a similarity to Newton's law of attraction or Coulomb's law of electrostatics and magnetostatics, but the electromagnetic force has nevertheless a totally different character. For it does not act in the direction of the connecting line but perpendicular to it. The three directions J, r, H are perpendicular to each other in pairs. From this we see that electrodynamic effects are intimately connected with the structure of Euclidean space; in a certain sense they furnish us with a natural rectilinear coordinate system.

The factor of proportionality c introduced in formula (54) is completely determined since the distance r, the current strength J, and the magnetic field H are measurable quantities. It clearly denotes the strength of that current which, flowing through a piece of conductor of unit length, produces a unit of magnetic field at a unit distance. It is customary and often convenient to choose in place of the unit of current that we have introduced (namely, the quantity of static electricity that flows through the cross-section per unit of time and is called the electrostatic unit), this current of strength c (in electrostatic measure) as the unit of current; it is then called the electromagnetic unit of current. Its use has an advantage in that equation (54) assumes the simple form $H=\frac{Jl}{r^2}$ or $J=\frac{Hr^2}{l}$, so that measurement of the strength of a current is reduced to that of two lengths and of a magnetic field. Most practical instruments for measuring currents depend on the deflection of magnets by currents, or the converse, and hence give the current strength in electromagnetic measure. To express this in terms of the electrostatic measure of current first introduced, the constant c must be known; for this, however, only one measurement is necessary.

Before we speak of the experimental determination of the quantity c, we shall get an insight into its nature by means of a simple dimensional consideration. According to (54) it is defined by $c=\frac{Jl}{Hr^2}$.

Now the following dimensional formulae hold:

$$[J] = \left[\frac{e}{t}\right], \qquad [H] = \left[\frac{p}{l^2}\right],$$

hence the dimensions of c become

$$[c] = \left[\frac{el}{pt}\right].$$

But we know that the electric charge e and the magnetic strength of pole p have the same dimensions because Coulomb's law for electric and magnetic force is exactly the same. Hence we get

$$[c] = \left[\frac{l}{t}\right],$$

that is, c has the dimensions of a velocity.

The first exact measurement of c was carried out by Weber and Kohlrausch (1856). These experiments belong to the most memorable achievements of precise physical measurement, not only on account of their difficulty but also on account of the far-reaching consequences of the result. *For the value obtained for* c *was* 3×10^{10} *cm./sec., which is exactly the velocity of light.*

This equality could not be accidental. Numerous thinkers, including Weber himself and many other mathematicians and physicists, felt the close relationship that the number $c = 3 \times 10^{10}$ cm./sec. established between two great realms of science, and they sought to discover the bridge that ought to connect electromagnetism and optics. This was accomplished by Maxwell after Faraday's wonderful and ingenious method of experimenting had brought to light new facts and new views. We shall next pursue this development.

5. Faraday's Lines of Force

Faraday came from no learned academy; his mind was not burdened with traditional ideas and theories. His sensational rise from a bookbinder's apprentice to the world-famous physicist of the Royal Institution of London is well known. The world of his ideas, which arose directly and exclusively from the abundance of his experiments, was just as free from conventional schemes as his life. We discussed already his researches on electrolytic dissociation. His method of trying all conceivable changes in the experimental conditions led him (1837) to insert nonconductors like petroleum and turpentine between the two metal plates (electrodes) of the electrolytic cell in place of a conducting fluid (acid or solution of a salt). These nonconductors did not dissociate, but they were not without influence on the electrical process. For it was found that when the two metal plates were charged by a voltaic battery with a definite potential difference, they took up different charges according to the substance that happened to be between them (Fig. 85). The non-conducting substance thus influences the power of taking up electricity or the *capacity* of the system of conductors composed of the two plates, which is called a *condenser*.

The discovery impressed Faraday so much that from that time on he gave up the usual idea that electrostatics was based on the direct action of electric charges at a distance, and developed a peculiar new interpretation of electric and magnetic phenomena, a theory of contiguous action. What he learned from the experiment

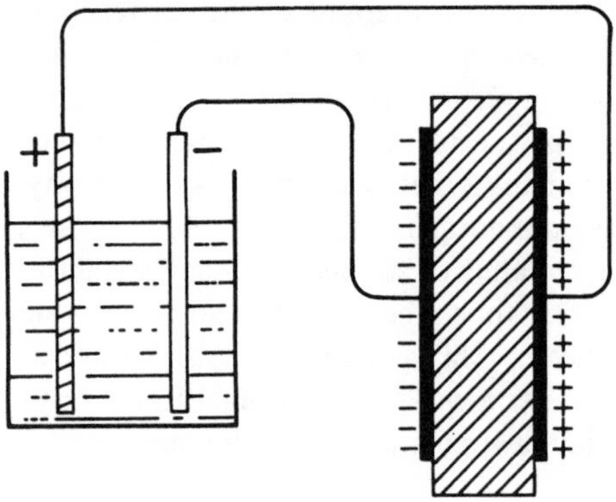

Fig. 85 *A condenser is charged up by a voltaic cell.*

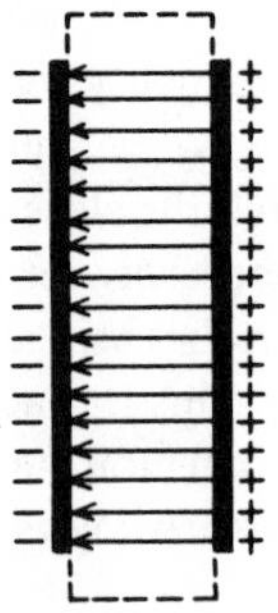

Fig. 86 *The lines of force in a condenser.*

described above was the fact that the charges on the two metal plates do not simply act on each other through the intervening space, but that this intervening space plays an essential part in the action. From this he concluded that the action of this medium is propagated from point to point and is therefore an action by contact, or a contiguous action.

We are familiar with the contiguous action of elastic forces in deformed rigid bodies. Faraday, who always kept to empirical facts, did indeed compare the electric contiguous action in non-conductors with elastic tensions, but he took care not to apply the laws of the latter to electrical phenomena. He used the graphical picture of "lines of force" that run in the direction of the electric field from the positive charges through the insulator to the negative charges. In the case of a plate condenser, the lines of force are straight lines perpendicular to the planes of the plate (Fig. 86). Faraday regarded the lines of force as the true substratum of electric phenomena; for him they are actually material configurations that move about, deform themselves, and thereby bring about electrical effects. For Faraday the charges play a quite subordinate part, as the places at which the lines of force start out or end. He was confirmed in this view by those experiments which proved that in conductors the total electric charge resides on the surface while the interior remains quite free. To give a dramatic proof of this, he built a large cage fitted all around with metal, into which he entered with sensitive electrical measuring instruments. He then had the cage very strongly charged and found that in the interior not the slightest influence of the charges was to be detected. We used this fact earlier (V,1) to derive Coulomb's law of action at a distance. But Faraday concluded from it that the charge was not the primary element of electrical phenomena and that it must not be imagined as a fluid which had the power of exerting forces at a distance. Rather, the primary element is the state of tension of the electric field in the nonconductors which is represented by the picture of the lines of force. The conductors are in a certain sense holes in the electric field, and the charges in them are only fictions invented to explain the pressures and tensions arising through the strains in the field as actions at a distance. Among the nonconductors or dielectric substances there is also the *vacuum*, the *ether*, which we here again encounter in a new form.

This strange view of Faraday's at first found no favor among the physicists and mathematicians of his own time. The view of action at a distance was maintained; this was possible even when the "dielectric" action of nonconductors discovered by Faraday was taken into account. Coulomb's law only needed to be altered a little: to every nonconductor there is assigned a peculiar constant ϵ, its *dielectric constant*, which is defined by the fact that the force acting between two charges e_1, e_2 embedded in the nonconductor is smaller in the ratio $1:\epsilon$ than that acting *in vacuo*:

$$K = \frac{1}{\epsilon}\frac{e_1 e_2}{r^2}. \tag{55}$$

For a vacuum $\epsilon = 1$, for every other body $\epsilon > 1$.

With this addition the phenomena of electrostatics could all be explained even when the dielectric properties of nonconductors were taken into account. We have already mentioned that electrostatics had previously passed over into a theory of pseudocontiguous action, the so-called theory of potential. This likewise easily succeeded in assimilating the dielectric constant ϵ. Nowadays we know that this actually was already equivalent to a mathematical formulation of

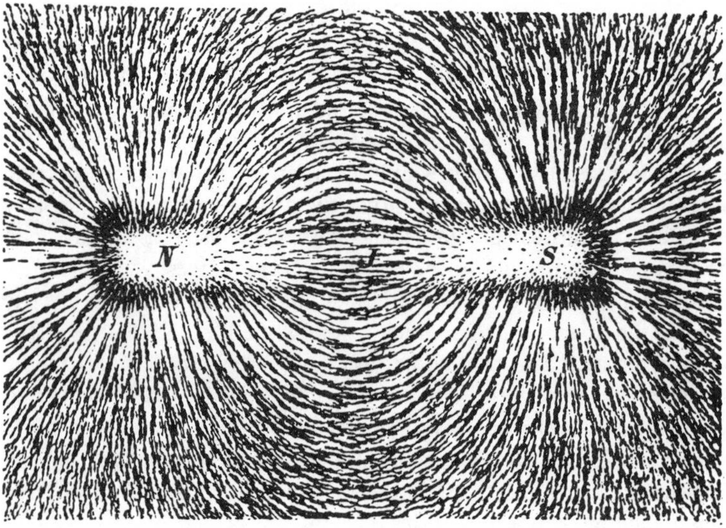

Fig. 87 *The magnetic field of a magnetized bar is made visible by iron filings on a paper above the bar.*

Faraday's concept of lines of force. But as this method of potential was then regarded only as a mathematical artifice, the antithesis between the classical theory of action at a distance and Faraday's idea of contiguous action still remained.

Faraday developed similar views about magnetism. He discovered that the forces between two magnetic poles likewise depend on the medium that happens to lie between them, and this again led him to the view that the magnetic forces, just as with the electric forces, are produced by a peculiar state of tension in the intervening media. The lines of force served to represent these tensions. They can, as it were, be made visible by scattering iron filings over a sheet of paper and holding the latter closely over a magnet (Fig. 87).

The theory of action at a distance leads to the formal introduction of a constant characteristic of the substance, the magnetic penetrability or *permeability* μ, and gives Coulomb's law in the altered form:

$$K = \frac{1}{\mu}\frac{p_1 p_2}{r^2}. \tag{55a}$$

Physicists have not, however, remained satisfied with this formal procedure, but have devised a molecular mechanism that makes the magnetic and dielectric power of polarization intelligible. We have already seen that the properties of magnets lead us to regard their molecules as small elementary magnets made to point in parallel directions by the process of magnetization. It is assumed that they retain this parallelism by themselves, say, through frictional resistances. Now it may be assumed that in the case of most bodies that do not occur as permanent magnets this friction is wanting. The parallel position is then indeed produced by an external magnetic field, but will at once disappear if the field is removed. Such a substance will then be a magnet only as long as an external field is present. But it need not even be assumed that the molecules are permanent magnets that are forced into parallel positions. If each molecule contains the two magnetic fluids, then they will separate under the action of the field and the molecule will become a magnet of itself. But this induced magnetism must have exactly the effect that the formal theory describes by introducing the permeability. Between the two magnetic poles (N, S) in such a medium there are formed chains of molecular magnets called magnetic dipoles, whose opposite poles everywhere compensate each other in the interior but end with opposite poles at N and S and hence weaken the actions of N and S (Fig. 88). (The converse effect, strengthening, also occurs, but we shall not enter into its interpretation.)

Exactly the same as has been illustrated for magnetism may be imagined for electricity. A dielectric, in this view, is composed of molecules that are either electric dipoles of themselves and assume a parallel position in an external field or that become dipoles through the separation of the positive and negative electricity under the action of the field. Between two plates of a condenser (Fig. 89) chains of molecules again form whose charges compensate each other in the interior but not on the plates. Through this a part of the charge on the plates is itself neutralized, and a new charge has to be imparted to the plates to charge them up to a definite tension or potential. This explains how the polarizable dielectric increases the receptivity or capacity of the condenser.

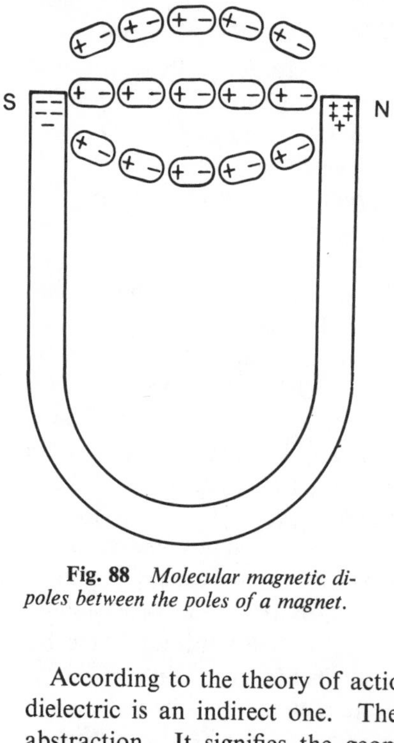

Fig. 88 *Molecular magnetic dipoles between the poles of a magnet.*

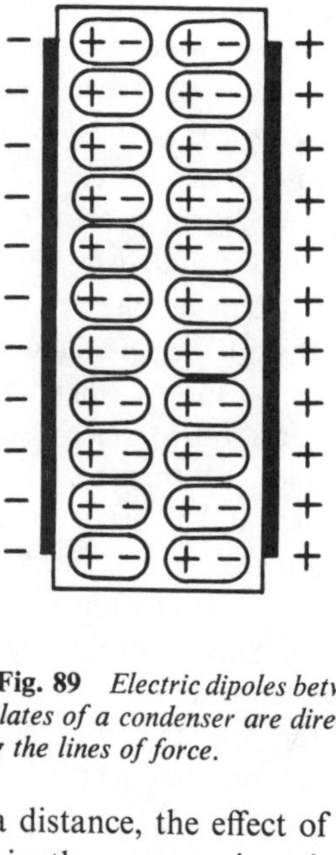

Fig. 89 *Electric dipoles between the plates of a condenser are directed along the lines of force.*

According to the theory of action at a distance, the effect of the dielectric is an indirect one. The field in the vacuum is only an abstraction. It signifies the geometrical distribution of the force that is exerted on an electric test body carrying a unit charge. But the field in the dielectric represents a real physical change of the substance consisting of the molecular displacement of the two kinds of electricity.

Faraday's theory of contiguous action knows no such difference between the field in the ether and in insulating matter. Both are dielectrics. For the ether the dielectric constant $\epsilon=1$, for other insulators ϵ differs from 1. If the graphical picture of electric displacement is correct for matter, it must also hold for the ether. This idea plays a great part in the theory of Maxwell, which is essentially the translation of Faraday's idea of lines of force into the exact language of mathematics. Maxwell assumes that in the ether, too, the production of an electric or a magnetic field is accompanied by "displacements" of the fluids. It is not necessary for this purpose to imagine the ether to have an atomic structure, yet Maxwell's idea comes out most clearly if we imagine ether molecules which become

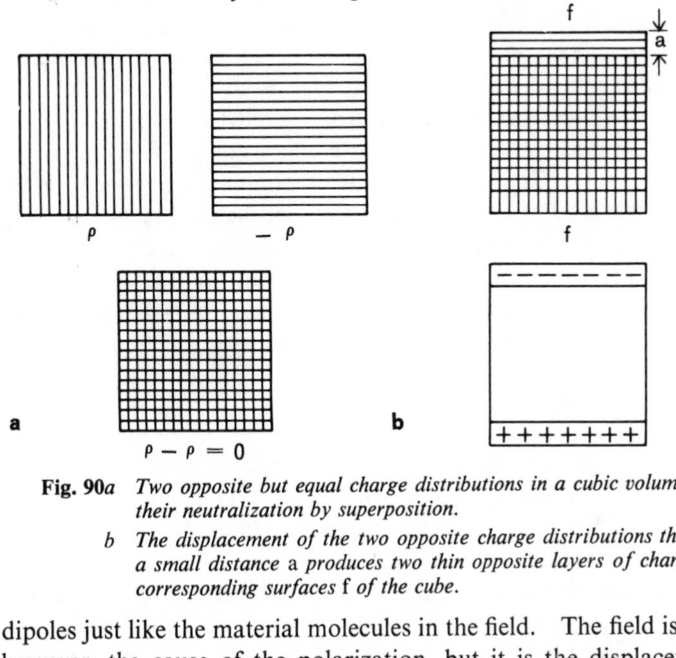

Fig. 90*a* *Two opposite but equal charge distributions in a cubic volume and their neutralization by superposition.*

b *The displacement of the two opposite charge distributions through a small distance* a *produces two thin opposite layers of charge on corresponding surfaces* f *of the cube.*

dipoles just like the material molecules in the field. The field is not, however, the cause of the polarization, but it is the displacement which is the essence of the state of tension that we call electric field. The chains of ether molecules are the lines of force, and the charges at the surface of the conductors are nothing but the end charges of these chains. If there are material molecules present besides the ether particles, the polarization becomes strengthened and the charges at the ends become greater.

We shall now discuss these ideas in more detail. We have just explained how magnetization and electrification can be illustrated by chains of dipole molecules (Fig. 88, 89). However, the idea of molecules in the ether has no empirical foundation. Therefore it is preferable to represent the situation by a continuous model. Imagine a rectangular block of space filled with a continuous positive charge density, ρ, and then the same part of space filled with a negative charge density, $-\rho$. If both kinds of charges are present simultaneously the space is uncharged (Fig. 90*a*). The establishment of an electric field E is, according to Faraday and Maxwell, nothing but a displacement of the two blocks of charge (see Fig. 90*b*) through a small distance a. The whole interior remains uncharged, although there is a shift of charges at each point; only on two opposite faces there appear opposite equal charges, for if f is the area of the face, there are two rectangular sheets of volume fa which contain only one kind of charge. As a is small, one can speak of surface charges ρfa and $-\rho fa$. The surface charge per unit area liberated by the little shift a is ρa; it represents a measure of the electric displacement D. However, one does not simply equate these two quantities, but must add a numerical factor for the following reason.

Consider a point charge e in a dielectric (see Fig. 91). The law of force (55) requires that the field E produced by it is

$$E = \frac{e}{\epsilon r^2}. \tag{56}$$

If one describes the same situation in Faraday's language, one has to assume a displacement which is constant on spheres around the center and diminishes with the distance r (Fig. 92). If a spherical shell with the outer radius r and the inner radius r' is imagined to be filled with mutually cancelling charge densities ρ and $-\rho$, and if these are displaced in the radial direction by a, there appears the charge $-f'a\rho$ at the inner sphere and the charge $fa\rho$ at the outer sphere. Both of these must be equal to the given central point charge; for if the inner radius is contracted to nothing the corresponding charge must just cancel the central charge e. Therefore $e = fa\rho$. But the total area of a sphere of radius r is $f = 4\pi r^2$; hence one has $e = 4\pi r^2 a\rho$. Substituting this in the expression for E, one has

$$E = \frac{e}{\epsilon r^2} = \frac{4\pi a\rho}{\epsilon}.$$

Field E and displacement $a\rho$ are therefore proportional. In order to get rid of the factor 4π in the final formula, it is customary to define $D=4\pi a\rho$; then $E=D/\epsilon$, or

$$D = \epsilon E. \tag{57}$$

One can say that the displacement D diverges from the central charge e in all directions.

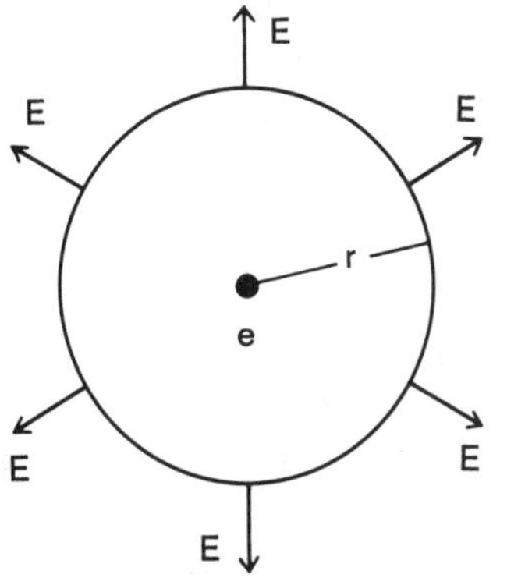

Fig. 91 *A point charge* e *produces a field* E *directed radially and having constant value on a concentric sphere.*

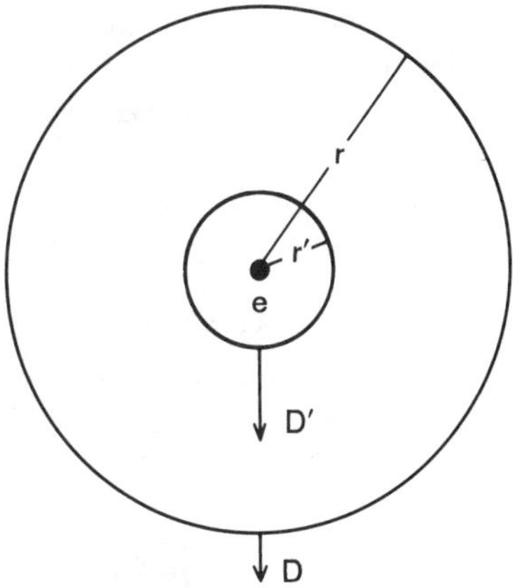

Fig. 92 *The displacement on two spheres with charge* e *in the center:*

$$4\pi r^2 \rho a = 4\pi r'^2 \rho a' = e$$

or

$$r^2 D = r'^2 D' = e.$$

This expression is also used in the general case where the true charge is not concentrated in one point but continuously distributed with a density ρ (which is not to be mistaken for the fictional density denoted by the same letter that we used to illustrate Maxwell's concept of displacement). One writes symbolically

$$\text{div } D = 4\pi\rho. \tag{58}$$

But this is more than a mnemonic help. Maxwell has managed to give the symbol div a definite meaning as a differential operation performed on the components of D. Thus to the mathematician (60) signifies a differential equation, a law of contiguous action.

Are Faraday's and Maxwell's ideas or those of the theory of action at a distance right?

So long as we confine ourselves to electrostatic and magnetostatic phenomena, both are equivalent. For the mathematical expression of Faraday's idea is what we have called a theory of pseudocontiguous action, because it does, indeed, operate with differential equations but recognizes no finite velocity of propagation of tensions. Faraday and Maxwell, however, themselves disclosed those phenomena which, in a way analogous to the inertial effects of mechanics, effect the delay in the transference of an electromagnetic state from point to point and hence bring about the finite velocity of propagation. These phenomena are the displacement current and the magnetic induction.

6. The Electrical Displacement Current

Suppose the poles of a galvanic cell to be connected with the plates of a condenser by means of two wires, one of them containing a switch (Fig. 93). If the switch is pressed down, a current flows

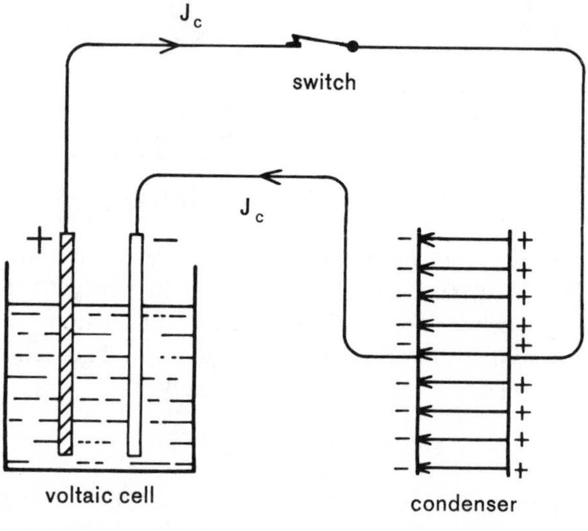

Fig. 93 *When the condenser is charged by a convection current* J_c, *the electric field inside the condenser changes and gives rise to a displacement current of the same amount as* J_c.

which charges the two plates of the condenser; an electric field E is thereby produced between them. Before Maxwell's time, this phenomenon was regarded as an "open circuit." Maxwell, however, realized that during the growth of the field E a *displacement current* flows between the condenser plates, and thus the circuit becomes closed. As soon as the condenser plates are completely charged, both currents, the conduction and the displacement current, cease.

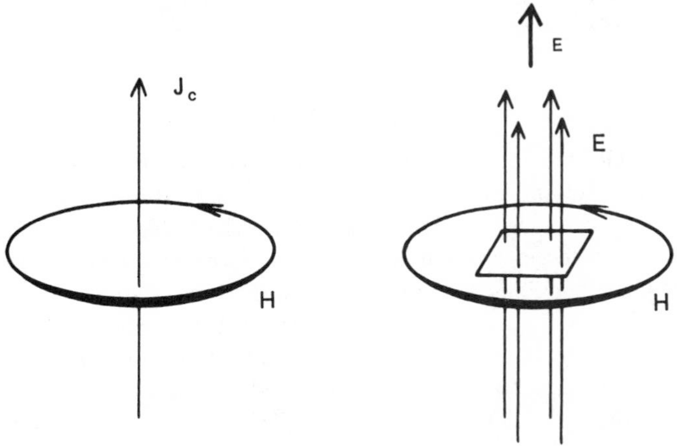

Fig. 94 *Both the convection current* J_c *and the displacement current* J_d *produce a surrounding magnetic field.*

Now, the essential point is Maxwell's affirmation that the displacement current, just like the conduction current, produces a magnetic field according to Biot and Savart's law. That this is actually so has not only been proved by the success of Maxwell's theory in predicting numerous phenomena but was also later confirmed directly by experiment.

The magnitude of the displacement current can easily be computed. We know that $\rho a f=\frac{D}{4\pi}f$ is the displacement of charge through a small area f perpendicular to the field. If $\mathfrak{D}$ is the change of D in the short time τ, then $\frac{D}{4\pi}\frac{f}{\tau}$ is the charge flowing in τ through f. Therefore $\frac{D}{4\pi}\frac{f}{\tau}$ is the current through f and $j_d = \frac{1}{4\pi}\frac{D}{\tau}$ is the density of the *displacement current*. With $D = \epsilon E$ hence $D = \epsilon E$, we write $j_d = \frac{\epsilon}{4\pi}\frac{E}{\tau}$.

Therefore, following Maxwell, the whole current density is the sum $j = j_c + j_d$, where j_c is the current density of the free movable charges and j_d is the displacement current. Both kinds of current are surrounded by a magnetic field in the usual way (Fig. 94).

7. Magnetic Induction

After Oersted had discovered that a conduction current produces a magnetic field and Biot and Savart had formulated this fact as an action at a distance, Ampère discovered (1820) that two voltaic currents exert forces on each other, and he succeeded in expressing the law underlying this phenomenon again in terms of an action at a distance. This discovery had far-reaching consequences, for it made it possible to regard magnetism as an effect of moving electricity. According to Ampère small closed currents are supposed to flow in the molecules of magnetized bodies. He showed that such currents behaved exactly like elementary magnets. This idea has stood the test of thorough examination; from his time on magnetic fluids became superfluous. Only electricity was left, which, when at rest, produced the electrostatic field, and when flowing, produced the magnetic field besides. Ampère's discovery may also be expressed in the following way: According to Oersted a wire in which the current J_1 is flowing produces a magnetic field in its neighborhood. A second wire in which the current J_2 is flowing is then pulled by forces due to this magnetic field. In other words, a field produced by one current tends to deflect or accelerate flowing electricity.

Hence the following question suggests itself: Can the magnetic field also set electricity that is at rest into motion? Can it produce or "induce" a current in the second wire which is initially without a current?

Faraday found the answer to this question (1831). He discovered that a static magnetic field is not able to produce an electric current but that a field which varies in time is able to. For example, when he quickly brought a magnet close to a loop of wire made of conducting material, a current flowed in the wire as long as the magnet moved. In particular, when he produced the magnetic field by means of a primary current, a short impulse of current occurred in the secondary wire whenever the first current was started or stopped.

From this it is clear that the induced electric force depends on the velocity of alteration of the magnetic field in time. Faraday succeeded in formulating the quantitative law of this phenomenon with the help of his concept of lines of force. Using Maxwell's ideas we shall give it such a form that its analogy with Biot and Savart's law comes out clearly.

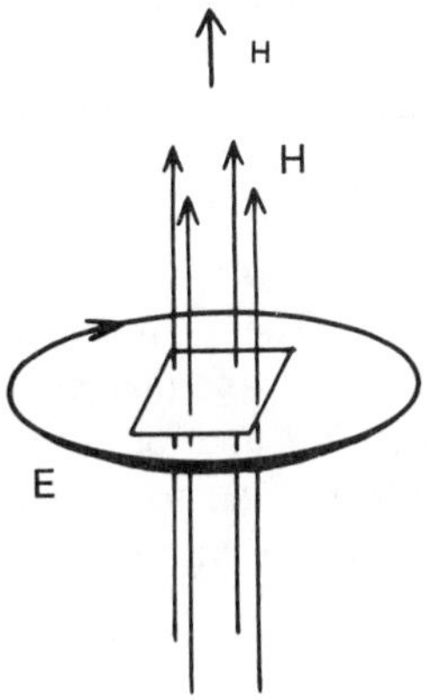

Fig. 95 *A changing magnetic field which represents a magnetic current* I *is surrounded by an electric field.*

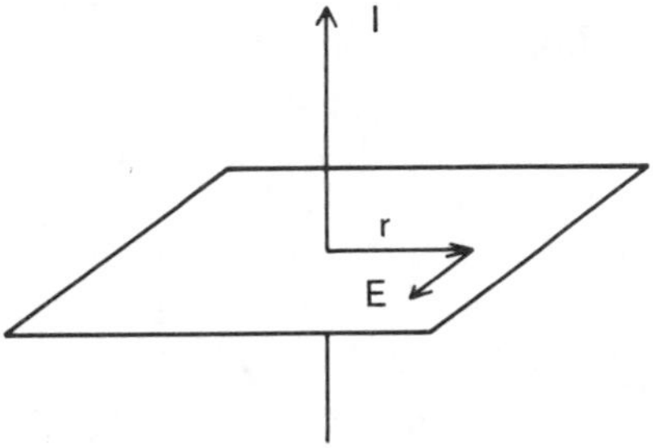

Fig. 96 *Direction of the electric field* E *induced by a magnetic current* I (*compare with Fig. 84*).

We imagine a bundle of parallel lines of magnetic force that constitute a magnetic field H. We suppose a circular conducting wire placed around this sheath (Fig. 95). If the intensity of field H changes in the small interval of time τ by the amount η, we call $\frac{\eta}{\tau}$ its velocity of change or the change in the number of lines of force. If in analogy to the electrical displacement we represent the lines of force as chains of magnetic dipoles (which, however, according to Ampère is actually not true) then with the change of H a displacement of the magnetic quantities will occur in every ether molecule, or a "magnetic displacement current" will flow whose current strength per unit of area or current density is given by $i=\frac{H}{4\pi\tau}\cdot$ If the field H is not in the ether but in a substance of permeability μ, the density of the magnetic displacement current is $i=\frac{\mu}{4\pi}\frac{H}{\tau}\cdot$ Thus the magnetic current $I=fi=f\frac{\mu}{4\pi}\frac{H}{\tau}$ passes through the cross-section f, that is, through the surface of the circle formed by the conducting wire.

Now, according to Faraday, this magnetic current produces all around it an electric field E, which encircles the magnetic current exactly as the magnetic field H encircles the electric current in Oersted's experiment, but in the reverse direction. It is this electric field E that drives the induced current around in the conducting wire; it is also present even if there is no conducting wire in which the current can form.

We see that the magnetic induction of Faraday is a perfect parallel to the electromagnetic discovery of Oersted. The quantitative law, too, is the same. According to Biot and Savart, the magnetic field H produced by a current element of length l and of strength J (compare Fig. 84) in the middle plane perpendicular to the element is perpendicular to the connecting line r and to the current direction, and has the value $H=\frac{Jl}{cr^2}$ (formula (54)).

Exactly the same holds when electric and magnetic quantities are exchanged and when the sense of rotation is reversed (Fig. 96). The induced electric intensity of field in the central plane is given by $E=\frac{Il}{cr^2}\cdot$

In it the same constant c, the ratio of the electromagnetic to the electrostatic unit of current, occurs which was found by Weber and Kohlrausch to be equal to the velocity of light. It can easily be seen from considerations about the energy involved that this must be so.

A great number of the physical and technical applications of electricity and magnetism depend on the law of induction. The transformer, the induction coil, the dynamo, and innumerable other apparatus and machines are appliances for inducing electric currents by means of changing magnetic fields. But however interesting these things may be, they do not lie on the road of our investigation, the final goal of which is to examine the relationship of the ether with the space problem. Hence we turn our attention at once to the theory of Maxwell, whose object was to combine all known electromagnetic phenomena into one uniform theory of contiguous action.

8. Maxwell's Theory of Action by Contact

We have already stated that soon after Coulomb's law had been established, electrostatics and magnetostatics were brought into the form of a theory of pseudocontiguous action. Maxwell now undertook to fuse this theory with Faraday's ideas, and to elaborate it so that it also included the newly discovered phenomena of dielectric and magnetic polarization, of electromagnetism, and magnetic induction.

Maxwell took as the starting point of his theory the idea already mentioned above that an electric field E is always accompanied by an electric displacement $D=\epsilon E$ not only in matter, for which ϵ is different from 1, but also in the ether, where $\epsilon=1$. We explained how the displacement can be visualized as the separation and flowing of electric fluids in the molecules. And we have found a differential law, which connects the charge density ρ in every space point with the divergence of $D=\epsilon E$:

$$\operatorname{div} \epsilon E = 4\pi\rho. \tag{58}$$

Exactly the same considerations apply to magnetism, but with one important difference: According to Ampère no real magnets exist, no magnetic quantities, but only electromagnets. The magnetic field is always to be produced by electric currents, whether they be conduction currents in wires or molecular currents in the molecules. From this it follows that the magnetic lines of force never end, that is, they are either closed or stretch to infinity. This is so in the case of an electromagnet, a coil through which a current is flowing (Fig. 97*a*, *b*); the magnetic lines of force are straight lines in the interior of the coil, but outside they are partly closed and partly going off to infinity. If we consider the coil between two planes A and B, then just as much "magnetic displacement" μH will enter through A as goes out through B. Therefore we must write

$$\operatorname{div} \mu H = 0. \tag{59}$$

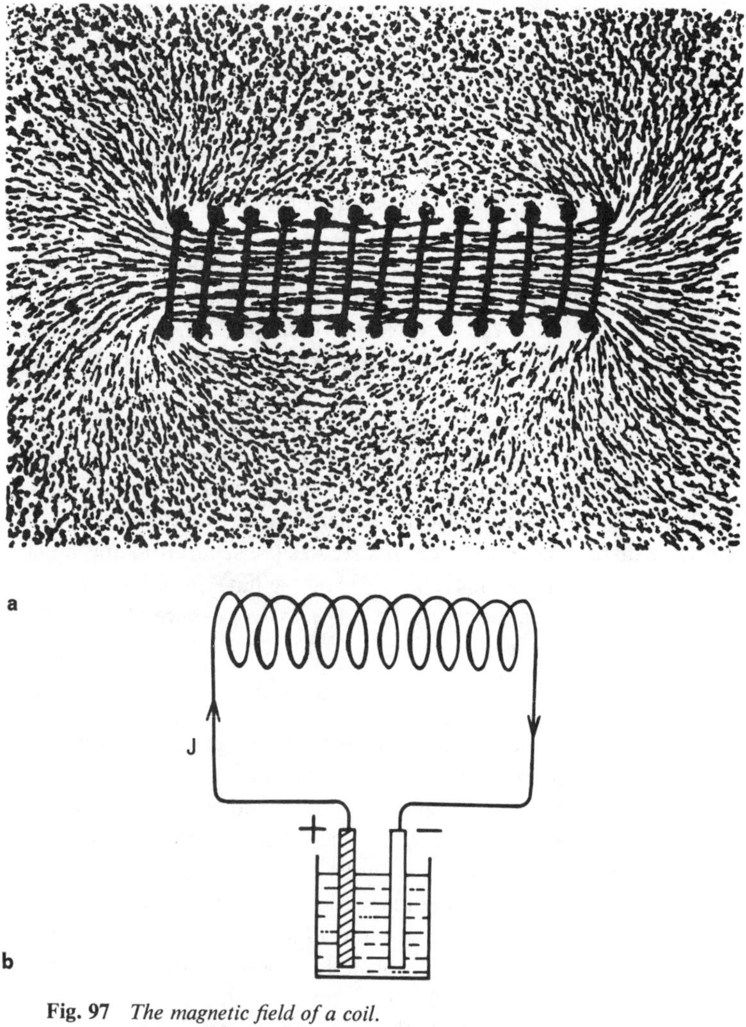

Fig. 97 *The magnetic field of a coil.*
a Lines of force of the coil made visible by iron filings.
b Current J *flowing through the coil.*

This is Maxwell's formula of contiguous action for magnetism. It may be remarked that the expression *magnetic induction* is used instead of displacement.

We now come to Biot and Savart's law of electromagnetism. To convert this into a law of contiguous action, we suppose the electric current not to be flowing in a thin wire, but to be distributed uniformly with the density $j=\frac{J}{f}$ over a circular cross-section f, and we then ask what is the magnetic intensity of field H at the edge of the cross-section. By Biot and Savart's law this is everywhere in the direction of the tangent to the circle and, according to formula (54), it has the value $H=\frac{Jl}{cr^2}$, where r is the radius of the circle and l the length of the current element. Now the cross-section, being circular, is $f=\pi r^2$; hence we may write formula (54) thus: $\frac{cH}{\pi l}=\frac{J}{\pi r^2}=\frac{J}{f}=j$, and this holds for every cross-section, however small, and for every length, however short. On the left, then, there is a certain differential quantity of the magnetic field, and the law states that this quantity is proportional to the current density. We cannot here carry out the mathematical investigation of how this differential quantity is formed. It has to take into account not only the intensity but also the direction of the magnetic field, and since this encircles or curls round the direction of the current, the differential operation is called "curl" of the field H (written *curl H*). Accordingly we write symbolically

$$c \operatorname{curl} H = 4\pi j \tag{60}$$

and again regard this formula only as a mnemonic for the relationships between the intensity and direction of the magnetic field H and the density of current j. To the mathematician, however, it is a differential equation of the same kind as the law (58).

Now exactly the same holds for magnetic induction, but we shall write the opposite sign to indicate the opposite sense of rotation:

$$c \operatorname{curl} E = -4\pi i. \tag{61}$$

The four symbolic formulae (58) to (61) show wonderful symmetry.

Formal agreement of this kind is by no means a matter of indifference. It exhibits the underlying simplicity of phenomena in nature, which remains hidden from direct perception because of the limitations of our senses and reveals itself only to our analytical faculty.

In general a conduction and a displacement current will be present simultaneously. For the former, Ohm's law, $j_c = \sigma E$, holds (52) (p. 162); for the latter, Maxwell's law, $j_d = \frac{\epsilon}{4\pi}\frac{E}{\tau}$. If both are present simultaneously we thus have

$$j = \frac{\epsilon}{4\pi}\frac{E}{\tau} + \sigma E.$$

There is no conduction current for magnetism, so we always have $i = \frac{\mu}{4\pi}\frac{H}{\tau}$. If we insert this in our symbolic equations (58) to (61), we get:

$$
\begin{aligned}
&(a) \qquad & \operatorname{div} \epsilon E &= 4\pi\rho, \\
&(b) \qquad & \operatorname{div} \mu H &= 0, \\
&(c) \qquad & c \operatorname{curl} H &= \epsilon \frac{E}{\tau} + 4\pi\sigma E, \qquad (62) \\
&(d) \qquad & c \operatorname{curl} E &= -\mu \frac{H}{\tau}.
\end{aligned}
$$

These are Maxwell's laws, which have remained the foundation of all electromagnetic and optical theories up to our own time. To the mathematician they are precise differential equations. To us they are mnemonics which state:

(a) Wherever an electric charge occurs, an electric field arises of such a kind that in every volume the charge is exactly compensated by the displacement.

(b) Through every closed surface just as much magnetic displacement passes outwards as comes inwards (there are no free magnetic charges).

(c) Every electric current, be it a conduction or a displacement current, is surrounded by a magnetic field.

(d) A magnetic displacement current is surrounded by an electric field in the reverse sense.

Maxwell's "field equations," as they are called, constitute a true theory of contiguous action or action by contact, for, as we shall presently see, they give a finite velocity of propagation for electromagnetic forces.

At the time they were set up, however, faith in direct action at a distance, according to the model of Newtonian attraction, was still so deeply rooted that a considerable interval elapsed before they were accepted, for the theory of action at a distance had also succeeded in mastering the phenomena of induction by means of formulae. This was done by assuming that moving charges exert, in addition to the Coulomb attraction, certain actions at a distance that depend on the amount and direction of the velocity. The first hypotheses of this kind were due to Neumann (1845). Another famous law is that set up by Wilhelm Weber (1846); similar formulae were given by Riemann (1858) and Clausius (1877). These theories have in common the idea that all electrical and magnetic actions are to be explained by means of forces between elementary electrical charges, or, as we say nowadays, "electrons." They were thus precursors of the present-day theory of electrons, with, however, an essential factor omitted: the finite velocity of propagation of the forces. These theories of electrodynamics, based on action at a distance, gave a complete explanation of the electromotive forces and induction currents that occur in the case of closed conduction currents. But in the case of "open" circuits that is, condensor charges and discharges, they were doomed to failure, for here the displacement currents come into play, of which the theories of action at a distance know nothing. It is to Helmholtz that we are indebted for appropriate experimental devices allowing us to decide between the theories of action at a distance and action by contact. He succeeded in carrying out the experiment with a certain measure of success, and he himself became one of the most zealous pioneers of Maxwell's theory. But it was his pupil Hertz who secured the victory for Maxwell's theory by discovering electromagnetic waves.

9. The Electromagnetic Theory of Light

We have already mentioned (V, 4, p. 163) the impression which the coincidence, established by Weber and Kohlrausch, of the electromagnetic constant c with the velocity of light made upon the physicists of the day. And there were still further indications that there is an intimate relation between light and electromagnetic phenomena. This was shown most strikingly by Faraday's discovery (1834) that a polarized ray of light which passes through a magnetized transparent substance is influenced by it. When the beam is parallel to the magnetic lines of force, its plane of polarization becomes turned. Faraday himself concluded from this that the luminiferous ether and the carrier of electromagnetic lines of force must be identical. Although his mathematical powers were not sufficient to allow him to convert his ideas into quantitative laws and formulae, his ideas were of a most abstract type and far surpassed the trivial view which accepted as known what was familiar. Faraday's ether was no elastic medium. He derived its properties, not by analogy from the apparently known material world, but from exact experiments and systematic deductions from them. Maxwell's talents were akin to those of Faraday, but they were supplemented by a complete mastery of the mathematical means available at the time.

We shall now show how the propagation of electromagnetic forces with finite velocity arises out of Maxwell's field laws (62). In doing so we shall confine ourselves to events that occur *in vacuo* or in the ether. The latter has no conductivity, that is, $\sigma=0$, and no true charges, that is, $\rho=0$; and its dielectric constant and permeability are equal to 1, that is, $\epsilon=1$, $\mu=1$. The first two field equations (62) then assert that

$$\operatorname{div} E = 0, \qquad \operatorname{div} H = 0 \tag{63}$$

or that all lines of force are either closed or run off to infinity. To obtain a rough picture of the processes we shall imagine individual, closed lines of force.

The other two field equations are then

$$(a)\ \frac{\mathsf{E}}{\tau} = c \operatorname{curl} H, \qquad (b)\ \frac{\mathsf{H}}{\tau} = -c \operatorname{curl} E. \tag{64}$$

We now assume that, somewhere in a limited space, there is an electric field E which alters by the amount E in the small interval of time τ; then $\frac{\mathsf{E}}{\tau}$ is its rate of change. According to the first equation, a magnetic field immediately coils itself around this electric field, and its strength is proportional to $\frac{E}{\tau}$. The magnetic field, too, will alter in time, say by H during each successive small interval τ. Again, in accordance with the second equation, its rate of change $\frac{H}{\tau}$ immediately induces an interwoven electric field. In the following interval of time the latter again induces an encircling magnetic field, according to the first equation, and so this chainlike process continues with finite velocity (Fig. 98).

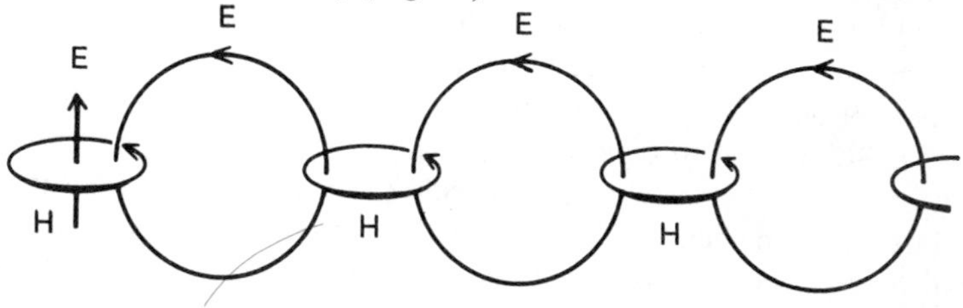

Fig. 98 *Electric and magnetic fields linked by induction.*

This is, of course, only a rough description of the process, which actually propagates itself in all directions continuously. Later we shall sketch a better picture.

What particularly interests us here is the following: We know from mechanics that the finite velocity of propagation of elastic waves is due to the delays that occur as a result of the inertia which comes into play when the forces are transmitted in the body from point to point. We have formulated this in equation (36) $\rho b = pf$ and with (37) $c^2 = \frac{p}{\rho}$ we get

$$b = c^2 f. \tag{36 a}$$

Here c^2 means the square of the velocity of the elastic waves, b the acceleration of the mass particles in the elastic body (i.e., the second-order differential coefficient with respect to time), and f is the second-order differential coefficient with respect to space.

Now, in the electromagnetic field, the case is nearly the same. The only difference is that instead of the dependency of the displacement on space and time in the elastic case we have two quantities

E and H depending on space and time. The rate of change of the electric field $\frac{E}{\tau}$ first determines the magnetic field H, and then the rate of change $\frac{H}{\tau}$ of the latter determines the electric field E at a neighboring point. The equations (64) contain only differential quantities of the first order, for instance $\frac{E}{\tau}$, a first-order differential coefficient with respect to time, and curl H, a first-order differential coefficient with respect to space. One gets an equation similar to (36) by the following procedure: To begin with, take the first-order differential coefficient of equation (64*a*) with respect to time. Then we have on the left-hand side the second-order differential coefficient of E with respect to time which is analogous to b in (36*a*) and which we will call b_E. On the right-hand side we have a mixed second-order differential coefficient (forming first the difference in space and then in time or vice versa). One gets the same mixed coefficient from (64*b*) by forming the first-order differential coefficient with respect to space. Then one sees that the mixed coefficient is equal to the product of c into the second-order differential coefficient in space of E, which is analogous to f in (36*a*) and may therefore be called f_E. Now one can eliminate the mixed coefficient in the equations and one gets

$$b_E = c^2 f_E. \tag{65}$$

This equation is in complete analogy with (36*a*) and shows the existence of electric waves with velocity c. By the same method one may derive a corresponding equation for the magnetic field $H (b_H = c^2 f_H)$. If one of the two partial effects would happen without loss of time, no propagation of the electric force in the form of waves would occur. This helps us to realize the importance of Maxwell's displacement current, for it provides just this rate of change $\frac{E}{\tau}$ of the electric field.

We shall now give a description of the propagation of an electromagnetic wave which is somewhat nearer to the actual process. Let two metal spheres have large, opposite, equal charges $+e$ and $-e$ so that a strong electric field exists between them. Next let a spark occur between the spheres. The charges then neutralize each other; the field collapses at a great rate of change $\frac{E}{T}$. The figure shows how the magnetic and electric lines of force then encircle each other alternately (Fig. 99). In our diagram the magnetic lines of

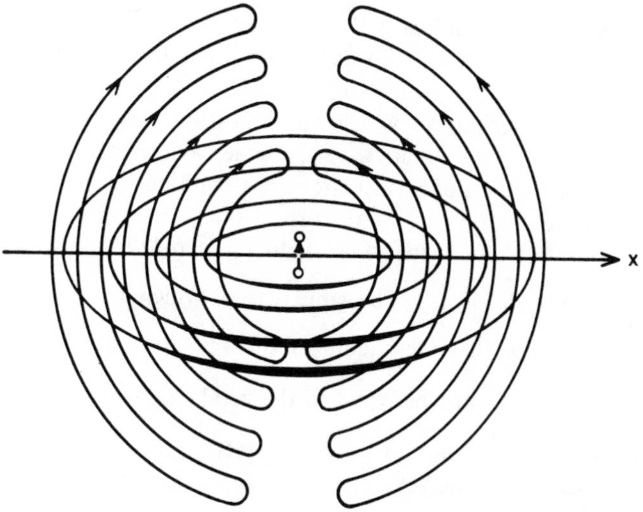

Fig. 99 *The electromagnetic field surrounding a discharge spark between two spheres. This field expands with velocity* c *of light in all directions.*

force are drawn only in the median plane between the spheres, the electric lines of force only in the plane of the paper, perpendicular to the median plane. The whole figure is, of course, radially symmetrical about the line connecting the centers of the spheres. Each successive loop of the lines of force is weaker than its immediate predecessor because it lies farther outwards and has a larger circumference. Accordingly, the inner part of a loop of electric force does not quite counterbalance the outer part of its predecessor, especially since it enters into action a little late.

If we pursue the process along a straight line which is perpendicular to the line connecting the centers of the spheres, say along the x-axis, then we see that the electric and magnetic forces are always perpendicular to this axis; moreover, they are perpendicular to each other.

This is true of any direction of propagation. Thus, the electromagnetic wave is precisely transversal. Furthermore, it is polarized, but we still have the choice of regarding either the electric or the magnetic intensity of field as the determining factor of the vibration.

Thus we have shown that the velocity of the waves is equal to the constant c and the waves are transversal. Further, since according to Weber and Kohlrausch the value of c is equal to that of the velocity of light c, Maxwell was able to conclude that *light waves are nothing other than electromagnetic waves.*

One of the inferences which Maxwell drew was soon confirmed experimentally to a certain extent, for he calculated the velocity of light c_1 for the case of an insulator ($\sigma=0$) and no free charge ($\rho=0$). Maxwell's equations (62c and d) show that we get nearly the same equations as (64) but with other c-values. In (64a) c is to be replaced by $\frac{c}{\epsilon}$ and in (64b) by $\frac{c}{\mu}$. The same reasons which led us to equation (65) show now that the square of the velocity c_1^2 of the electromagnetic waves must be equal to the product of $\frac{c}{\epsilon}$ and $\frac{c}{\mu}$: $c_1^2=\frac{c^2}{\epsilon\mu}$. Many materials are not noticeably magnetizable, so we can set $\mu=1$, which means that the velocity of light in an insulator with dielectric constant ϵ is given by $c_1=\frac{c}{\sqrt{\epsilon}}$. This leads to the value $n=\frac{c}{c_1}=\sqrt{\epsilon}$ for the refractive index.

Thus it should be possible to determine the refrangibility of light from the dielectric constant as given by purely electrical measurements. For some gases—for example, hydrogen, carbon dioxide, air—this is actually the case, as was shown by L. Boltzmann. For other substances Maxwell's relation $n=\sqrt{\epsilon}$ is not correct, but in all these cases the refractive index is not constant but depends on the color (frequency) of the light. This shows that dispersion of the light introduces a disturbing effect. We shall return to this fact later and deal with it from the point of view of the theory of electrons. At any rate, it is clear that the slower the vibrations or the longer the waves of the light that is used, the more closely the dielectric constant, as determined statically, agrees with the square of the refractive index. Waves of an infinite time of vibration are, of course, identical with a stationary state. Researches into the region of long waves (length of the order of centimeters) have completely confirmed Maxwell's formula.

Concerning the more geometrical laws of optics, reflection, refraction, double refraction, and polarization in crystals and so forth, the electromagnetic theory of light resolves all the difficulties that were quite insuperable for the theories of the elastic ether. In the latter, the greatest obstacle was the existence of longitudinal waves which appeared when light crossed the boundary between two media and which could be removed only by making quite improbable hypotheses about the constitution of the ether. The electromagnetic waves are always strictly transversal. Thus this difficulty vanishes. Maxwell's theory is almost identical formally with the ether theory of MacCullagh, as we mentioned above (IV, 6, p. 117); without repeating the calculations we can take over most of his deductions.

We cannot here enter into the later development of electrodynamics. The bond between light and electromagnetism became ever closer. New phenomena were continually being discovered which showed that electric and magnetic fields exerted an influence on light. Everything proved to be in accordance with Maxwell's laws, the certainty of which continued to grow.

But the striking proof of the oneness of optics and electrodynamics was given by Heinrich Hertz (1888) when he showed that the velocity of propagation of electromagnetic force was finite and when he actually produced electromagnetic waves. He made sparks jump across the gaps between two charged spheres and by this means generated waves such as are represented by our diagram (Fig. 99). When they encountered a circular wire with a small gap in it, they produced in it currents which manifested themselves by small sparks at the gap. Hertz succeeded in reflecting these waves and in making them interfere. This enabled him to measure their wave length. He knew the frequency of the oscillations and thus could calculate the velocity of the waves, which came out equal to c, that of light. This directly confirmed Maxwell's hypothesis. Nowadays the Hertzian waves of wireless stations travel over the earth without cessation and bear their tribute to the two great scientists, Maxwell and Hertz, one of whom predicted the existence of electromagnetic waves while the other actually produced them.

10. The Electromagnetic Ether

From this time on there was only one ether, which was the carrier of all electric, magnetic, and optical phenomena. We know its laws, Maxwell's field equations, but we know little of its constitution. Of what do the electromagnetic fields actually consist, and what is it that executes vibrations in the waves of light?

We recall that Maxwell took the concept of displacement as the foundation of his argument, and we interpreted this visually as meaning that in the smallest parts or molecules of the ether, just as in the molecules of matter, an actual displacement and separation of the electric (or magnetic) fluid occur. So far as this idea concerns the process of electric polarization of matter, it is well founded; it is also adopted in the modern modification of Maxwell's theory, the theory of electrons, for numerous experiments have rendered certain that matter has a molecular structure and that every molecule carries displaceable charges. But this is by no means the case for the free ether; here Maxwell's idea of displacement is purely hypothetical, and its only value is that it provides a visualizable image for the abstract laws of the field.

These laws state that with every change of displacement in time there is associated an electromagnetic field of force. Can we form a mechanical picture of this relationship?

Maxwell himself designed mechanical models for the constitution of the ether and applied them with some success. Lord Kelvin was particularly inventive in this direction, and strove unceasingly to comprehend electromagnetic phenomena as actions of concealed mechanisms and forces.

The rotational character of the relationship between electric currents and magnetic fields, and its reciprocal character, suggests that we regard the electric state of the ether as a linear displacement, the magnetic state as a rotation about an axis, or conversely. In this way we arrive at ideas that are related to MacCullagh's ether theory. According to this the ether was not to generate elastic resistances against distortions in the ordinary sense, but resistances against the absolute rotation of its elements of volume. It would take us too far to count the numerous and sometimes very fantastic hypotheses that have been put forward about the constitution of the ether. If we were to accept them literally, the ether would be a monstrous mechanism of invisible cogwheels, gyroscopes, and gears intergripping in the most complicated fashion, and of all this confused mass nothing would be observable but a few relatively simple features which would present themselves as an electromagnetic field.

There are also less cumbersome, and, in some cases, ingenious theories in which the ether is a fluid whose rate of flow represents, say, the electric field, and whose vortices represent the magnetic field. Bjerknes has sketched a theory in which the electric charges are imagined as pulsating spheres in the ether fluid, and he has shown that such spheres exert forces on one another which exhibit considerable similarity with the electromagnetic forces.

If we inquire into the meaning and value of such theories, we must grant them the credit of having suggested (though rather seldom) new experiments and of having led to the discovery of new phenomena. More often, however, elaborate and laborious experimental researches have been carried out to decide between two ether theories equally improbable and fantastic. In this way much effort has been wasted. Even nowadays there are people who regard a mechanical explanation of the electromagnetic ether as something demanded by reason. Such theories continue to crop up, and naturally they become more and more abstruse as the abundance of facts to be explained grows; hence the difficulty of the task increases without cessation.

Heinrich Hertz deliberately turned away from all mechanistic speculations. We give the substance of his own words: "The interior of all bodies, including the free ether, can, from an initial state of rest, experience some disturbances which we call electrical and others which we call magnetic. We do not know the nature of these changes of state, but only the phenomena which their presence calls up." This definite renunciation of a mechanical explanation is of great importance from the methodical point of view. It opens up the avenue for the great advances which have been made by Einstein's researches. The mechanical properties of solid and fluid bodies are known to us from experience, but this experience concerns only their behavior in a crude sense. Modern molecular researches have shown that these visible, crude properties are a sort of appearance, an illusion, due to our clumsy methods of observation, whereas the actual behavior of the smallest elements of structure, the atoms, molecules, and electrons, follows quite different laws. It is, therefore, naïve to assume that every continuous medium, like the ether, must behave like the apparently continuous fluids and solids of the crude world accessible to us through our coarse senses. Rather, the properties of the ether must be ascertained by studying the events that occur in it independent of all other experiences. The result of these researches may be expressed as follows: The state of the ether may be described by two directed magnitudes, which bear the names *electric* and *magnetic strength of field*, E and H, and whose changes in space and time are connected by Maxwell's equations. Under certain circumstances such an ether phenomenon produces mechanical, thermal, and chemical actions in matter that are capable of being observed.

Everything that goes beyond these assertions is superfluous hypothesis and fancy. It may be objected that such an abstract view undermines the inventive power of the investigator, which is stimulated by visual pictures and analogies, but Hertz's own example contradicts this opinion, for rarely has a physicist been possessed of such wonderful ingenuity in experiment, although as a theorist he recognized only pure abstraction as valid.

11. Hertz's Theory of Moving Bodies

A more important question than the pseudo problem of the mechanical interpretation of the ether is that concerning the influence of the motions of bodies (among which must be counted, besides matter, the ether) on electromagnetic phenomena. This brings us back, but from a more general standpoint, to the investigations which we made earlier (IV, 7) into the optics of moving bodies. Optics is now a part of electrodynamics, and the luminiferous ether is identical with the electromagnetic ether. All the inferences that we made earlier from the optical observations with regard to the behavior of the luminiferous ether must retain their validity since they are obviously quite independent of the mechanism of light vibrations; for our investigation concerned only the geometrical characteristics of a light wave, namely, frequency (Doppler effect), velocity (convection), and direction of propagation (aberration).

We have seen that up to the time when the electromagnetic theory of light was developed only quantities of the first order in $\beta=\frac{v}{c}$ were open to measurement. The result of these observations could be expressed briefly as the "optical principle of relativity": Optical events depend only on the relative motions of the involved material bodies that emit, transmit, or receive the light. In a system of reference moving with constant velocity relative to the ether all inner optical events occur just as if it were at rest.

Two theories were proposed to account for this fact. That of Stokes assumed that the ether inside matter was completely carried along by the latter; the second, that of Fresnel, assumed only a partial convection, the amount of which could be derived from experiments. We have seen that Stokes's theory, when carried to its logical conclusion, became involved in difficulties, but that Fresnel's represented all the phenomena satisfactorily.

In the electromagnetic theory the same two positions are possible, either complete convection, as advocated by Stokes, or the partial convection of Fresnel. The question is whether purely electromagnetic observations will allow us to come to a decision about these two hypotheses.

Hertz was the first to apply the hypothesis of complete convection to Maxwell's field equations. In doing so, he was fully conscious that such a procedure could be only provisional, because the application to optical events would lead to the same difficulties as those which brought Stokes's theory to grief. But the simplicity of a theory which required no distinction between the motion of ether and that of matter led him to develop and to discuss it in detail. This brought to light the fact that the induction phenomena in moving *conductors*, which are by far the most important for experimental physics and technical science, are correctly represented by Hertz's theory. Disagreements with experimental results occur only in finer experiments in which the displacements in *nonconductors* play a part. We shall investigate all possibilities in succession:

1. Moving conductors (*a*) in the electrical field.
   (*b*) in the magnetic field.
2. Moving insulators (*a*) in the electrical field.
   (*b*) in the magnetic field.

1*a*. A conductor acquires surface charges in an electric field. If it is moved, it carries them along with itself. But moving charges must be equivalent to a current, and hence must produce a surrounding magnetic field according to the law of Biot and Savart. To picture this to ourselves we imagine a plate condenser whose plates are parallel to the xz-plane (Fig. 100). Let them be oppositely charged

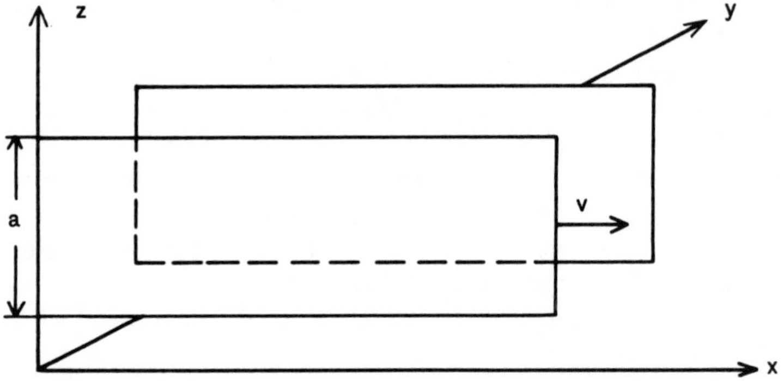

Fig. 100 *A charged plate of a condenser moving with velocity* v *perpendicular to the electric field.*

with density of charge σ on the surface. This means $e=\sigma f$ is the amount of electricity on an area f of the plate. Now let one plate be moved with respect to the ether in the direction of the x-axis with the velocity v. Then a *convection current* arises. The moving plate is displaced with velocity v, that is, by a length $v\tau$ in the time τ. If its width in z-direction is a, then an amount of electricity $e=\sigma av\tau$ passes in time τ through a plane that is parallel to the yz-plane, hence a current $J=\frac{e}{\tau}=\sigma av$ flows. This must exert exactly the same magnetic action as a conduction current of magnitude J flowing through the plate when it is at rest.

This was confirmed experimentally in Helmholtz's laboratory by H. A. Rowland (1875), and later, more accurately, by A. Eichenwald. Instead of a plate moving rectilinearly, a rotating metal disk was used.

1*b*. When conductors are moved about in a magnetic field, electric fields arise in them, and hence currents are produced. This is the phenomenon of induction by motion, already discovered by Faraday and investigated quantitatively by him. The simplest case is this: Let the magnetic field H produced, say, by a horseshoe magnet be parallel to the z-axis (Fig. 101). Let there be a straight piece of wire of length l parallel to the y-axis, and let this be moved with the velocity v in the direction of the x-axis. If the wire is now made part of

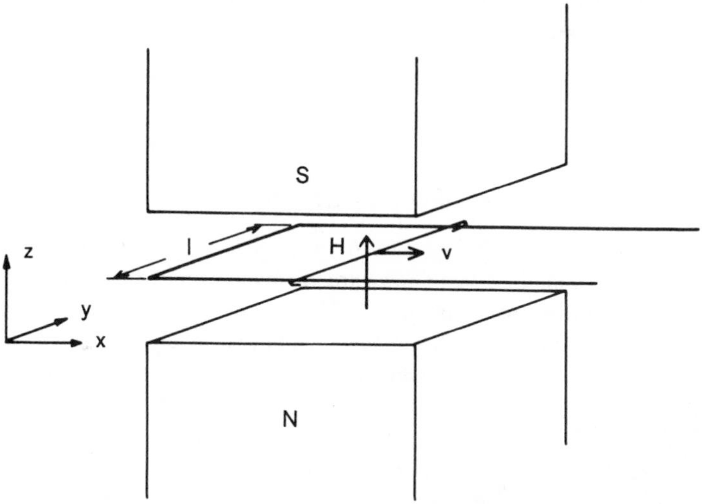

Fig. 101 *Motion of a wire of length* l, *which is part of a closed circuit, in a magnetic field between the ends of a large horseshoe magnet.*

a closed circuit by sliding it on the two opposite arms of a U-shaped piece of wire in such a way that the U takes no part in the motion (see figure), then an induction current J flows in the wire. This is given most simply by stating Faraday's law of induction thus: The current induced in a wire which forms part of a closed circuit is proportional to the change per second of the number of lines of force enclosed by the wire loop. This number is measured by the magnetic displacement per unit area μH multiplied by the area f of the loop, $f\mu H$. In the section on magnetic induction (p. 176) the change of this quantity was considered to be due to a change of H by H in the short time interval τ. Here it is due to a change of the area f produced by the movement of the wire. If its length is l and its velocity perpendicular to its extension is v, then it sweeps out the area lv each second, and this is the change of f. The change of the number of lines of force per second is therefore $vl\mu H$. According to Faraday's induction law an electric current J is induced in the wire. Instead of speaking of the current J it is better to express the effect in terms of the potential difference V produced between the ends of the wire. The experiment gives V proportional to the quantity just discussed, $vl\mu H$. Concerning the factor of proportionality, a remarkable law of symmetry has been revealed. If one measures all quantities in units here used, this factor turns out to be $\frac{1}{c}$, so that one has the equation $V=\frac{1}{c}\,vl\mu H$. Seen from the wire this corresponds to an electric field $E=\frac{V}{l}=\frac{v}{c}\,\mu H$. If the same piece of wire were to move without being part of a closed circuit, there would appear charges at the end of the wire corresponding to this field as long as the movement went on.

This law is the basis of all machines and apparatus of physics and electrotechnical science in which energy of motion is transformed by induction into electromagnetic energy; these include, for example, the telephone, and dynamo machines of every kind. Hence the law may be regarded as having been confirmed by countless experiments.

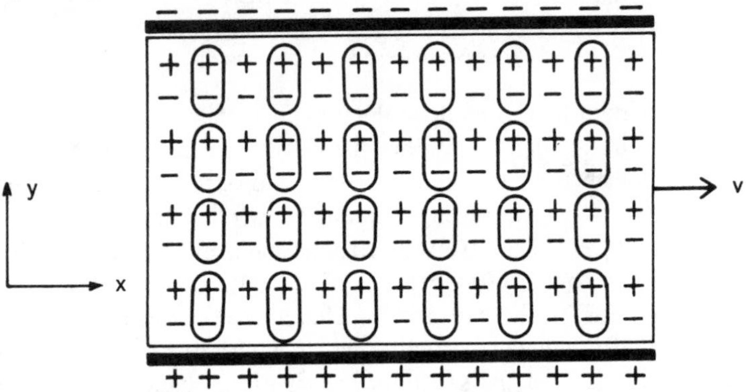

Fig. 102 *A charged condenser is filled with a disk-shaped insulator. In the insulator the displacement has induced charges on the surface of the disk. One part of the displacement (dipoles ±) is caused by the ether, the other part (dipoles* ⊞*) by the insulator. If the insulator is moved, only the insulator dipoles are moved with it.*

2*a*. We suppose the motion of a nonconductor in an electric field to be realized thus: A movable disk composed of the substance of the nonconductor is placed between the two plates of the condenser of Fig. 100 (see Fig. 102). The disk shall fill the space between the condenser plates so that the distance a marked in Fig. 100 measures also the corresponding width of the disk. If the condenser is now charged, an electric field E arises in the disk, and a displacement ϵE is induced which is perpendicular to the plane of the plates, that is, parallel to the y-direction. This causes the two boundary faces of the insulating disk to be charged equally and oppositely to the metal plates facing them respectively. The surface charge has a density σ which is proportional to the displacement D in the insulator: $4\pi\sigma = D = \epsilon E$. D consists of two parts, $D_e = E$, the displacement of the ether, and $D_m = D - D_e$, the displacement of matter alone.

If the insulating layer is now moved in the direction of the x-axis with the velocity v, then, according to Hertz, the ether in the layer will be carried along completely. Hence the field E and the charges of density $\sigma = \frac{\epsilon E}{4\pi}$ produced by it on the bounding planes will also be carried along.

Therefore the moving charge of a bounding surface again represents a current $\frac{\epsilon E}{4\pi} av$ and must generate, according to Biot and Savart's law, a magnetic field.

W. C. Röntgen proved experimentally (1885) that this was the case, but the deflection of the magnet needle that he observed was much smaller than it should have been from Hertz's theory. Röntgen's experiments show that only the excess of the charge density over the displacement of the ether alone (i.e., $D - D_e = E(\epsilon - 1) = D_m$, the displacement of the insulator alone) participates in the motion of the matter. We shall interpret this result later in a simple way. Here we merely establish that, as was to be expected according to the well-known facts of optics, Hertz's theory of complete convection also fails to explain purely electromagnetic phenomena.

Eichenwald (in 1903) confirmed Röntgen's result very strikingly by allowing the charged metal plates to take part in the motion. These give a convection current of the amount $\sigma av = \frac{\epsilon E}{4\pi} av$; according to Hertz this insulating layer ought, on account of the opposite and equal charges, exactly to compensate this current. But Eichenwald found that this was not the case. Rather, he obtained a current which was entirely independent of the material of the insulator. This is exactly what is to be expected according to Röntgen's results described above. For the current due to the insulator is $\left(\frac{\epsilon E}{4\pi} - \frac{E}{4\pi}\right)av$, of which the first term is compensated by the convection current of the plates, and so we are left with the current $\frac{E}{4\pi}\,av$, which is independent of the dielectric constant ϵ.

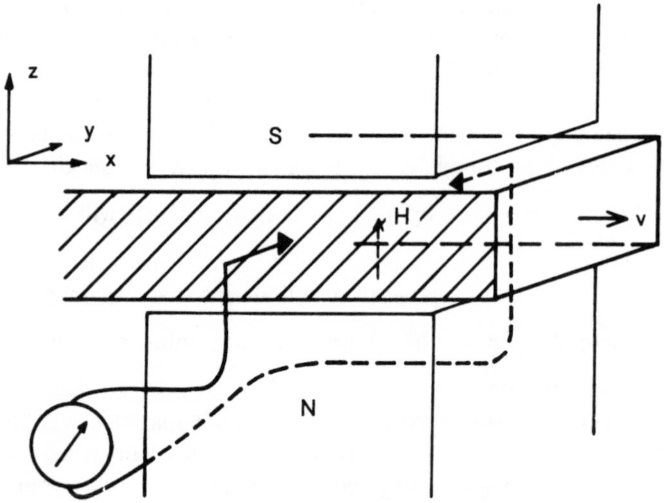

Fig. 103 *A piece of an insulator is moved in a magnetic field to measure the induced displacement charges at the surface of the disk.*

2*b*. We assume a magnetic field parallel to the z-axis, produced, say, by a horseshoe magnet, and a disk of nonconducting material moving through the field in the direction of the x-axis (Fig. 103). Let the insulator be not magnetizable ($\mu = 1$). Let the two bounding faces of the disk which are perpendicular to the y-axis be covered with metal, and let these surface layers be connected to an electrometer by means of sliding contacts so that the charges that arise on them can be measured.

This experiment corresponds exactly with the induction experiment discussed under (1*b*), except that a moving dielectric now takes the place of the moving conductor. The law of induction is applicable in the same way. It demands the existence of an electrical field $E = H\frac{v}{c}$, acting in the magnetic direction of the y-axis on the moving insulator. Hence, according to Hertz's theory, the two superficial layers must exhibit opposite charges of surface density $\frac{\epsilon E}{4\pi} = \frac{\epsilon H}{4\pi}\frac{v}{c}$ which cause a deflection of the electrometer. The experiment was carried out by H. F. Wilson, in 1905, with a rotating dielectric, and it did, indeed, confirm the existence of the charge produced, but again to a lesser extent, namely, corresponding to a surface density $(\epsilon - 1)\frac{H}{4\pi}\frac{v}{c}$. This means that there is only an effect of the moving matter and none of the ether. Here too, then, Hertz's theory fails.

In all these four typical phenomena what counts is clearly only the relative motion of the field-producing bodies with respect to the conductor or insulator investigated. Instead of moving this in the x-direction, as we have done, we could have kept it at rest and moved the remaining parts of the apparatus in the negative direction of the x-axis. The result would have been the same. For Hertz's theory recognizes only relative motions of bodies, the ether being also reckoned as a body. In a system moving with constant velocity everything happens, according to Hertz, as if it were at rest; that is, the classical principle of relativity holds.

But Hertz's theory is incompatible with the facts, and it soon had to make way for another which took exactly the opposite point of view with regard to relativity.

12. The Electron Theory of Lorentz

It is the theory of H. A. Lorentz (proposed in 1892) that signified the climax and the final step of the physics of the material ether.

It is a one-fluid theory of electricity that has been developed atomistically, and it is this feature which, as we shall presently see, determines the part allocated to the ether.

The fact that electric charges have an atomic structure, that is, occur in very small indivisible quantities, was first stated by Helmholtz (in 1881) in order to make intelligible Faraday's laws of electrolysis (p. 157). Actually, it was only necessary to assume that every atom in an electrolytic solution enters into a sort of chemical bond with an atom of electricity or an electron in order to make intelligible the fact that a definite amount of electricity always separates out equivalent amounts of substances.

The atomic structure of electricity proved of particular value for explaining the phenomena which are observed in the passage of the electric current through a rarefied gas. Here it was first discovered that positive and negative electricity behave quite differently. If two metal electrodes are introduced into a glass tube and if a current is made to pass between them (Fig. 104), very complicated phenomena are produced so long as gas is still present at an appreciable

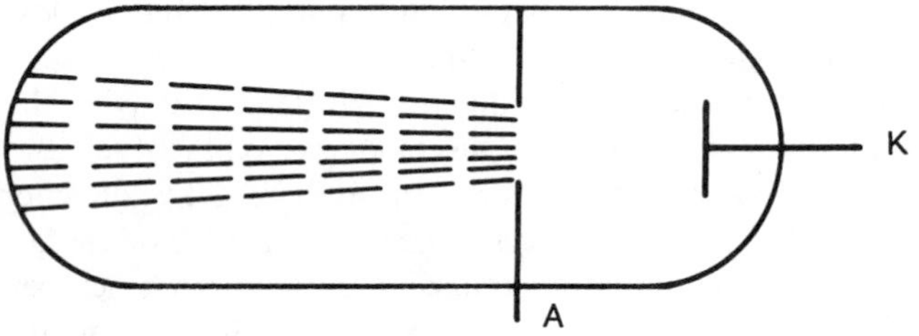

Fig. 104 *A tube to produce cathode rays.* K *cathode,* A *anode.*

pressure in the tube. But if the gas is pumped out more and more, the phenomena become increasingly simple. When the vacuum is very high, the negative electrode, the cathode K, emits rays which pass through a hole in the positive pole, the anode A, and are observed behind A through fluorescence produced on a screen (as known from any television set). These rays are called *cathode rays.* It was shown that they could be deflected by a magnet in the manner of a stream of negative electricity. The greatest share in investigating the nature of cathode rays was taken by Sir J. J. Thomson and P. L. Lenard. The negative charge of the rays could also be directly demonstrated by collecting it in a hollow conductor. Furthermore, the rays are deflected by an electric field applied perpendicular to their path, and this deflection is opposite to the direction of the field, which again proves the charge to be negative.

The conviction that the nature of cathode rays is corpuscular became a certainty when physicists succeeded in deducing quantitative conclusions concerning their velocity and their charge.

If we picture the cathode ray as a stream of small particles of mass m_{el}, then clearly it will be the less deflected by a definite electric or magnetic field the greater its velocity—just as the trajectory of a rifle bullet is straighter the greater its velocity. Now it is possible to produce cathode rays that can be strongly deflected—that is, slow cathode rays—by using a small difference of potential between cathode and anode. If one chooses a strong difference of potential between the two poles, the rays are strongly accelerated by the field from K to A. The velocity of the cathode rays behind the hole in A depends only on the acceleration from K to A, which may be calculated from the fundamental equation of mechanics

$$m_{el}b = K = eE,$$

where e is the charge and E is the field strength. We are here clearly dealing with a case analogous to that of "falling" bodies, in which the acceleration is not equal to that of gravity g but to $\frac{e}{m_{el}} E$. If the ratio $\frac{e}{m_{el}}$ were known, the velocity v could be found from the laws for falling bodies. But there are two unknowns, $\frac{e}{m_{el}}$ and v, and hence another measurement is necessary if they are to be determined. This is obtained by applying a lateral magnetic force. In discussing Hertz's theory (V, 11, 1b, p. 194) we saw that a magnetic field H sets up in a body moving perpendicular to H an electric field $E=\frac{v}{c} H$, which is perpendicular both to H and to v. Hence a deflecting force $eE=e\frac{v}{c} H$ will act on every cathode ray particle so that there will be an acceleration $b=\frac{e}{m_{el}} \frac{v}{c} H$ perpendicular to the original motion. This may be found by measuring the lateral deflection of the ray.

Hence we have a second equation for determining the two unknowns $\frac{e}{m_{el}}$ and v.

The determinations carried out by this or a similar method have led to the result that, for velocities that are not too great, $\frac{e}{m_{el}}$ has a definite constant value:

$$\frac{e}{m_{el}} = 5.31 \times 10^{17} \text{ electrostatic units per gm.} \tag{66}$$

On the other hand, in dealing with electrolysis (V, 2, formula (48), p. 159), we stated that hydrogen carries an amount of electricity $C_0 = 2.90 \times 10^{14}$ electrostatic units per gm. If we now make the readily suggested assumption that the charge of a particle is in each case the same, namely, an atom of electricity or an electron, we must conclude that the mass of the cathode ray particle m_{el} must bear the following ratio to that of the hydrogen atom m_H:

$$\frac{m_{el}}{m_H} = \frac{e}{m_H} : \frac{e}{m_{el}} = \frac{2.90 \times 10^{14}}{5.31 \times 10^{17}} = \frac{1}{1830}.$$

Thus, the cathode ray particles are nearly 2000 times lighter than hydrogen atoms, which are the lightest of all chemical atoms. This result leads us to conclude that cathode rays are a current of pure atoms of electricity.

This view has stood the test of innumerable researches. Negative electricity consists of freely moving electrons, but positive electricity is bound to matter and never occurs without it. Thus recent experimental researches have confirmed and given a precise form to the old hypothesis of the one-fluid theory. The amount of the charge e of the individual electron has also been successfully determined. The first experiments of this type were carried out by Sir J. J. Thomson (1898). The underlying idea is: Little drops of oil or water, or tiny spheres of metal of microscopic or submicroscopic dimensions, which are produced by condensation of a vapor or by spraying of a liquid in air, fall with constant velocity, since the friction of the air prevents acceleration. By measuring the rate of fall the size of the particles can be determined, and then their mass M is obtained by multiplying their size by the density. The weight of such a particle is then Mg, where $g = 981$ cm./sec.^{2} is the acceleration due to gravity. Now such particles may be charged electrically by subjecting the air to the action of x-rays or the rays of radioactive substances. If an electric field E, which is directed vertically upwards, is then applied, a sphere carrying the positive charge e is pulled upwards by it, and if the electric force eE is equal to the weight Mg, the sphere will remain poised in the air. The charge e may then be calculated from the equation $eE = Mg$. Millikan (1910), who carried out the most accurate experiments of this sort, found that the charge of the small drops is always an exact multiple of a definite minimum charge. Thus we shall call this the *elementary electrical quantum*. Its value is

$$e = 4.77 \times 10^{-10} \text{ electrostatic units.} \tag{67}$$

The absolute value of the elementary charge plays no essential part in Lorentz's theory of electrons. We shall now describe the physical world as suggested by Lorentz.

The material atoms are the carriers of positive electricity, which is indissolubly connected with them. In addition they also contain a number of negative electrons, so that they appear to be electrically neutral with respect to their surroundings. In nonconductors the electrons are tightly bound to the atoms; they may only be displaced slightly out of the positions of equilibrium so that the atom becomes a dipole. In electrolytes and conducting gases it may occur that an atom has one or more electrons too many or too few; it is then called an *ion* or a *carrier*, and it wanders in the electric field carrying electricity and matter simultaneously. In metals the electrons move about freely and experience resistance only when they collide with the atoms of the substance. Magnetism comes about when the electrons in certain atoms move in closed orbits and hence represent Amperian molecular currents.

The electrons and the positive atomic charges swim about in the sea of ether in which an electromagnetic field exists in accordance with Maxwell's equations. But we must set $\epsilon = 1$, $\mu = 1$ in them, and, in place of the density of the conduction current, we have the convection current ρv of the electrons. The equations thus become

$$\left.\begin{aligned} \operatorname{div} E &= 4\pi\rho; & \operatorname{curl} H - \frac{1}{c}\frac{E}{\tau} &= 4\pi\rho\frac{v}{c}, \\ \operatorname{div} H &= 0; & \operatorname{curl} E + \frac{1}{c}\frac{H}{\tau} &= 0, \end{aligned}\right\} \qquad (68)$$

and include the laws of Coulomb, Biot and Savart, and Faraday in the usual way.

Thus *all* electromagnetic events consist fundamentally of the motions of electrons and of the fields accompanying them. All matter is an electrical phenomenon. The various properties of matter depend on the various possibilities of motion of the electrons with respect to atoms, in the manner just described. The problem of the theory of electrons is to derive the ordinary equations of Maxwell from the fundamental laws (68) for the individual, invisible electrons and atoms, that is, to show that material bodies appear to have, according to their nature, respectively, a conductivity σ, a dielectric constant ϵ, and a permeability μ.

Lorentz has solved this problem and has shown that the theory of electrons not only gives Maxwell's laws in the simplest case, but, more than this, also explains numerous facts which were inexplicable for the descriptive theory or could be accounted for only with the aid of artificial hypotheses. These facts comprise, above all, the more refined phenomena of optics, color dispersion, the magnetic rotation of the plane of polarization (p. 184) discovered by Faraday, and similar interactions between light waves and electric or magnetic fields. We shall not enter further into this extensive and mathematically complicated theory, but shall restrict ourselves to the question which is of primary interest to us: What part does the ether play in this concept of matter?

Lorentz proclaimed the very radical thesis which had never before been asserted with such definiteness:

The ether is at rest in absolute space.

In principle this identifies the ether with absolute space. Absolute space is no vacuum, but something with definite properties whose state is described with the help of two directed quantities, the electrical field E and the magnetic field H, and, as such, it is called the ether.

This assumption goes still further than the theory of Fresnel. In the latter the ether of astronomic space was at rest in a special inertial system, which we might regard as being absolute rest. But the ether inside material bodies is partly carried along by them.

Lorentz dispenses with even this partial convection and arrives at practically the same result. To see this, we consider the phenomenon that occurs in a dielectric between the plates of a condenser. When the latter is charged, a field perpendicular to the plate arises (Fig. 89), and this displaces the electrons in the atoms of the dielectric substance and transforms them into dipoles, as we explained earlier (pp. 197 and 198). The dielectric displacement in Maxwell's sense is ϵE, but only a part of it is due to the actual displacement of the

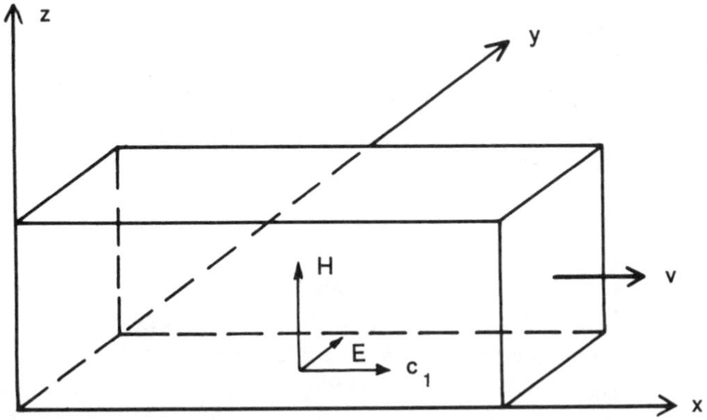

Fig. 105 *A light ray with its electric vibration* E *and its magnetic vibration* H *travelling in an insulator with velocity* c_1. *The insulator moves with velocity* v.

electrons. For a vacuum has the dielectric constant $\epsilon=1$, and hence the displacement E; consequently the true value of the electronic displacement is $\epsilon E-E=(\epsilon-1)E$. Now we have seen that the experiments of Röntgen and Wilson on the phenomena in moving insulators affirm that actually only this part of the displacement takes part in the motion. Thus Lorentz's theory gives a correct account of electromagnetic facts without having recourse to hypotheses about the ether carried along by matter.

The fact that the convection of light comes out in exact agreement with Fresnel's formula (44), (p. 134) is made plausible by the following argument:

As in Wilson's experiment, we consider a dielectric body which moves in the x-direction with the velocity v and in which a light ray travels in the same direction (Fig. 105). Let this ray consist of an electrical vibration E parallel to the y-axis and a magnetic vibration parallel to the z-axis. Now we know from Wilson's experiment that such a magnetic field in the moving body produces a corresponding displacement of the value $(\epsilon - 1)\frac{v}{c}H$ in the y-direction. From this we get a superposed electrical field if we divide by ϵ. Thus the total electrical field is

$$E + \frac{\epsilon - 1}{\epsilon}\frac{v}{c}H.$$

If the convection were complete, as is assumed in Hertz's theory, we should have only ϵ in place of $\epsilon - 1$, thus the total field would have the value $E + \frac{v}{c}H$. We see that in our formula v is replaced by

$$\frac{\epsilon - 1}{\epsilon}v.$$

Therefore, this value should correspond to the absolute velocity of the ether within matter according to Fresnel's theory, that is, to the convection coefficient called ϕ in optics (compare formula (44)). This is precisely the case, for, according to Maxwell's electromagnetic theory (p. 188), the dielectric constant is equal to the square of the index of refraction n, i.e., $\epsilon = n^2$. If we insert this value, we get

$$\frac{\epsilon - 1}{\epsilon}v = \frac{n^2 - 1}{n^2}v = \left(1 - \frac{1}{n^2}\right)v = \phi$$

in agreement with formula (44).

We recall that Fresnel's theory encountered difficulties through color dispersion; for if the refractive index n depends on the frequency (color) of the light, so also will the convection coefficient ϕ. But the ether can be carried along in only one definite way, not differently for each color. This difficulty vanishes for the theory of electrons, since the ether remains at rest and it is the electrons situated in the matter that are carried along; color dispersion is due to their being forced into vibration by light and reacting, in turn, on the velocity of light.

We cannot enter further into the details of this theory and its many ramifications, but we shall summarize the result as follows: Lorentz's theory presupposes the existence of an ether that is absolutely at rest. It then proves that, in spite of this, all electromagnetic and optical phenomena depend only on the relative motions of translation of material bodies so far as terms of the first order in β come into account. Hence it accounts for all known phenomena, above all for the fact that the absolute motion of the earth through the ether cannot be demonstrated by experiments on the earth involving only quantities of the first order (this is the optical, or rather the electromagnetic, principle of relativity).

There is, however, one experiment of the first order which can no more be explained by Lorentz's theory than by any of the other theories previously discussed: this would be a failure of the experiment to find an absolute motion of the whole solar system with Römer's method (see pp. 91 and 129).

The deciding point for Lorentz's theory is whether it stands the test of experiments that allow quantities of the second order in β to be measured. For they should make it possible to establish the absolute motion of the earth through the ether. Before we enter into this question, we have yet to discuss an achievement of Lorentz's theory of electrons through which its range became greatly extended, namely, the electrodynamic interpretation of inertia.

13. Electromagnetic Mass

The reader will have remarked that from the moment when we left the elastic ether and turned our attention to the electrodynamic ether, we have had little to say about mechanics. Mechanical and electrodynamic phenomena each form a realm for themselves. The former take place in absolute Newtonian space, which is defined by the law of inertia and which betrays its existence through centrifugal forces; the latter are states of the ether which is at rest in absolute space. A comprehensive theory, such as Lorentz's aims at being, cannot allow these two realms to exist side by side unassociated.

We have seen that physicists in spite of incredible effort and ingenuity were not able to reduce electrodynamics to terms of mechanics. The converse then boldly suggests itself: Can mechanics be reduced to terms of electrodynamics?

If this could be carried out successfully the absolute abstract space of Newton would be transformed into the concrete ether. The inertial resistances and centrifugal forces would appear as physical actions of the ether, say, as electromagnetic fields of particular form, but the principle of relativity of mechanics would lose its strict validity and would be true, like that of electrodynamics, only approximately, for quantities of the first order in $\beta=\frac{v}{c}$.

Science has not hesitated to take this step, which entirely reverses the order of rank of concepts. And, although the doctrine of an

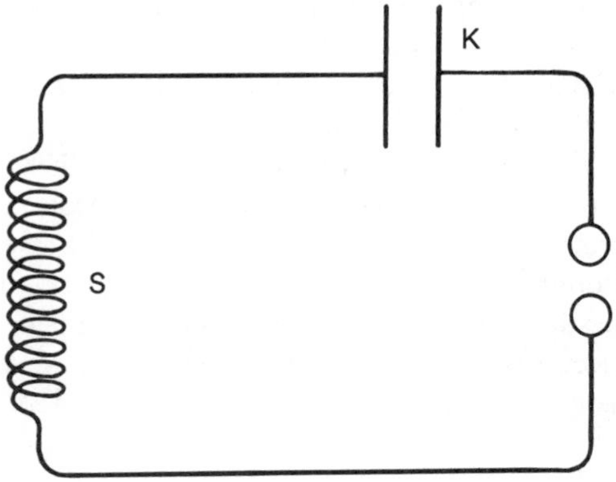

Fig. 106 *A circuit with condenser* K, *coil* S, *and spark gap* F *used to demonstrate the oscillations of electricity.*

ether absolutely at rest later had to be dropped, this revolution, which ejected mechanics from its throne and raised electrodynamics to sovereign power in physics, was not in vain. Its results have retained their validity in a somewhat altered form.

We saw already (p. 185) that the propagation of electromagnetic waves comes about through the mutual action of electrical and magnetic fields producing an effect analogous to that of mechanical inertia. An electromagnetic field has a power of persistence quite similar to that of matter. To generate the field, work must be performed, and when it is destroyed, this work again appears. This is observed in all phenomena that are connected with electromagnetic vibrations—for example, in the various forms of wireless transmitters. An old-fashioned wireless transmitter of the Marconi type contains an electric oscillator, consisting essentially (Fig. 106) of a spark gap *F*, a coil *S*, and a condenser *K* (two metal plates that are separated from each other) connected by wires to form an "open" circuit. The condenser is charged until a spark jumps across the gap at *F*. This causes the condenser to become discharged and the quantities of electricity that have been stored flow away. They do not simply neutralize each other, but shoot beyond the state of equilibrium and again become collected on the condenser plates, but with reversed signs, just as a pendulum swings past the position of equilibrium to the opposite side. When the condenser has thus

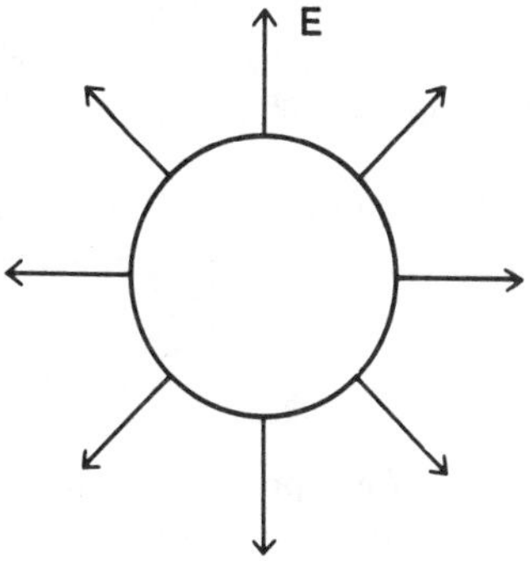

Fig. 107 *The electric field around a charge at rest.*

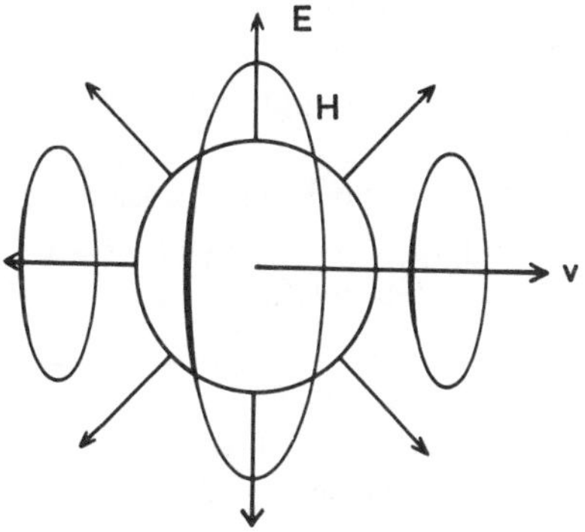

Fig. 108 *The electric field around a charge is supplemented by a magnetic field when it is moved.*

been charged up afresh, the electricity again flows back causing another spark to jump the gap, and thus the system oscillates to and fro until its energy has been used up in warming the conducting wires or in being passed on to other parts of the apparatus, for example, the emitting antenna. Thus the oscillation of the electricity proves the inertial property of the field, which exactly corresponds to the inertia of mass of the pendulum bob. Maxwell's theory represents this fact correctly in all details. The electromagnetic vibrations that occur in a definite apparatus can be predicted by calculations from the equations of the field.

This led J. J. Thomson to infer that the inertia of a body must be increased by an electric charge which is imparted to it. Let us consider a charged sphere first at rest and then moving with the velocity v. The stationary sphere has an electrostatic field with lines of forces directed radially outwards (Fig. 107); the moving sphere has, in addition, a magnetic field with circular lines and forces that encircle the path of the sphere (Fig. 108). For a moving charge is a convection current (combined with a displacement current) and produces a magnetic field in accordance with Biot and Savart's law. Both states have the inertial property above described. The one can be transformed into the other only by the application of work. The force that is necessary to set the stationary sphere into motion is thus greater for the charged than for the uncharged sphere. To accelerate the moving charged sphere still further, the magnetic field H must clearly be strengthened. Thus, again, an increase of force is necessary.

We remember that a force K that acts for a short time τ represents an impulse $J=K\tau$ which produces a change of velocity w in a mass m in accordance with the formula (7) (II, 9, p. 34):

$$mw = J.$$

If the mass carries a charge, a definite impulse J will produce a smaller change of velocity and the remainder J' will be used to change the magnetic field. Thus we have

$$mw = J-J'.$$

Now, calculation gives the rather obvious result that the impulse J' necessary to increase the magnetic field is greater, the greater the change of velocity w, and, indeed, it is approximately proportional to this change of velocity. Thus we may set $J'=m'w$, where m' is a factor of proportionality which, moreover, may depend on the state, that is, the velocity v, before the change of velocity occurs. We then have

$$mw = J-m'w$$

or

$$(m+m')w = J.$$

Thus, it is as if the mass m were augmented by an amount m', which is to be calculated from the electromagnetic field equations and which may be dependent on the velocity v. The exact value of m' for any velocity v may be calculated only if assumptions are made about the distribution of the electric charge over the moving body. But the limiting value for velocities that are small compared with that of light c, that is, for small values of β, is obtained independently of such assumptions as

$$m' = \frac{4}{3}\frac{S}{c^2}, \tag{69}$$

where S is the electrostatic energy of the charges on the body.

We have seen that the mass of the electron is about 2000 times smaller than that of the hydrogen atom. Hence the idea occurs that the electron has, perhaps, no "ordinary" mass at all, but is nothing other than an "atom of electricity," and that its mass is entirely electromagnetic in origin. Is such an assumption reconcilable with the knowledge that we have of the size, charge, and mass of the electron?

Since the electrons are to be the structural elements of atoms, they must at any rate be small compared with the size of atoms. Now we know from atomic physics that the radius of atoms is of the order 10^{-8} cm. Thus the radius of the electron must be smaller than 10^{-8} cm. If we imagine the electron as a sphere of radius a with charge e distributed over its surface, then, as may be derived from Coulomb's law, the electrostatic energy is $S = \frac{1}{2}\frac{e^2}{a}$. Hence, by (69), the electromagnetic mass becomes

$$m_{el} = \frac{4}{3}\frac{S}{c^2} = \frac{2}{3}\frac{e^2}{ac^2}.$$

From this we can calculate the radius a:

$$a = \frac{2}{3}\frac{e}{c^2}\frac{e}{m_{el}}.$$

On the right-hand side we know all the quantities, $\frac{e}{m_{el}}$ from the deflection of the cathode rays (formula (66), p. 202), e from Millikan's measurements (formula (67), p. 203), and c the velocity of light. If we insert the values given, we get

$$a = 1.88 \times 10^{-13} \text{ cm}.$$

a length which is about 100,000 times smaller than the radius of an atom.

Thus the hypothesis that the mass of the electron is electromagnetic in origin does not conflict with the known facts. But this does not prove the hypothesis.

At this stage the theory found strong support in refined observations of cathode rays and of the β-rays of radioactive substances, which are also ejected electrons. We explained above how electric and magnetic action on these rays allows us to determine the ratio of the charge to the mass, $\frac{e}{m_{el}}$, and also their velocity v, and that at first a definite value for $\frac{e}{m_{el}}$ was obtained, which was independent of v. But, on proceeding to higher velocities, a decrease of $\frac{e}{m_{el}}$ was found. This effect was particularly clear and could be measured quantitatively in the case of the β-rays of radium, which are only slightly slower than light. The assumption that an electric charge should depend on the velocity is incompatible with the ideas of the electron theory. But that the mass should depend on the velocity was certainly to be expected if the mass was to be electromagnetic in origin. To arrive at a quantitative theory, it is true, definite assumptions had to be made about the form of the electron and the distribution of the charge on it. M. Abraham (1903) regarded the electron as a rigid sphere, with a charge distributed on the one hand, uniformly over the interior, or, on the other, over the surface, and he showed that both assumptions lead to the same dependence of the electromagnetic mass on the velocity, namely, to an increase of mass with increasing velocity. The faster the electron travels, the more the electromagnetic field resists a further increase of velocity. The increase of m_{el} explains the observed decrease of $\frac{e}{m_{el}}$, and Abraham's theory agrees quantitatively very well with the results of measurement of Kaufmann (1901) if it is assumed that there is no "ordinary" mass present.

Thus, the object of tracing the inertia of electrons back to electromagnetic fields in the ether was attained. At the same time a further perspective presented itself. Since atoms are the carriers of positive electricity, and also contain numerous electrons, perhaps their mass is also electromagnetic in origin? In that case, mass as the measure of the inertial persistence would no longer be a primary phenomenon, as it is in elementary mechanics, but a secondary consequence of the structure of the ether. Therefore, Newton's absolute space, which is defined only by the mechanical law of inertia, becomes superfluous; its part is taken over by the ether whose electromagnetic properties are well known.

We shall see (V, 15, p. 221) that new facts contradict this view. But the relationship between mass and electromagnetic energy, which was first discovered in this way, constitutes a fundamental discovery the deep significance of which was brought into prominence only when Einstein proposed his theory of relativity.

We have yet to add that, besides Abraham's theory of the rigid electron, other hypotheses were set up and worked out mathematically. The most important is that of Lorentz (1904) which is closely connected with the theory of relativity. Lorentz assumed that every moving electron contracts in the direction of motion, so that from a sphere it becomes a flattened spheroid of revolution, the amount of flattening depending in a definite way on the velocity. This hypothesis seems at first sight strange. It certainly gives a simpler formula for the way electromagnetic mass depends on velocity than does Abraham's theory, but this in itself does not justify it. The actual confirmation came from the course Lorentz's theory of electrons took when it had to consider quantities of the second order in the discussion of experimental researches, to which we shall presently direct our attention. Lorentz's formula then turned out to have a universal significance in the theory of relativity. The experimental decision between it and Abraham's theory will be discussed later (VI, 7, p. 278).

At the beginning of the new century, after the theory of electrons had reached the stage above described, the possibility of forming a uniform physical picture of the world seemed at hand—a picture which would reduce all forms of energy, including mechanical inertia, to the same root, to the electromagnetic field in the ether. Only one form of energy—gravitation—seemed still to remain outside the system; yet it could be hoped that that, too, would allow itself to be interpreted as an action of the ether.

14. Michelson and Morley's Experiment

Twenty years before this period, however, the base of the whole structure had already cracked and, while the building was going on above, the foundations needed repairing and strengthening.

We have several times emphasized that any decisive experiment in regard to the theory of the stationary ether had to be precise enough to determine quantities of the second order in β. Only then could it be ascertained whether or not a fast-moving body is swept by an ether wind which blows away the light waves as is demanded by theory.

Michelson and Morley (1881) were the first successfully to carry out the most important experiment of this type. They used Michelson's interferometer (IV, 4, p. 102) which they had refined to a precision instrument of unheard-of efficiency.

In investigating the influence of the earth's motion on the velocity of light (IV, 9, p. 130), it has been found that the time taken by a ray of light to pass back and forth along a distance l parallel to the earth's motion differs only by a quantity of the second order from the value which it has when the earth is at rest. We found earlier that this time was

$$t_1 = l\left(\frac{1}{c+v}+\frac{1}{c-v}\right) = \frac{2lc}{c^2-v^2},$$

for which we may also write

$$t_1 = \frac{2l}{c}\frac{1}{1-\beta^2}.$$

If this time could be so accurately measured that the fraction $\frac{1}{1-\beta^2}$ could be distinguished from 1 in spite of the extremely small value of the quantity β^2, we should have means of proving the existence of an ether wind.

It is by no means possible, however, to measure the short time taken by a light ray to traverse a certain distance. Interferometric methods give us rather only differences of the times taken by light to traverse various different routes between two given points. But they give these with amazing accuracy.

For this reason Michelson and Morley caused a second ray of light to traverse a path AB of the same length l backwards and forwards, but perpendicular to the earth's orbit (Fig. 109). While

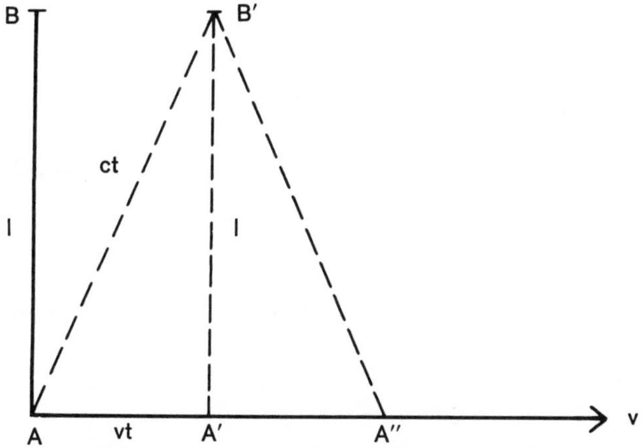

Fig. 109 *The path of the light in Michelson's experiment.*

the light passes from A to B, the earth moves a short distance forward so that the point B arrives at the point B' of the ether. Thus the true path of the light in the ether is AB', and if it takes a time t to cover this distance, then $AB' = ct$. During the same time A has moved on to the point A' with the velocity v, thus $AA' = vt$. If we now apply Pythagoras' theorem to the right-angled triangle $AA'B$, we get

$$c^2t^2 = l^2 + v^2t^2$$

or

$$t^2(c^2 - v^2) = l^2, \qquad t^2 = \frac{l^2}{c^2 - v^2} = \frac{l^2}{c^2}\frac{1}{1-\beta^2},$$

$$t = \frac{l}{c}\frac{1}{\sqrt{1-\beta^2}}.$$

The light requires exactly the same time to make the return journey, for the earth shifts by the same amount, so that the initial point A moves from A' to A''.

Thus the light takes the following time for the journey backwards and forwards:

$$t_2 = \frac{2l}{c} \frac{1}{\sqrt{1-\beta^2}}.$$

The difference between the times taken to cover the same distance parallel and perpendicular to the earth's motion is thus

$$t_1 - t_2 = \frac{2l}{c}\left(\frac{1}{1-\beta^2} - \frac{1}{\sqrt{1-\beta^2}}\right).$$

Now, by neglecting terms of higher order than the second in β (similar to what was done on p. 126) we may approximate by replacing $\frac{1}{1-\beta^2}$ by $1+\beta^2$, and $\frac{1}{\sqrt{1-\beta^2}}$ by $1+\frac{\beta^2}{2}$.*

Hence we may write to a sufficient degree of approximation

$$t_1 - t_2 = \frac{2l}{c}\left[(1+\beta^2) - \left(1+\frac{\beta^2}{2}\right)\right] = \frac{2l}{c}\frac{\beta^2}{2} = \frac{l}{c}\beta^2.$$

The retardation of the one light wave compared with the other is thus a quantity of the second order.

This retardation may be measured with the help of Michelson's interferometer (Fig. 110). In this the light coming from the source Q is divided at the half-silvered plate P into two rays which run in perpendicular directions to the mirrors S_1 and S_2, at which they are reflected and sent back to the plate P. From P onwards they run parallel into the telescope F where they interfere. If the distances S_1P and S_2P are equal and if one arm of the apparatus is placed in the direction of the earth's motion, one duplicates the

* To show the approximate validity of the equation $\frac{1}{1-\beta^2} = 1+\beta^2$ write it $1 = (1-\beta^2)(1+\beta^2) = 1-\beta^4$, which is correct if β^4 is neglected. In the same way, squaring $\frac{1}{\sqrt{1-\beta^2}} = 1+\frac{\beta^2}{2}$, one obtains $\frac{1}{1-\beta^2} = 1+\beta^2+\frac{\beta^4}{4}$; if the last term is neglected one has the same formula as above.

case just discussed. Thus the two rays reach the field of vision with a difference of time $\frac{l}{c}\beta^2$. Hence the interference fringes are not situated precisely where they would be if the earth were at rest. But if we now turn the apparatus through 90° until the other arm is parallel to the direction of the earth's motion, the interference fringes will be displaced by the same amount but in the opposite direction.

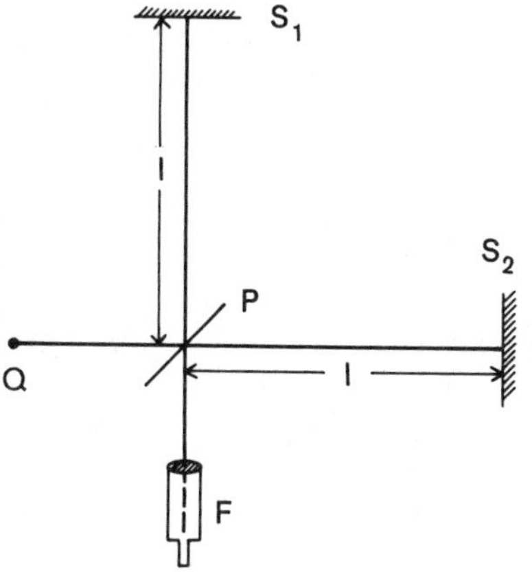

Fig. 110 *Michelson's interferometer.*

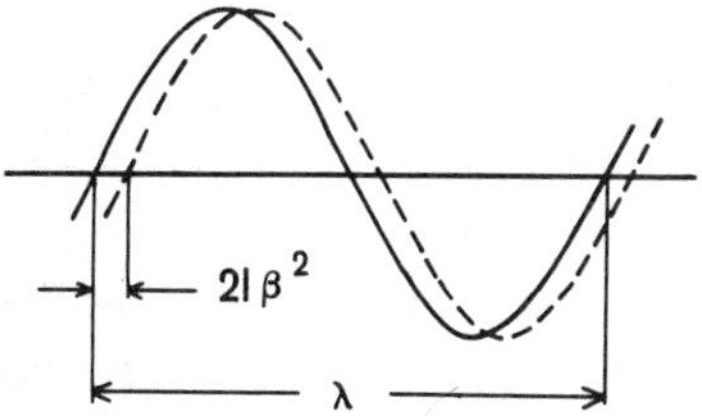

Fig. 111 *Two waves of the same wave length λ, one shifted against the other by 2 lβ².*

Hence if we observe the position of the interference fringes while the apparatus is being rotated, a displacement should be measured which corresponds to the double retardation $2\frac{l}{c}\beta^2$.

If T is the period of vibration of the light used, the ratio of the retardation to the period is $\frac{2l}{cT}\beta^2$, and since by formula (35) (p. 99) the wave length $\lambda = cT$, we may write this ratio as $2\frac{l}{\lambda}\beta^2$.

Hence, when the apparatus is rotated, the two interfering trains of waves experience a relative displacement whose ratio to the wave length is given by $\frac{2l\beta^2}{\lambda}$ (Fig. 111). The interference fringes themselves arise because the rays which leave the source in different directions have to traverse somewhat different paths. The distance between two fringes corresponds to a path difference of one wave length, hence the observable displacement of the fringes is the fraction $\frac{2l\beta^2}{\lambda}$ of the width of the fringe.

Now Michelson, in a repetition of his experiment with Morley (1887) carried out on a larger scale, extended the length of the path traversed by the light by means of several reflections forward and back to 11 m. $= 1.1 \times 10^3$ cm. The wave length of the light used was about $\lambda = 5.9 \times 10^{-5}$ cm. We know that β is approximately equal to 10^{-4}, and hence $\beta^2 = 10^{-8}$. So we get

$$\frac{2l\beta^2}{\lambda} = \frac{2 \times 1.1 \times 10^3 \times 10^{-8}}{5.9 \times 10^{-5}} = 0.37,$$

that is, the interference fringes must be displaced by more than one-third of their distance apart when the apparatus is turned through 90°. Michelson was certain that the one-hundredth part of this displacement would still be observable.

When the experiment was carried out, however, not the slightest sign of the expected displacement manifested itself, and later repetitions with still more refined means led to no other result. From this we must conclude that the ether wind *does not exist. The velocity of light is not influenced by the motion of the earth even to the extent involving quantities of the second order.*

15. The Contraction Hypothesis

Michelson and Morley concluded from their experiment that the ether is carried along completely by the moving earth, as is maintained in the elastic theory of Stokes and in the electromagnetic theory of Hertz. But this conclusion contradicts the numerous experiments which prove partial convection. Michelson then investigated whether it was possible to establish a difference in the velocity of light at different heights above the earth's surface, but without a positive result. He concluded from this that the motion of the ether that is carried along by the earth must extend to very great heights above the earth's surface. Thus, then, the ether would be influenced by a moving body at considerable distances. But this is in fact not the case, for Oliver Lodge showed (1892) that the velocity of light in the neighborhood of rapidly moving bodies is not influenced in the slightest, not even when the light passes through a strong electric or magnetic field, carried along by the body. But all these efforts seem superfluous, for even if they led to an unobjectionable explanation of Michelson's experiment, the rest of electrodynamics and optics of moving bodies which speaks in favor of partial convection would remain unexplained.

We see now Lorentz's theory of electrons placed in a very difficult position by Michelson and Morley's experiment. The doctrine of the stationary ether seems to demand that an ether wind exist on the earth, and hence stands in contradiction to the results of Michelson and Morley's experiment. The fact that it did not at once succumb to this challenge proves the inherent strength of the theory, a strength deriving from the consistency and completeness of its physical picture of the world. Finally, it overcame even this difficulty to a certain extent, although by a very strange hypothesis, which was proposed by Fitzgerald (1892) and at once taken up and elaborated by Lorentz.

Let us recall the reflections on which Michelson and Morley's experiment were based. We found that the time taken by a light ray to travel to and fro along a distance l differs according to whether the ray travels parallel or perpendicular to the earth's motion. In the former case

$$t_1 = \frac{2l}{c}\frac{1}{1-\beta^2},$$

in the second,

$$t_2 = \frac{2l}{c}\frac{1}{\sqrt{1-\beta^2}}.$$

If we now assume that the arm of the interferometer which is directed parallel to the direction of the earth's motion is shortened in the ratio $\sqrt{1-\beta^2}:1$, the time t_1 would become reduced in the same ratio, namely,

$$t_1 = \frac{2l\sqrt{1-\beta^2}}{c(1-\beta^2)} = \frac{2l}{c}\frac{1}{\sqrt{1-\beta^2}}.$$

Thus we should have $t_1 = t_2$.

This suggests the following general hypothesis, the crudeness and boldness of which is startling indeed: *every body which has the velocity* v *with respect to the ether contracts in the direction of motion by the fraction*

$$\sqrt{1-\beta^2} = \sqrt{1-\frac{v^2}{c^2}}.$$

Michelson and Morley's experiment must actually, then, give a negative result, since for both positions of the interferometer t_1 equals t_2. Furthermore—and this is the important point—such a contraction could not be ascertained by any means on earth, for every earthly measuring rod would be contracted in just the same way. An observer who was at rest in the ether outside the earth would, it is true, observe the contraction. The whole earth would be flattened in the direction of motion and likewise all things on it.

The contraction hypothesis seems so remarkable—indeed, almost absurd—because the contraction is not a consequence of any forces but appears only as a companion circumstance to motion. Lorentz, however, did not allow this objection to keep him from absorbing this hypothesis into his theory, particularly as *new* experiments confirmed that no second-order effect of the earth's motion through the ether could be detected.

We cannot describe all these experiments or even outline them. They are partly optical and concern the events involved in reflection and refraction, double refraction, rotation of the plane of polarization, and so forth; and they are partly electromagnetic and concern induction phenomena, the distribution of the current in wires, and the like. The improved technique of physics allows us nowadays to establish unambiguously the existence or absence of second-order effects in these phenomena. A particularly noteworthy experiment is that of Trouton and Noble (1903), which was intended to detect a torsional force which should occur in a suspended plate condenser in consequence of the ether wind.

These experiments produced without exception a negative result. There could no longer be any doubt that a motion of translation through the ether cannot be detected by an observer sharing in the motion. Thus the principle of relativity which holds for mechanics is also valid for all electromagnetic phenomena.

Lorentz next proceeded to bring this fact into harmony with his ether theory. To do this there seemed no other way than to assume the contraction hypothesis and to fuse it into the laws of the electron theory so as to form a consistent whole free from inner contradictions. He first observed that a system of electric charges which keep in equilibrium only through the action of their electrostatic forces contracts of itself as soon as it is set into motion; or, more accurately, the electromagnetic forces that arise when the system is moving uniformly change the configuration of equilibrium in such a way that every length is contracted in the direction of its motion by the factor $\sqrt{1-\beta^2}$.

Now, this mathematical theorem leads, if we assume that all physical forces are ultimately electrical in origin or that they at least follow the same laws of equilibrium in uniformly moving systems, to an explanation of Fitzgerald's contraction. The difficulty of regarding all forces as electrical is due to the circumstance that they lead, in accordance with old and well-known theorems deriving from Gauss and Green, to charges being in equilibrium but never in *stable* equilibrium. The forces which bind the atoms to form molecules and the latter together to form solid bodies cannot, therefore, simply be electrical. The necessity for assuming nonelectric forces emerges most clearly if we inquire into the dynamical constitution of an individual electron. This is supposed to be an accumulation of negative charge which we must assume to be of finite amount, for, as we have seen (p. 211), the energy of a spherically shaped charge of radius a is equal to $\frac{1}{2}\frac{e^2}{a}$, and it becomes infinitely great if a is set equal to zero. But the component parts of the electron strive to separate, since similar charges repel. Consequently, there must be a new force which keeps them together. In Abraham's theory of the electron it is assumed that an electron is a rigid sphere, that is, that the nonelectric forces are to be so great that they admit of no deformation whatsoever. But it is, of course, possible to make other assumptions.

Now, it suggested itself to Lorentz that the electron also experiences the contraction $\sqrt{1-\beta^2}$. We have already stated (p. 213) that a much simpler formula results for the mass of the electron than that arising from Abraham's hypothesis. But, in addition to electromagnetic energy, Lorentz's electrons have also an energy of deformation of foreign origin, which is wanting in the rigid electron of Abraham.

Lorentz next investigated the question whether the contraction hypothesis is sufficient for deriving the principle of relativity. After laborious calculations he established that this was not the case, but he also found (1899) what assumption had to be added in order that all electromagnetic phenomena in moving systems occur just as in the resting ether. His result is at least as remarkable as the contraction hypothesis. It is: *A new time measure must be used in a system which is moving uniformly.* He called this time, which differs from system to system, "local time." The contraction hypothesis may clearly be expressed thus: The measure of length in moving systems is different from that in the ether. Both hypotheses together state that space and time must be measured differently in moving systems and in the resting ether. Lorentz enunciated the laws according to which the measured quantities in various systems may be transformed into one another, and he proved that these transformations leave the field equations of the electron theory unchanged. This is the mathematical content of his discovery. Larmor (1900) and Poincaré (1905) arrived at similar results about the same time.* We shall examine these transformation formulae presently from Einstein's point of view, and so we shall not enter into them here. But we shall consider what consequences the new turn in Lorentz's theory had for the idea of the ether.

In the new theory of Lorentz the principle of relativity holds, in conformity with the results of experiment, for all electrodynamic events. Thus, an observer perceives the same phenomena in his system no matter whether it is at rest in the ether or moving uniformly and rectilinearly. He has no means at all of distinguishing the one from the other. For even the motion of other bodies in the world, which are moving independently of him, always informs him only of relative motion with respect to them and never of absolute motion with respect to the ether. Thus he can assert that he himself is at

* It is interesting historically that the formula of transformation to a moving system, which we nowadays call Lorentz's transformation (see VI, 2, p. 236, formula (70*a*)), was set up by Voigt as early as 1877 in an investigation which was still founded on the elastic theory of light.

rest in the ether, and no one can contradict him. It is true that a second observer on another body moving relative to the first can assert the same with equal right. There is no empirical and no theoretical means of deciding whether one or the other of them is right.

Consequently, we arrive at the same position with respect to the ether as the principle of relativity of classical mechanics did with respect to the absolute space of Newton (III, 6, p. 69). In the latter case it had to be admitted that it is meaningless to regard a definite place in absolute space as something real in the sense of physics. For there is no physical means of fixing a place in absolute space or of finding it a second time. In precisely the same way we must now admit that a definite position in the ether is nothing real in the physical sense, and for this reason the ether itself entirely loses the character of a substance. Indeed, we may say: If each of two observers who are moving relative to each other can assert with equal right that he is at rest in the ether, there can be no ether.

Thus, the extreme development of the ether theory leads to its dissolution as a fundamental concept. But it has required a great effort to admit the failure of the ether idea. Even Lorentz, whose ingenious suggestions and laborious efforts led the ether theory to this crisis, hesitated for a long time before taking this step. The reason is this: The ether was conceived for the express purpose of being a carrier of light vibrations, or, more generally, of the electromagnetic forces in empty space. Vibrations without something which vibrates seemed to be unthinkable. On the other hand, the assertion that in empty space there are observable vibrations goes beyond all possible experience. Light or electromagnetic forces are never observable except in connection with bodies. Empty space free of all matter is no object of observation at all. All that we can ascertain is that an action starts from one material body and arrives at another material body some time later. What occurs in the interval is purely hypothetical, or, more precisely, a matter of suitable assumption. Theorists may use their own judgment to attribute properties to the vacuum, with the one restriction that these serve to correlate changes of material things.

This view is a step in the direction of higher abstraction, releasing us from ideas that previously were considered to be necessary components of our thinking. At the same time, it is an approach to the ideal of allowing only that which is directly given by experience to be valid as a constructive part of the physical world, all superfluous pictures and analogies originating from more primitive and unrefined experience being eliminated.

From now on ether as a substance vanishes from theory. In its place we have the electromagnetic field as a mathematical device for conveniently describing processes in matter and their relationships.* There remains the task of building up a description of the physical world afresh on these more abstract but empirically sound foundations. As mentioned already, Lorentz and Poincaré have succeeded in doing this by careful analysis of the properties of Maxwell's equations. They were indeed in possession of a great deal of mathematical theory. Lorentz, however, was so attached to his assumption of an ether absolutely at rest that he did not acknowledge the physical significance of the equivalence of the infinite numbers of systems of reference which he had proved. He continued to believe that one of them represented the ether at rest. Poincaré went a step further. It was quite clear to him that Lorentz's viewpoint was not tenable and that the mathematical equivalence of systems of reference meant the validity of the principle of relativity. He also was quite clear about the consequences of his theory. What he missed was a simple physical—or should we say philosophical—point, which would make the theory of relativity independent of its derivation from Maxwell's equations, even though by rather tedious calculations.

This important step was to come from Einstein. He noticed that to overcome the difficulties met in relativistic considerations one had to go back to the fundamental concepts of space and time. He found that in the current concepts there was an assumption not based on facts and succeeded in rebuilding the theory by eliminating this preconceived notion.

* Einstein in later years proposed calling empty space equipped with gravitational and electromagnetic fields the "ether," whereby, however, this word is not to denote a substance with its traditional attributes. Thus, in the "ether," there are to be no determinable points, and it is meaningless to speak of motion relative to the "ether." Such a use of the word "ether" is of course admissible, and when once it has been sanctioned by usage in this way, probably quite convenient.

CHAPTER VI

EINSTEIN'S SPECIAL PRINCIPLE OF RELATIVITY

1. The Concept of Simultaneity

The difficulties which had to be overcome by applying the principle of relativity to electrodynamical events consisted of bringing into harmony the following two apparently inconsistent statements:

1. According to classical mechanics the velocity of any motion has different values for two observers moving relative to each other.
2. Experiment informs us that the velocity of light is independent of the state of motion of the observer and has always the same value c.

The older ether theory endeavored to get rid of the contradiction between these two laws by dividing the velocity of light into two components: (*a*) the velocity of the luminiferous ether, and (*b*) the velocity of light with respect to the ether. Of these two (*a*) can be appropriately described with the help of convection coefficients. This theory, however, was successful in eliminating the contradiction only with regard to quantities of the first order. To maintain the law of constancy of the propagation of light, Lorentz's theory had to introduce a special measure of length and time for every moving system. The compatibility of the statements (1) and (2) then appears as produced by a sort of "physical illusion."

In 1905 Einstein recognized that Lorentz contractions and local times were not mathematical devices and physical illusions but involved the very concepts of space and time.

Of the two statements (1) and (2), the first is purely theoretical and conceptual in character whereas the second is founded on fact.

Now, since the second statement, that of the constancy of the velocity of light, must be regarded as being experimentally established with certainty, nothing remains but to give up the first law and hence the ideas about space and time as hitherto accepted. Thus there must be an error in these ideas, or at least a fallacy, due to a confusion of habits of thought with logical consistency, a tendency we all realize to be an obstacle to progress.

Now, the *concept of simultaneity* is a fallacy of this type. It is regarded as self-evident that there is sense in the statement that an event at the point A, say the earth, and an event at the point B, say the sun, are simultaneous. It is assumed that concepts like "moment of time," "simultaneity," "earlier," "later," and so forth, have a meaning in themselves a priori which is valid for the whole universe. This was Newton's view, too, when he postulated the existence of an absolute time or duration of time (III, 1, p. 57) which was to pass "equably without regard to any thing external."

But there is certainly no such time for the quantitative physicist. He sees no meaning in the statement that an event at A and an event at B are simultaneous, since he has no means of deciding the truth or falsity of this assertion. To be able to decide whether two events at different points are simultaneous we must have clocks at every point which we can be certain will go at the same rate or beat "synchronously." Thus the question resolves into this: Can we define a means of testing the equal rate of beating of two clocks situated at different points?

Let us imagine the two clocks at A and B a distance l apart at rest in a system of reference S. Now there are two methods of regulating the clocks so that they go at the same rate:

1. We may take them to the same point, regulate them there so that they go in unison, and then restore them to A and B, respectively.
2. We may use time signals to compare the clocks.

Both processes are adopted in practice. A ship at sea carries with it a chronometer which beats accurately and which has been regulated in accordance with the normal clock in the home port; moreover, it receives time signals by wireless telegraphy.

The fact that these signals are regarded as necessary proves our lack of confidence in "transported" time. The practical weakness in the method of transporting clocks is that the smallest error in the beating increases continually. But even if the assumption is made that there are ideal clocks free from error (such as the physicist is convinced he has in the atomic vibrations that lead to the emission of light), it is logically inadmissible to base on them the definition of time in systems moving relative to each other; for the equal beating of two clocks, however good they may be, cannot be tested directly, that is, without the intervention of signals, unless they are close to and at rest relative to each other. It cannot be established without signals that they maintain the same rate when in relative motion. The contrary is the kind of pure hypothesis which we should avoid if we wish to adhere to the principles of physical research. Thus we are obliged to adopt the method of time signals to define time in systems moving relative to each other. If this allows us to arrive at a method of measuring times which is free from contradictions, we shall have to investigate subsequently how an ideal clock has to be constituted in order that it should always show the "right" time in systems moving arbitrarily (see VI, 5, p. 251).

Let us picture a long series of barges *B*, *C*, *D*, pulled by a steam tug *A* over the sea. Suppose there is no wind but that the fog is so thick that each ship is invisible to the others. Now, if the clocks on the barges and the tug are to be compared, sound signals will be used. The tug *A* sends out a shot at 12 o'clock, and when the sound is audible on the barges, the bargemen will set their clocks at 12. But it is clear that in doing so they commit a small error, since the sound requires a short time to arrive from *A* at *B*, *C*, *D*. If the velocity of sound *c* is known this error can be eliminated. The symbol *c* represents about 340 m./sec. If the barge *B* is at a distance $l=170$ m. behind *A*, the sound will take $t=\frac{l}{c}=\frac{170}{340}=\frac{1}{2}$ sec. to travel from *A* to *B*, and hence the clock at *B* must be set $\frac{1}{2}$ sec. after 12 at the moment when the sound reaches it. But this correction again is right only if the tug and barges are at rest. As soon as they are in motion the sound clearly requires less time to get from *A* to *B*, because the barge *B* is moving towards the sound wave. If we now wish to apply the true correction we must know the absolute velocity of the ships with respect to the air. If this is unknown it is impossible to compare times absolutely with the help of sound. In clear weather we can use light in place of sound. Since light travels enormously faster, the error will at any rate be small, but in a question of principle the absolute magnitude is of no account. If we imagine in place of the tug and the barges a heavenly body in the sea of ether, and light signals in place of sound signals, all our reflections remain valid and unchanged. There is no more rapid messenger than light in the universe. We see that the theory of the absolutely stationary ether leads to the conclusion that an absolute comparison of times can be carried out in moving systems only if we know the motion with respect to the ether.

But the result of the experimental researches was that it is impossible to detect motion with respect to the ether by physical means. From this it follows that absolute simultaneity can likewise be ascertained in no way whatever.

The paradox contained in this assertion vanishes if we remind ourselves that to compare times by means of light signals we must know the exact value of the velocity of light, while the measurement of the latter again entails the determination of a length of time. Thus we are obviously moving in a vicious circle.

Even if we cannot attain absolute simultaneity, however, it is possible, as Einstein has shown, to define a *relative simultaneity* for all clocks that are relatively at rest with respect to one another without knowing the value of the velocity of the signals.

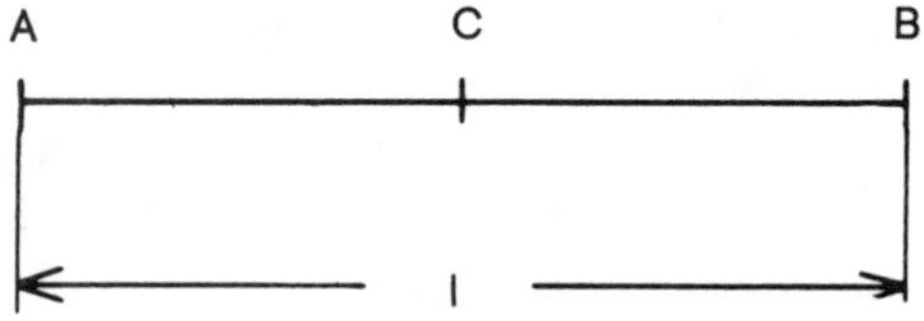

Fig. 112 *On the definition of simultaneity.*

We shall first show this for the case of the tug and the barges (Fig. 112). When they are at rest we can make the clocks in the boats A and B go at the same rate by the following means: We place a boat C exactly halfway between the boats A and B and send off a shot from C. Then it must be heard simultaneously at A and B.

Now if the series of boats, which we shall call S, are in motion, we may clearly apply the same method. If it does not occur to the bargemen that they are moving relative to the air they will be convinced that the clocks at A and B are synchronized.

Suppose a second series of boats S', whose barges A', B', C' are at exactly the same distances apart as the corresponding boats of the first series S, compare their clocks in exactly the same way. If now the one series overtakes the other, whether the latter be at rest or not, the ships will pass each other, and at a certain moment A will coincide with A', and B with B', and the bargemen can test whether their clocks agree. Of course they will find that they do not. Even if A and A' should accidentally be beating synchronously, B and B' will not do so.

This will bring the error to light. When the boats are in motion the signal from the middle point C evidently takes more time to arrive at the preceding ship A, and less to arrive at the following ship B, than when the boats are at rest, because A is moving away from the sound wave while B is moving towards it, and this difference varies with the velocities of the two series of barges.

Now in the case of sound, *one* system has the correct time, namely, that which is at rest relative to the air. In the case of light, however, it is not possible to assert this because absolute motion with respect to the luminiferous ether is a concept which, according to all experience, has no physical reality. The method of comparing clocks which we just illustrated for sound is, of course, also possible with light. The clocks at A and B are set so that every flash of light sent out from the middle point C of the distance AB reaches the clocks A and B just as their hands are in the same position. In this way every system can have the synchronism of its clocks adjusted. But when two such systems meet each other, even if the clocks A, A' agree in time, then B, B' will exhibit different positions of the hands. Each system may claim with equal right that it has the correct time, for each can assert that it is at rest, since all physical laws are the same in each. But when two claim what, by its very meaning, can belong to only one, it must be concluded that the claim itself is meaningless. *There is no such thing as absolute simultaneity.*

Whoever has once grasped this will find it difficult to understand why it took many years of exact research for this simple fact to be recognized. It is a repetition of the old story of Columbus' egg.

The next question is whether the method of comparing clocks which we have introduced leads to a consistent relative concept of time. This is actually the case. To see it we shall use Minkowski's representation of events or world points in an *xt*-plane; i.e., we restrict ourselves to motions in the *x*-direction and omit those in the *y*- and *z*-directions (Fig. 113*a*).

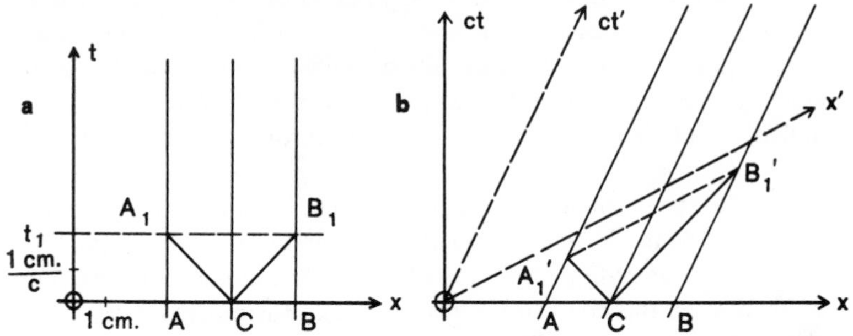

Fig. 113*a* *The points* A, B, C *resting in the system* S(x, t) *have world lines parallel to the* t-*axis. The world lines of a light signal starting at* t=*0 in* C *reaches* A *and* B *in the world points* A_1 *and* B_1 *at the time* t_1 *simultaneously.*

b *The points* A, B, C *are moving in the system* S(x, ct) *with velocity* v. *They are at rest in the system* S′(x′, ct′). A'_1 *and* B'_1 *are the world points where the signal from* C *reaches* A *and* B. *The equations of the* ct′-*axis are:*

$$in\ S\ x = vt = \frac{v}{c}\cdot ct,\ in\ S'\ x' = 0.$$

The slope of the x′-*axis has to be parallel to the straight line* $A'_1B'_1$. *A calculation using analytical geometry in the system* S *leads to equations of the* x′-*axis:*

$$in\ S\ x = \frac{c^2}{v}t = \frac{c}{v}\cdot ct,\ in\ S'\ ct' = 0.$$

The points *A*, *B*, and *C*, at rest on the *x*-axis, are represented in the *xt*-coordinate system as three parallels to the *t*-axis. Let the point *C* lie midway between *A* and *B*. At the moment $t=0$ a light signal is to be sent out from it in both directions.

We assume that the system *S* is "at rest," that is, that the velocity of light is the same in both directions. Then the light signals moving to the right and left are represented by straight lines which are equally inclined to the *x*-axis and which we call "*light lines.*" We shall make their inclination 45°; this evidently amounts to saying that the same distance which represents the unit length 1 cm. on the *x*-axis in the figure signifies the very small time $\frac{1\text{ cm.}}{c}$ on the *t*-axis which the light takes to traverse the distance 1 cm. For sake of simplicity it is better to use ct instead of t as a measure of time. This means that on the "time axis" the time is measured by the light path. That is the length of path ct traversed by the light during the time t.

The t-values of the points of intersection A_1, B_1 of the light lines with the world lines of the points A and B give the times at which the two light signals arrive. We see that A_1 and B_1 lie on a parallel to the x-axis and have the same t-value, that is, they are simultaneous.

Next, let three points A, B, C move uniformly with the same velocity. Their world lines are then again parallel, but inclined to the x-axis (Fig. 113*b*). The light signals are represented by the same light lines, proceeding from C, as above, but their points of intersection A_1', B_1' with the world lines A, B do *not* now lie on a parallel to the x-axis; thus they are *not* simultaneous in the x, ct-coordinate system, and B_1' is later than A_1'. On the other hand an observer moving with the system can with equal right assert that A_1', B_1' are simultaneous events (world points). He will use an x', ct'-coordinate system S' in which the points A_1', B_1' lie on a parallel to the x'-axis. The world lines of the points A, B, C are, of course, parallel to the ct'-axis, since A, B, C are at rest in the system S' and hence their x'-coordinates have the same values for all t's.

It follows that the moving system S' is represented in the x, ct-plane by an oblique coordinate system x', ct', in which *both* axes are inclined to the original axes.

We now recall that in ordinary mechanics the inertial systems in the x, ct-plane are likewise represented by oblique coordinates with ct-axes arbitrarily directed, the x-axis, however, always remaining the same (III, 7, p. 75). We have already pointed out that from the mathematical point of view this is a flaw which is eliminated by the theory of relativity. We now see clearly how this happens as a result of the new definition of simultaneity. At the same time a glance at the figure convinces us that this definition must be consistent in itself, for it signifies nothing other than that we use oblique instead of rectangular coordinates.

The units of length and time in the oblique system are not yet determined by the construction, for this makes use only of the fact that in a system S light is propagated with equal velocity in all directions, but not of the law that the velocity of light has the same value c in all inertial systems. If we accept the latter, too, we arrive at the complete kinematics of Einstein.

2. Einstein's Kinematics and Lorentz's Transformations

We once more state the hypotheses of Einstein's kinematics:

1. *The principle of relativity.* There are an infinite number of systems of reference (inertial systems) moving uniformly and rectilinearly with respect to each other, in which all physical laws assume the simplest form (originally derived for absolute space or the stationary ether).

2. *The principle of the constancy of the velocity of light.* In all inertial systems the velocity of light has the same value when measured with rods and clocks of the same kind.

Our problem is to derive the relations between lengths and times in the various inertial systems. In doing so we shall again restrict ourselves to motions parallel to a definite direction in space, the x-direction.

We will use two methods. The first will start from our diagrams at the end of the previous section; the second will give a more algebraic derivation of the relations between two systems S and S' with the relative velocity v.

To obtain a numerical relation between the two systems, we have to know the units and their relations in S and S'. For this purpose we have to find the images for the units on the x'- and ct'-axis of the system S' in Fig. 113*b* which represent the same length and time interval in S' as those chosen as units in the system S. Let us assume the stretch $\overline{OE}$ from O to E (Fig. 114*a*) represents a rod of unit length at rest in S. The world lines of the ends of this rod are the ct-axis and the parallel line through E. This line cuts the x'-axis in e'.

The world lines of the same rod at rest in S' will be the ct'-axis and a parallel line through a point E' on the x'-axis. The segment $\overline{OE'}$ represents the unit of the length in the S'-system. The world line through E' cuts the x-axis in e.

For the sake of brevity we shall now let E, e, etc., stand for the segments $\overline{OE}$, $\overline{Oe}$, etc., as well as for the end points of these segments.

The meaning of e' is as follows: An observer at rest in S' who wants to measure the length of the unit rod at rest in S will find as a result of a simultaneous observation O and e' for its end points. Simultaneous observation in the S'-system is essential because the S-unit is moving with regard to the observer in S'. Since the unit in S' is given by E' the result of the S-measurement is the $\frac{e'}{E'}$ part of the S'-unit. When E corresponds to 1 cm., the observer in S' would find $\frac{e'}{E'}$ cm. The same applies to e where now $\frac{e}{E}$ is the factor relating the two measurements.

Now, according to the principle of relativity, the two systems are equivalent, i.e., the relative changes $\frac{e'}{E'}$ and $\frac{e}{E}$ have to be equal:

$$\frac{e'}{E'} = \frac{e}{E} \qquad \text{or} \qquad Ee' = E'e. \tag{α}$$

This relation allows us to construct the point E'.

The unit of time in the S'-system $E_{ct'}$ can be constructed in a corresponding way from E_{ct}, the point defining the unit of time in S.

According to Fig. 114*a* one now gets two relations*

* The first relation (β) is Pythagoras' theorem for the triangle OEe' $\left(\overline{Ee'}=E\frac{v}{c}\right)$. The second relation may be proved by using Fig. 114*b*. From Pythagoras' theorem we get $E'^2 = D^2\left(1+\frac{v^2}{c^2}\right)$. Now $e = D - \overline{De} = D\left(1-\frac{v^2}{c^2}\right)$. It follows

$$E'^2 = e^2 \frac{1+\frac{v^2}{c^2}}{\left(1-\frac{v^2}{c^2}\right)^2}.$$

Taking the root of the last equation and of (β),

$$E' = e\,\frac{\sqrt{1+\frac{v^2}{c^2}}}{1-\frac{v^2}{c^2}}, \quad e' = E\,\sqrt{1+\frac{v^2}{c^2}}$$

and substituting this in (α): $Ee' = E'e$, one gets, by cancelling $\sqrt{1+\frac{v^2}{c^2}}$, the result (γ).

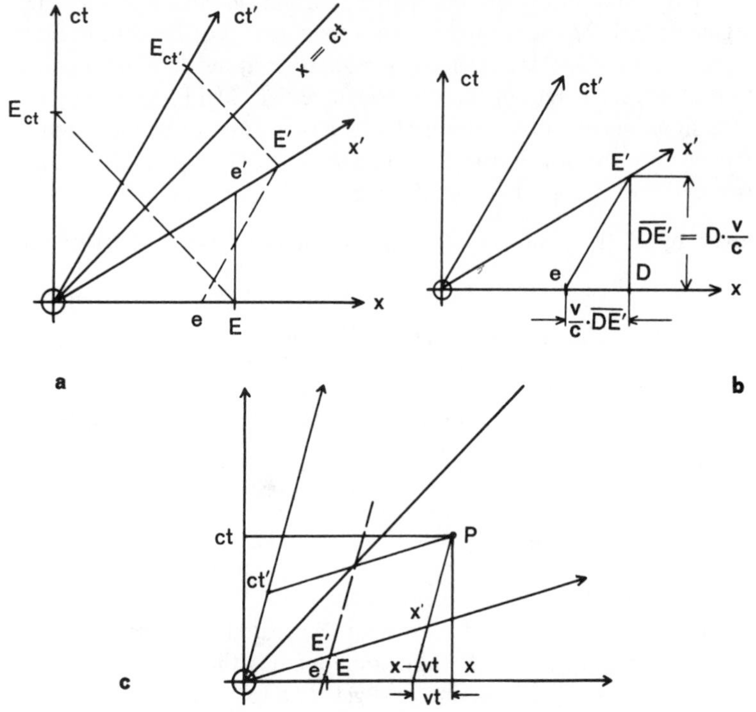

Fig. 114*a* *Units of space and time in* S (E, E_{ct}) *and in* S′ (E′, $E'_{ct'}$). $\overline{Oe}$ *is the representation in* S *of the unit rod resting in* S′, *whereas* $\overline{Oe'}$ *represents in* S′ *the unit rod resting in* S.
b *On the calculation of the ratio* E′/e.
c *Lorentz transformation of the coordinates of a world point* P.

$$e'^2 = E^2\left(1+\frac{v^2}{c^2}\right), \qquad (\beta)$$

$$e^2 = E^2\left(1-\frac{v^2}{c^2}\right). \qquad (\gamma)$$

We are now able to transform the coordinates x and t of any world point P in the system S into the coordinates x' and t' of P in the system S'.

In Fig. 114c there are represented two systems S and S' and the units of length E and E' in the systems and the segment e, known from Fig. 114*a*. A point P with the coordinates x, ct in S has the coordinates x', ct' in S'. Now one may measure lengths on the figure in units U (for example, centimeters), but the coordinates are defined in units E in S or E' in S'; that means

$$x' = \frac{\overline{Ox'}}{E'},$$

where $\overline{Ox'}$ is the length measured in units U and x' is the coordinate, and

$$x = \frac{\overline{Ox}}{E} \quad \text{or} \quad (x-vt) = \frac{\overline{O(x-vt)}}{E},$$

respectively. From Fig. 114*c* one gets two proportions:

$$\frac{x'}{x-vt} = \frac{\overline{Ox'}}{\overline{O(x-vt)}} \frac{E}{E'},$$

and

$$\frac{\overline{Ox'}}{\overline{O(x-vt)}} = \frac{E'}{e}.$$

Substituting the second into the first and using (γ) we obtain the result:

$$\frac{x'}{(x-vt)} = \frac{E}{e} = \frac{1}{\sqrt{1-\frac{v^2}{c^2}}}.$$

The correspondent relation for the time coordinates is

$$\frac{ct'}{\left(ct-\frac{v}{c}x\right)} = \frac{E}{e} = \frac{1}{\sqrt{1-\frac{v^2}{c^2}}}.$$

The last two formulae, completed with $y'=y$ and $z'=z$ (for y and z are perpendicular to the direction of motion and do not change) constitute the so-called *Lorentz transformation*, which allows one to calculate the coordinates of a world point in S' from those in S. We write them in the usual form:

$$x' = \frac{x-vt}{\sqrt{1-\frac{v^2}{c^2}}}, \quad y' = y, \quad z' = z, \quad t' = \frac{t-\frac{v}{c^2}x}{\sqrt{1-\frac{v^2}{c^2}}}. \tag{70a}$$

They are exactly the formulae that Lorentz found by analyzing Maxwell's field equations (see V, 15, p. 222).

Let us now consider an algebraic method of deriving the same transformation formulae. A world point P (coordinates x, t in S and x', t' in S') may lie on a world line $x'=C'$, which represents a point resting at the space position C' in S'. This world line has the equation $x-vt=C$ in S (Fig. 114c). The two equations represent the same world line. Dividing one by the other one obtains $(x-vt)/x'=C/C'=\alpha$, where α, like C and C', is constant along the line. Therefore

$$\alpha x' = x-vt. \tag{δ}$$

According to the principle of relativity, however, both systems are equivalent. Thus we may equally well apply the same arguments to a world line of a point resting in S, except that now the relative velocity v has the reverse sign. Therefore $x'+vt'$ must be proportional to x, and, on account of the equivalence of both systems, the factor of proportionality α will be the same in each case:

$$\alpha x = x'+vt'. \tag{ϵ}$$

From this and the preceding equation, t' may be expressed in terms of x and t. We get

$$vt' = \alpha x - x' = \alpha x - \frac{x-vt}{\alpha} = \frac{1}{\alpha}[(\alpha^2-1)x+vt],$$

thus

$$\alpha t' = \frac{\alpha^2-1}{v}x+t.$$

From this equation combined with the first, one can obtain x' and t' when x and t are known. The factor of proportionality α is as yet not determined, but it must be chosen consistent with the principle of the constancy of the velocity of light.

To use this principle, we assume a light signal to be emitted from the origin of our systems. According to the principle of the constancy of the velocity of light, the world line of the light signal has the equation $x = ct$ and $x' = ct'$ in both systems. This we put into (δ) and (ϵ) and get

$$\alpha ct' = ct - vt = (c-v)t,$$

$$\alpha ct = ct' + vt = (c+v)t'.$$

Multiplying these we have

$$\alpha^2 c^2 t' t = (c-v)(c+v)t' t$$

or

$$\alpha^2 = \frac{c^2 - v^2}{c^2} = 1 - \frac{v^2}{c^2}.$$

The transformation formulae now become

$$\alpha x' = x - vt, \qquad \alpha t' = -\frac{v}{c^2} x + t.$$

This is the same result which we have derived by the more geometric method above.

If we wish to express x, y, z, t by means of x', y', z', t' we must solve the equations (70a) with respect to x, y, z, t. But we can deduce from the equivalence of both systems S and S' without calculation that the formulae given by solution must have the same forms except that v becomes changed into $-v$. Thus

$$x = \frac{x' + vt'}{\sqrt{1 - \frac{v^2}{c^2}}}, \quad y = y', \quad z = z', \quad t = \frac{t' + \frac{v}{c^2} x'}{\sqrt{1 - \frac{v^2}{c^2}}}, \qquad (70b)$$

and this can be checked by direct calculation.

Particular interest attaches to the limiting case in which the velocity v of the two systems becomes very small in comparison with the velocity of light. We then arrive directly at the Galileo transformation (formula (29), p. 74). For if $\frac{v}{c}$ can be neglected in comparison with 1, we get from (70)

$$x' = x - vt, \qquad y' = y, \qquad z = z', \qquad t' = t.$$

Thus we understand how, on account of the small value that $\frac{v}{c}$ has in most practical cases, Galilean and Newtonian mechanics satisfied all requirements for some centuries.

3. Geometrical Representation of Einstein's Mechanics

Before we discuss the content of these formulae we shall interpret, with the help of Minkowski's geometrical method of representation in the four-dimensional "world" x, y, z, t (or x, y, z, ct), the connections which they exhibit between two inertial systems. In doing so we may omit the coordinates y, z that remain unchanged, and restrict ourselves to a consideration of the x, ct-plane. All kinematical laws, then, appear as geometrical facts in the x, ct-plane. The reader is, however, strongly advised to practice translating the relationships obtained in geometrical form back into the ordinary language of kinematics. Thus, a world line is to be taken as denoting the motion of a point, the intersection of two world lines the meeting of the two moving points, and so forth. In order to visualize the processes represented in the figures, one should take a ruler and pass it along the t-axis parallel to the x-axis with constant velocity, concentrating attention on the intersections of the edge of the ruler with the world lines. These points, then, move to and fro on the edge and give an idea of the motion in space.

As we have seen, every inertial system S (VI, 1, p. 231) is represented by an oblique set of axes in the x, ct-plane. The fact that one among them is rectangular must be regarded as an accidental circumstance and plays no particular part in our considerations, as shown clearly by our second proof.

Every point in space may be regarded as the source of a light wave which spreads out spherically and uniformly in all directions. Of this spherical wave only two light signals which pass along the x-direction are represented in our figures. One of these moves to the left, the other to the right. Thus they correspond in the x, ct-plane to two intersecting straight lines which are, of course, entirely independent of the choice of the system of reference, since they connect events (world points) with one another, namely, the points of x, ct-space reached successively by the light signal.

We draw these *light lines* for a world point O, taken as the origin, of all the x, ct-coordinate systems considered, and draw them as two mutually perpendicular straight lines. We choose these as the axes of an XY-coordinate system (Fig. 115).

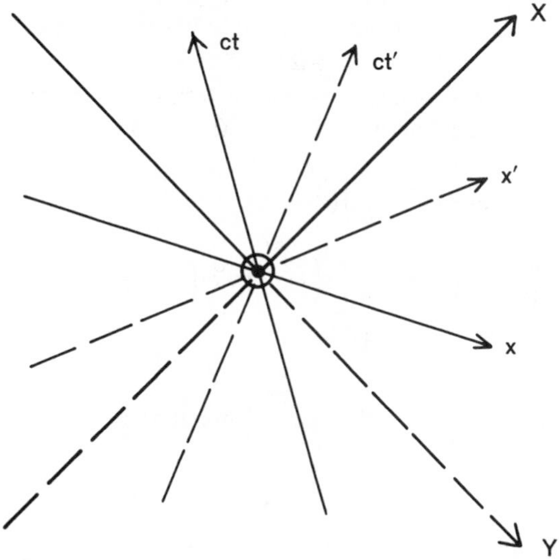

Fig. 115 *The invariant lines* X, Y *correspond to light signals passing through* O. *The solid lines represent light signals starting from* O*; the dotted lines represent light signals converging on* O.

Here we are at the roots of Einstein's theory. The XY-system is uniquely determined and fixed in the "world"; its axes are not lines in space but are formed by world points, namely, those corresponding to space points at the time when they are reached by a light signal emitted from the origin. This invariant or "absolute" coordinate system is thus highly abstract. We must accustom ourselves to seeing such abstractions in the modern theory replace the concrete idea of the ether. The strength of such abstractions lies in the fact that they contain nothing that goes beyond the concepts necessary to interpret the results of experience.

The *calibration curves* which cut off the units of length and time from the axes of an arbitrary inertial system x, ct must be rigidly connected with this absolute system of reference X, Y. These calibration curves must be represented by an invariant law; the problem is to find it.

The light lines themselves are invariant. The X-axis ($Y=0$) is represented in a system of reference S by the formula $x=ct$, and in another system of reference S' by the formula $x'=ct'$, for these formulae express that the velocity of light has the same value in both systems. Now we shall express the difference $x'-ct'$, which is equal to zero for the points of the Y-axis, in terms of the coordinates x, X, by means of the Lorentz transformation (70). Then

$$\begin{aligned} x'-ct' &= \frac{1}{\alpha}\left[(x-vt)-c\left(t-\frac{v}{c^2}x\right)\right] \\ &= \frac{1}{\alpha}\left[x\left(1+\frac{v}{c}\right)-ct\left(1+\frac{v}{c}\right)\right] \\ &= \frac{1+\beta}{\alpha}(x-ct). \end{aligned}$$

Here we have introduced the usual abbreviation

$$\beta = \frac{v}{c}. \tag{71}$$

From this we see that when $x-ct=0$, so is $x'-ct'=0$.

The Y-axis ($X=0$) is given by $x=-ct$, or $x'=-ct'$. If we carry out the corresponding transformation from x' and ct' to x and ct, we have only to change c into $-c$ and β into $-\beta$ (whereas $\alpha=\sqrt{1-\beta^2}$ remains unchanged) and we get

$$x'+ct' = \frac{1-\beta}{\alpha}(x+ct).$$

From these formulae we easily deduce an invariant expression. For $(1+\beta)(1-\beta)=(1-\beta^2)=\alpha^2$; hence if we multiply the two equations the constant factor becomes 1 and we find

$$(x'-ct')(x'+ct') = (x-ct)(x+ct)$$

or

$$x'^2-c^2t'^2 = x^2-c^2t^2;$$

that is, the *expression*

$$F = x^2-c^2t^2 \tag{72}$$

is an invariant. On account of its importance we call it the *fundamental invariant.* We see that F has the dimension $[l^2]$.

It serves in the first place for determining the units of length and time in an arbitrary system of reference S. These units, in different systems S, are marked on rods and clocks of identical physical magnitude and construction. Though they have the dimensions length and time, we use in the following simply the symbol 1 for lengths and areas because the actual choice of physical unit does not matter.

Which are the world points for which F has the value $+1$ or -1?

We have $F=1$ for the world point $x=1$, $ct=0$ (Fig. 116). But

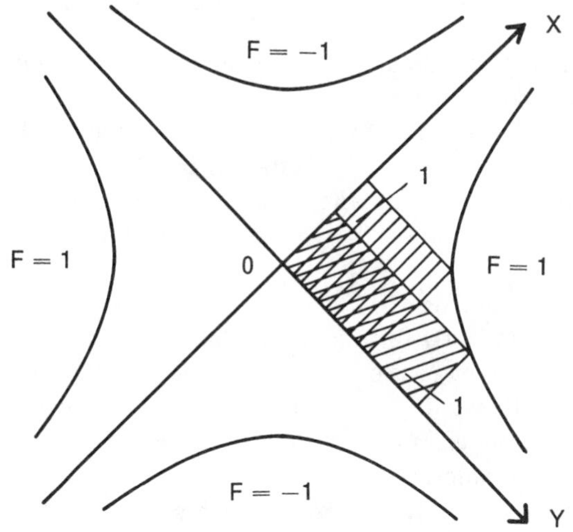

Fig. 116 *Calibration curves* F=*1 and* F=*−1.*

this is the end point of a segment of unit length whose other end is the origin at the moment $t=0$. As this holds in the same way for all systems of reference S, we recognize that the world points for which $F=1$ define the unit of length which is at rest in any arbitrary system of reference, as we shall presently show in greater detail.

In the same way $F=-1$ for the world point $x=0$, $ct=1$; here t is the small time light needs to run through the unit of length. Thus this world point is correspondingly connected with the unit of time of the clock which is at rest in the system S.

Now, it is very easy to construct the points $F=+1$ or $F=-1$ geometrically by starting from the invariant coordinate system XY. The X-axis is formed by the points for which $Y=0$. On the other hand, the same world points are characterized in any arbitrary inertial system S by the relation $x=ct$. Hence Y must be proportional to $x-ct$. By choosing the unit of Y appropriately, we may set

$$Y = x-ct.$$

In the same way by considering the Y-axis we find that we may set

$$X = x+ct.$$

Then we have

$$XY = (x-ct)(x+ct) = x^2-c^2t^2 = F. \qquad (73)$$

Now $F=XY$ is the area of a rectangle with the sides X and Y. The world points for which $F=XY=1$ are the free corners of the rectangles of unit area formed by the coordinates X, Y. These rectangles are shown in the diagram (Fig. 116). They include the square whose side is unity; the others are higher in proportion as they are narrower, and lower in proportion as they are wider, in agreement with the condition $Y=\frac{1}{X}$ (Fig. 116). These points X, Y for which $XY=1$ clearly form a curve which approaches the x- and y-axes more and more closely. This curve is called an *equilateral hyperbola*. If X and Y are both negative, XY is positive. Hence the construction gives us a second branch, the image of the first, in the opposite quadrant.

For $F=-1$ the same construction holds in the remaining two quadrants where the coordinates X and Y have different signs.

The four hyperbolae now form the calibration curves by which the units of length and time are fixed for all systems of reference.

Let the x-axis meet the branch $F=+1$ of the hyperbola at the points P and P', and the t-axis the branch $F=-1$ of the hyperbola at Q and Q' (Fig. 117).

Through P we draw a parallel to the ct-axis and we maintain that this does not cut the right branch $F=+1$ of the calibration curve in a second point but just touches it at P. In other words, we maintain that not a single point of this branch of the calibration lies to the left of the straight line but that the whole branch runs to the right of it, so that all its points have x-coordinates that are larger than the distance OP.

This is actually the case; since for every point of the calibration curve $F=x^2-c^2t^2=1$, we have $x^2=1+c^2t^2$. Thus $x^2=1$ for the point P of the calibration curve which at the same time lies on the x-axis $ct=0$, but for every other point on the calibration curve x^2 is larger than 1 by the positive amount c^2t^2. Accordingly $\overline{OP}=1$, and for every point of the right branch of the calibration curve x is larger than 1.

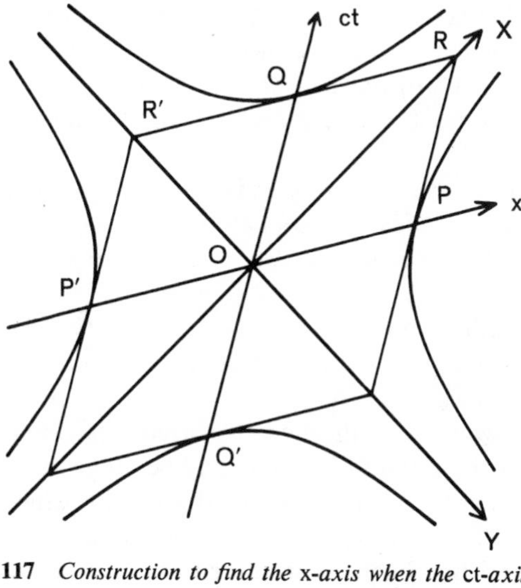

Fig. 117 *Construction to find the* x-*axis when the* ct-*axis is known, or vice versa.*

In just the same way, it follows that the parallel through P' to the ct-axis touches the left branch of the hyperbola $F=1$ at P' and that the parallels through Q and Q' to the x-axis touch the branches $F=-1$ of the hyperbola at Q and Q'. This clearly makes the distance $\overline{OQ}=1$. For the point Q lies on the calibration curve $F=x^2-c^2t^2=-1$ and $x=0$ is the t-axis, thus $c^2t^2=1$; hence $ct=1$ is the value of $\overline{OQ}$.

The two parallels to the ct-axis through P and P' meet the light lines X, Y in the points R and R'. But the parallels through Q and Q' to the x-axis pass through the same points. For the point R, for example, we have $x = ct$ because it lies on the X-axis, and $x = 1$ because it lies on the parallel to the ct-axis through P. From this it follows that $ct = 1$, that is, R lies on the parallel to the x-axis through Q.

Now we see that this construction of the x-axis agrees with that previously given (p. 230) defining simultaneous world points. For the ct-axis $\overline{OQ}$ and the two parallels $\overline{PR}$ and $\overline{P'R'}$ are the world lines of three points, one of which, O, lies midway between the other two, P, P'. Now if a light signal is sent out from O in both directions, it will be represented by the light lines $\overline{OX}$ and $\overline{OY}$, and these cut the two parallel world lines in R and R'. Consequently, these two world points are simultaneous, their connecting line is parallel to the x-axis, exactly as given by our new construction.

We condense the result of our reflections into the following short statement:

The axes x *and* ct *of a system of reference* S *are so situated with respect to each other that each is parallel to that straight line which is touched by the calibration curve at the point of intersection with the other axis.*

The unit of length is represented by the distance $\overline{OP}$. The unit of time is determined by the distance $\overline{OQ}$, which actually is also a length on our ct-scale.

Every straight world line through the origin that intersects the branches $F = 1$ of the calibration curve may be taken as the x-axis. The t-axis is then fixed as a parallel to the straight line which touches the hyperbola at P. In the same way the ct-axis may be chosen as an arbitrary world line intersecting the curves $F = -1$ of the calibration curves. The corresponding x-axis is uniquely determined by the analogous construction.

These rules take the place of the laws of classical kinematics. There the x-axis was the same for all inertial systems, the unit of length was fixed on it, and the unit of time was equal to the section cut off by a definite straight line parallel to the x-axis from the ct-axis, which is in general oblique (see Fig. 41, p. 75).

Now, how does it happen that these constructions apparently so different can in practice scarcely be distinguished?

This is due to the enormously great value of the velocity of light c, compared with the usual velocities of bodies in the world. In our figures the unit of, for instance, 1 cm. on the *ct*-scale corresponds to $\frac{1 \text{ cm.}}{c} = \frac{1}{3} \times 10^{-10}$ sec. If we wish to represent 1 sec. and 1 cm. in the figure by lines of the same length, we must obviously compress the diagram in the direction of *t* so that all distances parallel to the *t*-axis are shortened in the ratio $1 \frac{\text{cm.}}{\text{sec.}} : c$. If *c* were equal to 10 cm./sec., the picture given would be something like that depicted in Fig. 118. The two light lines would form a very small angle representing the limits for the directions of the *x*-axes, and, on the other

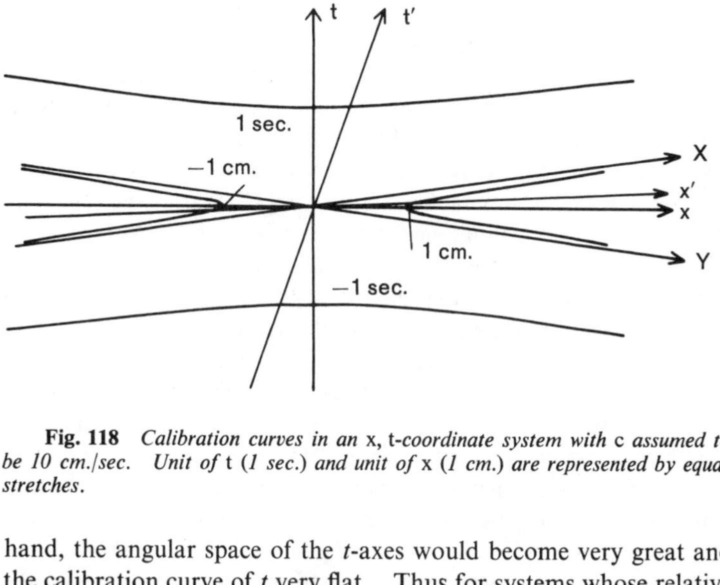

Fig. 118 *Calibration curves in an* x, t*-coordinate system with* c *assumed to be 10 cm./sec. Unit of* t *(1 sec.) and unit of* x *(1 cm.) are represented by equal stretches.*

hand, the angular space of the *t*-axes would become very great and the calibration curve of *t* very flat. Thus for systems whose relative velocities *v* are very small compared with *c*, the units of time differ imperceptibly from one another. The larger the value of *c*, the more conspicuous the quantitative difference between the ranges of free play of the *x*- and the *t*-direction will be. For the actual value of *c* ($c = 3 \times 10^{10}$ cm./sec.), the drawing could not be done on the paper at all: both light lines would coincide in practice, and the *x*-direction which always lies between them would thus be constant. This is exactly what ordinary kinematics assumes. We obtain

Fig. 41 (p. 75). Hence we see that Galilean kinematics is a special case or, rather, a limiting case of Einstein's, namely, that for which the velocity of light is infinitely great.

4. Moving Measuring Rods and Clocks

We shall now answer the simplest questions of kinematics, those concerning the measurement of the length of one and the same measuring rod and of the duration of one and the same time in different systems of reference.

Let a rod of unit length be placed at the origin of the system S along the x-axis. We inquire what its length is in the system S'. It is at once evident that it will differ from the unit length, for the observers moving with S' will, of course, measure the positions of the end points of the rod simultaneously—i.e., simultaneously in the system of reference S'. But this does not mean simultaneously in the system of reference S. Thus, even if the position of one end of the rod is read off simultaneously in S and S', that of the other end will not be read off simultaneously with respect to the S-time by the observers of the systems S and S'. In the meantime, the system S' has moved forward and the reading of the S'-observers concerns a displaced position of the second end of the rod.

At first sight this matter seems hopelessly complicated. There are opponents of the principle of relativity, simple minds who, when they have become acquainted with this difficulty in determining the length of a rod, indignantly exclaim, "Of course, everything can be derived if we use false clocks; here we see to what absurdities blind faith in the magic power of mathematical formulae leads us," and then condemn the theory of relativity at one stroke. Our readers will, it is hoped, have grasped that the formulae are by no means the essential feature, but that we are dealing with purely conceptual relationships which can be understood quite well without mathematics. Indeed, we might not only do without the formulae but also do without the geometrical figures and present the whole thing in ordinary words, but then this book would become so bulky and impossible of design that no one would publish it and none be found to read it.

We first use our figure in the x, ct-plane to solve the question of determining the length of the rod in the two systems S and S' (Fig. 119).

The rod is supposed to be at rest in the system S (x, ct). Accordingly the world line of its initial point is the ct-axis, and the world line of its end point is the straight line parallel to this at the distance 1; the latter touches the calibration curve at the point P. Hence the whole rod is represented for all times by the strip between these two straight lines.

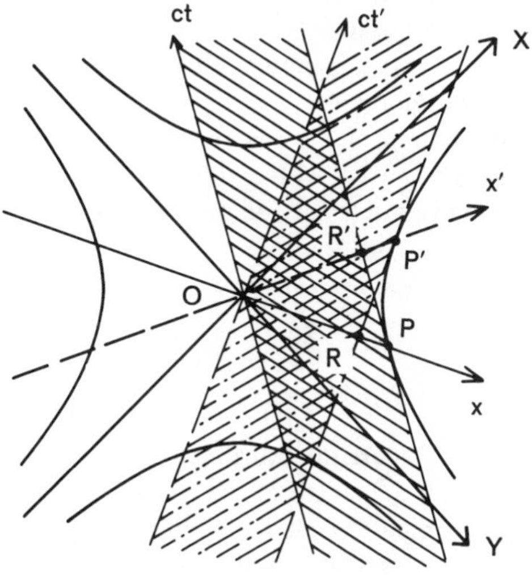

Fig. 119 *Lorentz contraction.*

Now its length is to be determined in the system S' (x', t') which is moving with respect to S. Thus its ct'-axis is inclined to the ct-axis. We find the corresponding x'-axis by drawing the tangent at the point of intersection Q' of the ct'-axis with the calibration curve and then draw the parallel $\overline{OP'}$ to this tangent through O. The distance $\overline{OP'}$ is the unit of length on the x'-axis. The length of the rod of unit length at rest in the system S as measured in the system S' is, however, determined by the distance $\overline{OR'}$ which the parallel strip representing the rod cuts out of the x'-axis. This is clearly shorter than $\overline{OP'}$, thus $\overline{OR'}$ is less than 1, and hence the rod appears shortened in the moving system S'.

This is exactly the contraction devised by Fitzgerald and Lorentz to explain Michelson and Morley's experiment. Here it appears as a natural consequence of Einstein's kinematics.

If, conversely, a rod at rest in the system S' is measured from the system S, it likewise appears contracted and not lengthened. For such a rod is represented by the strip which is bounded by the ct'-axis and the world line parallel to it through the point P'. But the latter meets the unit distance $\overline{OP}$ of the system S in an internal point R, so that $\overline{OR}$ is smaller than 1.

Thus the contraction is reciprocal, and this is what the theory of relativity demands. Its magnitude is best found with the help of the Lorentz transformation (70).

Let l_0 be the length of the rod in the system of reference S' in which it is at rest; l_0 is called the *statical length* or *proper length* of the rod. The two ends of the rod may be at the points x_1' and x_2', then $x_2' - x_1' = l_0$.

If the rod is observed in S, we have from the first formula of (70*a*)

$$x_1' = \frac{x_1 - ct_1}{\sqrt{1-\frac{v^2}{c^2}}}; \qquad x_2' = \frac{x_2 - ct_2}{\sqrt{1-\frac{v^2}{c^2}}},$$

where x_1, t_1 and x_2, t_2 are the coordinates of x_1' and x_2' in S. Now we wish to measure the length of the rod in S; that means we must determine the coordinates x_1 and x_2 simultaneously with regard to S: we must set $t_1 = t_2$. In doing this and subtracting the two equations, we obtain

$$x_2' - x_1' = \frac{x_2 - x_1}{\sqrt{1-\frac{v^2}{c^2}}}.$$

Setting $x_2 - x_1 = l$ this can be written

$$l = l_0\sqrt{1-\frac{v^2}{c^2}}. \tag{74}$$

This states that the length of the rod in the system S appears shortened in the ratio $\sqrt{1-\beta^2} : 1$ exactly in agreement with the contraction hypothesis of Fitzgerald and Lorentz (V, 15, p. 218).

The same arguments apply to the determination of an interval of time in two different systems S and S'.

We suppose clocks that go at the same rate to be placed at each of the space points of the system S. These have a definite position of the hands with respect to S at the same moment. The position $ct_1=0$ is represented by the world points of the x-axis, and the position $ct_2=1$ by the world points of the straight line which passes through the point Q and is parallel to the x-axis (Fig. 120).

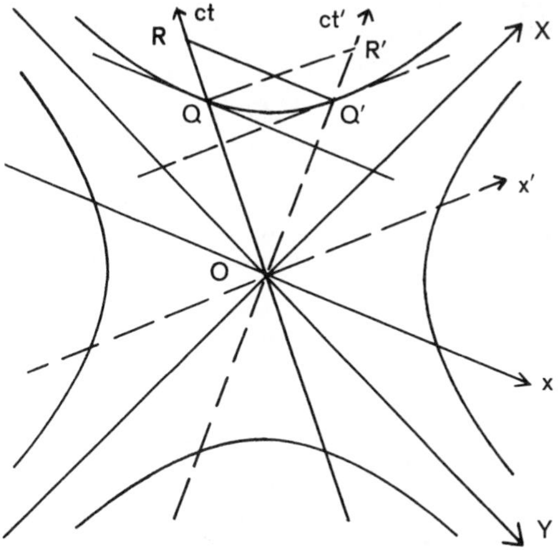

Fig. 120 *Time dilatation.*

Suppose a clock for which $t_1'=0$ when $t_1=0$ is placed at the origin of the system S'. We inquire what is the position of the hands of a clock of the system S which is at the point where the clock at rest in S' exactly indicates the time $ct_2'=1$. The required value of ct_2 is evidently determined by the intersection Q' of the ct'-axis with the calibration curve $F=-1$. The point R where the ct-axis is intersected by the parallel through Q' to the x-axis is simultaneous with Q' in S; hence $\overline{OR}$ represents ct_2. The figure shows that $\overline{OR} > \overline{OQ}$, or ct_2 is larger than $ct_1 = 1$. This means that a time interval in the system S' appears lengthened if measured in S.

Conversely, the unit of time $\overline{OQ}$ in S appears lengthened if measured in S'; for R' and Q are simultaneous in S' and $\overline{OR'} > \overline{OQ'}$.

To find the amount of the lengthening, we consider a time difference T_0 which begins at t_1' and ends at t_2' as shown by a clock at rest in the system S': $t_2'-t_1'=T_0$. From the second formula of (70*a*) we have

$$t_1' = \frac{t_1-\frac{v}{c^2}x_1}{\sqrt{1-\frac{v^2}{c^2}}}, \qquad t_2' = \frac{t_2-\frac{v}{c^2}x_2}{\sqrt{1-\frac{v^2}{c^2}}}.$$

We measure the whole time T_0 at the space point $x_1'=x_2'$ of the clock in S'. From the first formula of (70*a*) we deduce $x_2-x_1=v(t_2-t_1)$, since the clock has the velocity v in S. Now we subtract t_1' from t_2':

$$t_2'-t_1' = \frac{t_2-t_1-\frac{v}{c^2}(x_2-x_1)}{\sqrt{1-\frac{v^2}{c^2}}} = \frac{t_2-t_1-\frac{v^2}{c^2}(t_2-t_1)}{\sqrt{1-\frac{v^2}{c^2}}}$$

$$= (t_2-t_1)\sqrt{1-\frac{v^2}{c^2}}.$$

The time difference $T=t_2-t_1$ measured in S is therefore connected with the time difference T_0 in S' by

$$T = \frac{T_0}{\sqrt{1-\frac{v^2}{c^2}}}. \tag{75}$$

The time dilatation is reciprocal to the contraction in length.

Conversely, of course, the unit of time of a clock at rest in the system S appears increased in the system S'.

Or we may say that, viewed from any one system, the clocks of every other system moving with respect to it appear to be losing time. The course of events in time in the systems in relative motion are slower, so that all events in a moving system lag behind the corresponding events in the system regarded at rest. We shall return later to the consequences which arise from this fact and which are often regarded as being paradoxical.

The time as read from a clock in the system of reference in which it is at rest is called the *proper time* of the system. This is identical with the "local time" of Lorentz. The advance made by Einstein's theory is not in the formulation of laws but rather in a fundamental change of viewpoint towards the laws. Lorentz made local time appear a mathematical auxiliary quantity in contrast to true absolute time. Einstein established that there is no means of determining this absolute time or of distinguishing it from the infinite number of equivalent local times of the various systems of reference that are in motion. But this signifies that absolute time has no physical reality. Time data have a significance only relative to definite systems of reference. This completes the relativization of the concept of time.

5. Appearance and Reality

Now that we have become acquainted with the laws of Einstein's kinematics in the double form of figures and formulae, we shall discuss it from the point of view of the theory of knowledge.

It might be imagined that Einstein's theory furnishes no new knowledge about things of the physical world but is concerned only with definitions and conventions which are suggested by and in agreement with the facts but which might equally well be replaced by others. We are led to such an interpretation if we think of the starting point of our reflections, the example of the tug, in which the conventional and arbitrary nature of Einstein's definition of simultaneity attracted our attention. As a matter of fact, if we use sound signals to regulate the clocks, Einstein's kinematics can be applied in its entirety to ships that move through motionless air. The symbol c would then denote the velocity of sound in all formulae. Every moving ship would have its own units of length and time according to its velocity, and the Lorentz transformations would hold between the system of measurement of the various ships. We should have before us a consistent Einsteinian world on a small scale.

But this world would be consistent only so long as we admitted that the units of length and time are to be restricted by no postulate other than the two principles of relativity and the constancy of the velocity of sound or light. Is this the meaning of Einstein's theory?

Certainly not. Rather it is assumed as self-evident that a measuring rod which is brought into one system of reference S and then into another S' under exactly the same physical conditions would represent the same length in each (provided it is affected in neither system by external forces). A fixed rod that is at rest in the system S and is of length 1 cm. will, of course, also have the length 1 cm. when it is at rest in the system S', provided that the remaining physical conditions (gravitation, position, temperature, electric and magnetic field, and so forth) are the same in S' as in S. Exactly the same would be postulated for the clocks.

We might call this tacit assumption of Einstein's theory the "principle of the physical identity of the units of measure."

As soon as we are conscious of this principle, we see that to apply Einstein's kinematics to the case of the ships and to compare clocks with sound signals is incompatible with it. For if the units of length and time are determined according to Einstein's rule with the help of the velocity of sound, they will, of course, by no means be equal to the units of length and time measured with physically identical measuring rods and clocks; for the former are not only different on every moving ship according to its velocity, but the unit of length in the direction of motion is different from that perpendicular to (or athwart) it. Thus Einstein's kinematics would be a possible definition but in this case not even a useful one. Ordinary knowledge of measuring rods and clocks would without doubt be superior to it.

For just this reason it is hardly possible to illustrate Einstein's kinematics by means of models. These certainly give the relationships between lengths and times in the various systems correctly, but they are inconsistent with the principle of the identity of the units of measure; nothing can be done but choose two different scales of length in two systems S and S' of the model moving relative to each other.

According to Einstein, the state of affairs is quite different in the real world. In it the new kinematics is to be valid just when the *same* rod and the *same* clock are used first in the system S and then in the system S' to fix lengths and times. This is the feature of Einstein's theory by which it rises above the standpoint of a mere convention and asserts definite properties of real bodies. This gives it its fundamental importance for the whole physical view of nature,

We can illustrate this by considering Römer's method of measuring the velocity of light with the help of Jupiter's moons. The whole solar system moves relative to the fixed stars. We imagine a system of reference S rigidly connected with the latter and let the sun and its planets define another system S'. Jupiter and its satellites form an (ideally perfect) clock. It moves round in a circle so that at one time it has the direction of the relative motion of S with respect to S'—at another, the opposite direction. We can by no means arbitrarily determine the beating of the Jupiter clock in these positions in such a way that the time that the light takes to traverse the diameter of the earth's orbit is the same in all directions, but rather this is so *quite by itself*, thanks to the way the Jupiter clock is constructed. For it shows only the proper time of the solar system S', and not some absolute time or the foreign time of the system S of the fixed stars. In other words, the time of revolution of Jupiter's moons is constant relative to the solar system (the velocity of Jupiter itself relative to the solar system being left out of account).

Now it is asserted by some that this view denotes a *violation of the causal law*. For if one and the same measuring rod, as judged from the system S, has a different length according to its being at rest in S or moving relative to S, then, so these people say, there must be a cause for this change. But Einstein's theory gives no cause; rather it states that the contraction occurs by itself, that it is an accompanying circumstance of the fact of motion. In fact, this objection is not justified. It is due to a too limited view of the concept "change." In itself such a concept has no meaning. It denotes nothing absolute, just as data denoting distances or times have no absolute significance. For we do not mean to say that a body which is moving uniformly in a straight line with respect to an inertial system S "undergoes a change," although it actually *changes* its *situation* with respect to the system S. Nor is it clear a priori what "changes" physics counts as effects for which causes are to be found; rather, this is to be determined by experimental research itself.

The standpoint of Einstein's theory about the nature of the contraction is as follows: A material rod is physically not a spatial thing but a space-time configuration. Every point of the rod exists at this moment, at the next, and still at the next, and so on, at every moment of time. The adequate picture of the rod under consideration (one-dimensional in space) is thus not a section of the x-axis but rather a strip of the x, ct-plane (Fig. 121). The same rod, when at rest in various moving systems S and S', is represented by various strips. There is *no* a priori rule as to how these two-dimensional configurations of the x, ct-plane are to be drawn so that they may

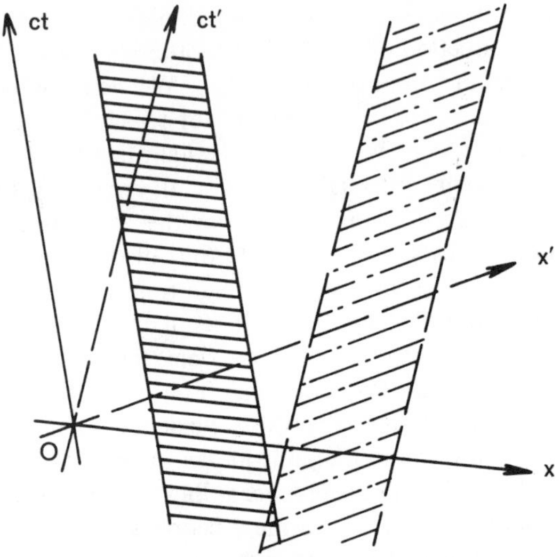

Fig. 121 *World lines of two measuring rods moving relative to each other. Each rod is represented by a strip of world lines parallel to their respective* t- *and* t'-*axis. The hatching lines represent the rods in their rest systems at different times.*

represent correctly the physical behavior of one and the same rod at different velocities. To achieve this a calibration curve in the x, ct-plane must first be fixed. Classical kinematics draws this differently from Einsteinian kinematics. It cannot be ascertained intuitively which is correct. In the classical theory both strips have the same width if this is measured parallel to a fixed x-axis. In Einstein's theory they have the same width if this is measured for each rod in the x-direction of the system of reference in which the rod is at rest. The "contraction" does not affect the strip at all but rather a section cut out of the x-axis. It is, however, only the strip as a manifold of world points (events) which has physical reality, and not the cross-section. Thus the contraction is only a consequence of our way of regarding things and is not a change of a physical reality. Hence it does not come within the scope of the concepts of cause and effect.

The view expounded in the preceding paragraph does away with the notorious controversy as to whether the contraction is "real" or only "apparent." If we slice a cucumber, the slices will be larger the more obliquely we cut them. It is meaningless to call the sizes of the various oblique slices "apparent" and call, say, the smallest which we get by slicing perpendicularly to the axis the "real" size.

In exactly the same way a rod in Einstein's theory has various lengths according to the point of view of the observer. One of these lengths, the statical or proper length, is the greatest, but this does not make it more real than the others. The application of the distinction between "apparent" and "real" in this naïve sense is no more reasonable than asking what is the real x-coordinate of a point x, y when it is not known which xy-coordinate system is meant.

The same remarks apply to the relativity of time. An ideal clock has always one and the same rate of beating in the system of reference in which it is at rest. It indicates the "proper time" of the system of reference. Regarded from another system, however, it goes more slowly. In such a system a definite interval of the proper time seems longer. Here, too, it is meaningless to ask what is the "real" duration of an event.

When understood in the right way, Einstein's kinematics contains no obscurities and no inconsistencies. But many of its results appear contrary to our customary forms of thought and to the doctrines of classical physics. When these antitheses occur in a particularly marked way they are often felt to be paradoxical, even unbearable. Later we shall draw numerous deductions from Einstein's theory which first encountered violent opposition until physicists succeeded in confirming them experimentally. One of the most striking examples is the so-called "clock paradox" which, though satisfactorily treated by Einstein almost half a century ago, is still hotly discussed in our day.

Let us consider an observer A at rest at the origin O of the inertial system S. A second observer B is at first to be at rest at the same point O, and is then to move off with uniform velocity along a straight line, say the x-axis, until he has reached a point C, when he is to turn around and return to O along the x-axis with the same velocity.

Let both observers carry with them ideal clocks which indicate their proper time. The time lost in getting started, in turning around, and in slowing down on arrival at B can be made as short as we please compared with the times of moving uniformly there and back by making these sufficiently great. If, say, the rate of the clocks should be influenced by the acceleration, this effect will be comparatively small if the duration of the journey is sufficiently great, so that this effect may be neglected. But then the clock of the observer *B* must have lost time compared with the clock of *A* after *B*'s return to *O*. For we know (VI, 4, p. 249) that during the periods of *B*'s uniform motion, which are the determining factors for the result, the proper time lags behind the time of any other inertial system. This is seen particularly vividly in the geometrical picture in the *x*, *ct*-plane (Fig. 122). In this we have for the sake of convenience drawn the axes of

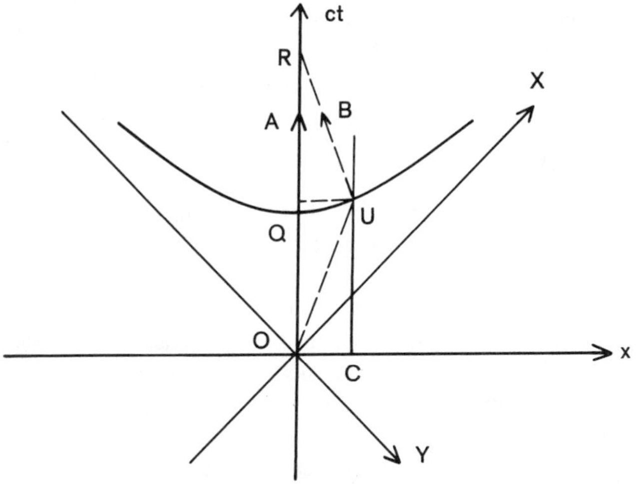

Fig. 122 *Illustration of the clock paradox.*

the *x*, *ct*-system perpendicular to each other. The world line of the point *A* is the *ct*-axis. The world line of the point *B* is the bent line $\overline{OUR}$ (drawn as a dotted line) whose corner *U* lies on the world line of the turning point *C* drawn parallel to the *ct*-axis.

Through *U* we draw the hyperbola $F=x^2-c^2t^2=-c^2t_U{}^2$, where t_U is the proper time of *B* in *U*. Let this meet the *ct*-axis in *Q*. Then clearly the length $\overline{OQ}$ of the proper time for the observer *A* is exactly equal to the length $\overline{OU}$ of the proper time for the observer *B*, as *Q* and *U* are on the same calibration curve. But the length of the proper time for *A* until the turning point *R* is reached is, as the figure tells us, more than twice as great as $\overline{OQ}$, whereas it is exactly twice as great as $\overline{OU}$ for B. Thus, at the moment of turning around, A's clock is in advance of B's clock. The amount of this advance can easily be calculated from formula (75), in which T is the proper time of A, and T_0 denotes the time measured in the system B.

$$T = \frac{T_0}{\sqrt{1-\frac{v^2}{c^2}}}$$

holds for every moment of the motion since the outward and the inward journey take place with the same velocity. Hence, in particular, it also holds for the moment of turning, where T denotes the whole time of the voyage according to the proper time of A, and T_0, the time of the voyage according to the proper time of B. For $v \ll c$ we write (75) approximately $T = T_0\left(1+\frac{1}{2}\frac{v^2}{c^2}\right)$. Hence the advance of A's clock with respect to B's clock is

$$T - T_0 = \frac{v^2}{2c^2} \times T_0. \tag{76}$$

The paradoxical feature of this result lies in the circumstance that *every* internal process in the system B must take place more slowly than the same process in the system A. All atomic vibrations—indeed, even the course of life itself—must behave just like the clocks. Thus, if A and B were twin brothers then B must be younger than A when the voyage is finished. This is truly a strange deduction, which can, however, be avoided by no twist of reasoning. We must put up with it just as, some centuries ago, it had to be accepted that our fellow creatures in the antipodes stood on their heads. As formula (76) shows, we have to do with an effect of the second order. To confirm it one needs very large velocities. Up to now the velocities obtainable by rockets or the like are much too small. We will show presently that the effect can really be observed by investigating small particles with velocities near c.

Nevertheless we shall illustrate equation (75) by the consideration of an imaginary journey to the stars. In Fig. 122 the ct-axis may be the world line of the earth, $\overline{CUR}$ that of the star. Then $\overline{OC}$ is the distance l of the star from the earth (as measured on the earth). The whole path $2l$ of the journey equals vT. We write the relation (75) between the times in the form

$$T_0 = \frac{2l}{v}\cdot\sqrt{1-\frac{v^2}{c^2}} = 2\frac{l}{c}\times\frac{c}{v}\sqrt{1-\frac{v^2}{c^2}}.$$

Here T_0 is the time the traveller spends in his space ship, and $\frac{l}{c}=T_l$ is the distance of the star measured by the time light needs to travel the distance l. It is well known that even the nearest fixed stars are very far, some light years, away. This means that light needs some years to travel from the star to the earth.

Now if one undertakes a journey to a distant star with a rocket of velocity v, the proper time spent by the passenger in the rocket during the journey to the star and back again is

$$T_0 = 2T_l\times\frac{c}{v}\sqrt{1-\frac{v^2}{c^2}}, \tag{75a}$$

whereas the corresponding time measured on the earth is

$$T = 2T_l\times\frac{c}{v}. \tag{75b}$$

The next larger fixed star is α-Centauri on the southern sky, with a distance of about 4.5 light years. One of the brightest stars on the northern sky is Sirius, with a distance of 9 light years. To illustrate formula (75*a*), imagine a journey to α-Centauri where $2T_l=9$ years. In Fig. 123 T_0 is plotted against $\frac{v}{c}$, with a value $2T_l=9$ years. One sees that with a velocity $v=c\times 0.67$ the time delay T_0 for the passenger will be 10 years (20 years for $v=c\times 0.41$, etc.). The time delay on the earth according to (75*b*) would be 13.5 years for the 10-year journey (22 years for the 20-year journey).

In Fig. 124 we have plotted T_l against $\frac{v}{c}$ for $T_0=10$ years. This gives the distance T_l which can be travelled during a time of 10 years with a velocity v. For instance one sees that with $v=c\times 0.9$, one can reach a star at a distance of 10 light years, or that the minimum velocity needed for a 10-year journey to α-Centauri is $0.67c$.

We may mention that the surprising fact expressed in equation

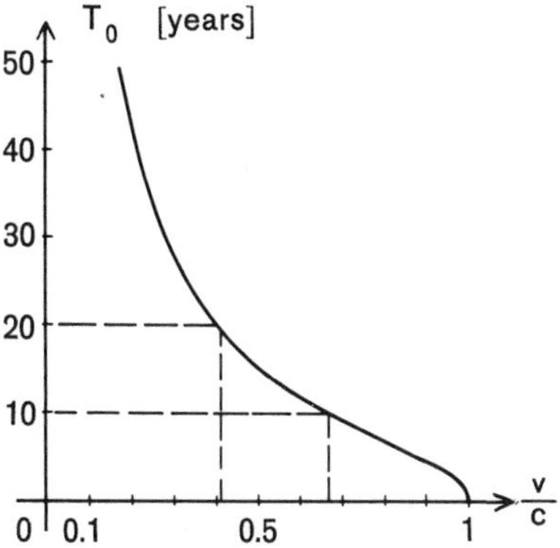

Fig. 123 *The amount of proper time* T_0 *of the space traveller plotted against* $\frac{v}{c}$ *for a journey to α-Centauri (distance 4.5 light years).*

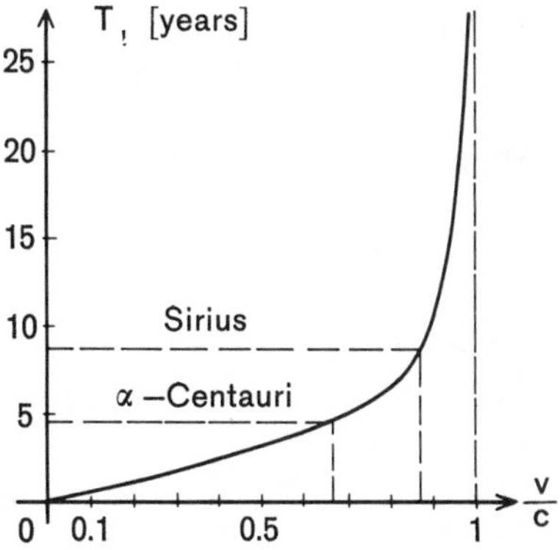

Fig. 124 *Traversed distance* T_l *in light years measured on earth for a space traveller within a proper time* $T_0 = 10$ *years plotted against* $\frac{v}{c}$.

(75) can be derived in quite another way with the help of the Lorentz contraction of the distance l as measured by the rocket passenger. According to (74) this distance will be $l\sqrt{1-\frac{v^2}{c^2}}$. The velocity of the star relative to the rocket is v. Therefore the time needed for this distance will be

$$\frac{T_0}{2} = \frac{l\sqrt{1-\frac{v^2}{c^2}}}{v},$$

which is the same as equation (75*a*).

As we have said, these space experiments cannot at present be performed. But there are phenomena due to small cosmic particles which can be observed and used for a perfectly convincing confirmation of the time dilation and the effect described in the clock paradox. The existence of cosmic rays has been known for the last fifty years. They come from outer space and consist of extremely small, fast-moving particles, mainly protons (i.e., nuclei of hydrogen atoms) but also nuclei of other atoms. They enter the earth's atmosphere from all sides and collide with particles of the air. When a cosmic particle hits the nucleus of an atmospheric atom (nitrogen or oxygen), new particles are produced which are called *mesons*; their mass is intermediate between that of the proton and the electron. These primary mesons, called π-mesons, are unstable and decay after a short lifetime into another, lighter type of mesons and electrons and other light particles. The π-mesons can also be produced artificially with the help of the large, modern accelerating machines (cyclotrons, etc.); these artificial mesons are relatively slow and their lifetime is practically the same as if they were at rest. Thus one knows the proper lifetime $T_0 = 10^{-8}$ sec. of the π-mesons. Now if the velocity of the cosmic mesons were as large as that of light, the distance travelled by them would be only $c \times T_0 = 3 \times 10^{10} \times 10^{-8} =$ 300 cm. But π-mesons of very high energy are observed on sea level. How is it possible that they penetrate the atmosphere, travelling a distance of about $h = 30$ km. $= 3 \times 10^6$ cm. during their lifetime? This paradox is resolved by taking into account the dilation of time; the lifetime as observed on the earth, T, is very much longer than T_0. Indeed, one has $T = T_0\sqrt{1-\frac{v^2}{c^2}}$; in order that the π-mesons reach the surface of the earth, this time must be larger than the height of the atmosphere divided by the velocity v; the minimum velocity must therefore satisfy the condition

$$\frac{T_0}{\sqrt{1-\frac{v^2}{c^2}}} = \frac{h}{v}$$

or

$$\frac{\frac{v}{c}}{\sqrt{1-\frac{v^2}{c^2}}} = \frac{h}{T_0 c} = \frac{3 \times 10^6 \text{ cm.}}{10^{-8} \times 3 \times 10^{10} \text{ cm.}} = 10^4.$$

From this one can compute $\frac{v}{c}$ and obtain

$$v = c(1 - \tfrac{1}{2}\, 10^{-8}) = 0.999999995c.$$

Velocities of this enormous magnitude frequently occur in cosmic radiation.

Thus the mesons illustrate the clock paradox; each meson carries its own clock which determines the proper time T_0 of decay. But the lifetime T observed by a terrestrial observer is much larger. As we mentioned above, one can express this also in a different manner: The moving meson "sees" the distances on the earth contracted and is able to pass through considerable distances depending on its velocity.

If we feel worried by this result and call it paradoxical, we simply mean that it is unusual, or "peculiar"; time will help us to conquer this strange feeling.

But there are also opponents to the theory of relativity who seek to make of these conclusions an objection against the logical consistency of the theory. Their argument is as follows: According to the theory of relativity two systems in relative motion are equivalent. We may therefore also regard B as at rest. The system A then performs a journey in exactly the same way as B previously, but in the opposite direction. We must therefore conclude that when A returns, B's clock will be in advance of A's. But previously we had come to exactly the opposite conclusion. Now since A's clock cannot be in advance of B's and at the same time B's in advance of A's, this argument discloses an inherent contradiction in the theory. This objection is raised again and again. But it is superficial reasoning and the error is obvious; the principle of relativity concerns *only* such systems as are moving uniformly and rectilinearly with respect to each other. In the form in which it has been so far developed it is *not* applicable to accelerated systems. But the system B *is* accelerated and it is *not*, therefore, equivalent to A. The system A is an inertial system, B is not. Later, it is true, we shall see that the general theory of relativity of Einstein also regards systems as equivalent which are accelerated with respect to each other, but in a sense which requires more detailed discussion. When dealing with this more general standpoint we shall return to the clock paradox and show that on close examination there are no difficulties in it. For in the considerations above we made the assumption that for sufficiently long journeys the short periods of acceleration exert no influence on the beating of the clocks. But this holds *only* when we are judging things from the inertial system A and *not* for the measurement of time in the accelerated system B. According to the principles of the general theory of relativity, gravitational fields occur in the latter which affect the beating of the clocks. When this influence is taken into account, it is found that under all circumstances B's clock goes in advance of A's, and thus the apparent contradiction vanishes (see VII, 11, p. 354).

The relativization of the concepts of length and intervals of time appears difficult to many, but probably only because it is strange. The relativization of the concepts "below" and "above" which occurred through the discovery of the spherical shape of the earth probably caused people of that period no less difficulty. In this case, too, the result of research contradicted a view that had its source in direct experience. Similarly, Einstein's relativization of time seems not to be in accord with the experience of the time of the individual. For the *feeling* of "now" stretches without limit over the world, linking all being with the ego. The fact that the same moments that the ego experiences as "simultaneous" are to be called "consecutive" by another ego cannot be comprehended in fact by the actual *experience of time*. But exact science has *other* criteria of truth. Since absolute "simultaneity" cannot be ascertained, science must remove this concept from its system.

6. The Addition of Velocities

We shall now enter more deeply into the laws of Einstein's kinematics. In doing so we shall for the most part restrict ourselves to considering the x, ct-plane. There is no essential difficulty in generalizing the theorems obtained for the case of the four-dimensional $xyzt$-space, and we shall therefore merely touch upon it.

The light lines that are characterized by $F = x^2 - c^2t^2 = 0$ divide the x, ct-plane into four quadrants (Fig. 116). Now F evidently retains the same values in each quadrant, F being >0 in the two opposite quadrants which contain the hyperbolic branches $F = +1$, and F being <0 in the two opposite quadrants which contain the hyperbolic branches $F = -1$. A straight world line passing through the origin 0 can be chosen as the x-axis or the ct-axis according to whether it lies in the quadrant $F > 0$ or $F < 0$. Corresponding to this we distinguish world lines as "spacelike" or "timelike" (Fig. 125).

In any inertial system the x-axis separates the world points of the "past" ($t < 0$) from those of the future ($t > 0$). But this separation is different for each inertial system, since for another position of the

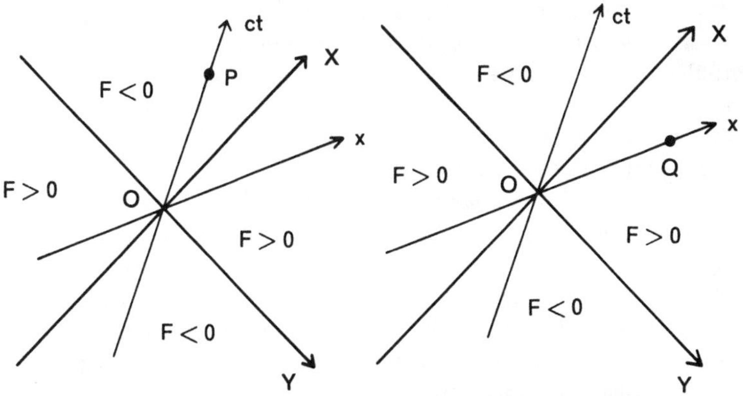

Fig. 125*a* *Timelike distance* $\overline{OP}$. *b* *Spacelike distance* $\overline{OQ}$.

x-axis world points which previously lay above the x-axis, that is, in the "future," now lie below the x-axis, or in the past, and conversely. Only the events represented by the world points within the quadrants $F<0$ are uniquely either "past" or "future" for every inertial system. For such a world point P (Fig. 125*a*) we have $c^2t^2>x^2$, that is, in every admissible system of reference the time distance of the two events O and P is greater than the time the light requires to pass from the one point to the other. We can then always introduce an inertial system S such that its ct-axis passes through P—that is, in which P represents an event which takes place at the spatial origin. Regarded from another inertial system this inertial system S will move rectilinearly and uniformly in such a way that its origin coincides exactly with the events O and P. Then, obviously, for the event P in the system S, we must have $x=0$, that is,

$$F = -c^2t^2 < 0.$$

In every inertial system the ct-axis represents the world points of events that occur at the spatial origin on the x-axis, and it separates (in the two-dimensional figure) the points to the left of the origin from those on its right. But for a different inertial system with a different ct-axis this demarcation will be a different one. It is uniquely determined only for the world points that lie within the quadrants $F>0$, whether they lie "before" or "behind" the spatial origin. For such a point Q (Fig. 125*b*) we have $c^2t^2<x^2$, that is, in

every admissible system of reference the time interval between two events O and Q is less than the time the light takes to pass from one to the other. Then we can introduce an appropriate moving inertial system S, whose x-axis passes through Q, in which both events O and Q are simultaneous. In this system we evidently have for the event Q that $t=0$, and thus $F=x^2>0$.

From this it follows that the invariant F is for every world point P a measurable quantity having a significance which we can visualize. By introducing a suitable reference system S the world point P may be transformed either "to the same place" as O, in which case $F=-c^2t^2$ (where t is the difference of time of the event P with respect to the event O occurring at the same point of space of the system S); or else P may be transformed "to simultaneous times" with O, and then $F=x^2$ (where x is the spatial distance between the two events that occur in the system S).

In every coordinate system the light lines $F=0$ represent motions which occur with the velocity of light. Accordingly, every timelike world line represents a motion of a velocity smaller than c. Or, the other way around, every motion which occurs with a velocity less than that of light can be "transformed to rest" because there is a timelike world line corresponding to it.

But what about motions that occur with a velocity greater than that of light? After the preceding discussion it would seem clear that Einstein's theory of relativity must declare such motions to be impossible. For the new kinematics would lose all meaning if there were signals which allowed us to control the simultaneity of clocks by means involving a velocity greater than that of light. Here a difficulty appears to arise.

Let us assume that a system S' has a velocity v with respect to another system S. Let a moving body K move relative to S' with the velocity u'. According to ordinary kinematics, the relative velocity of the body K with respect to S is then

$$u = v+u'.$$

Now, if v, as well as u', is greater than half the velocity of light, then $u=v+u'$ is greater than c, and this is to be impossible according to the theory of relativity.

This contradiction is due, of course, to the circumstance that velocities cannot simply be added in the kinematics of relativity theory, where every system of reference has its own units of length and time.

The necessity of taking this circumstance into account is obvious from the fact that in any two systems moving with respect to each other the velocity of light is supposed to have always the same value —the fact that we used earlier in deriving the Lorentz transformation (VI, 2, p. 237). The actual law for the composition of velocities can be derived from this transformation (formula (70)). Consider a moving body in the system S'. It may move in the x', y'-plane and thus have two components of velocity $u_{x'}$ and $u_{y'}$, and may start at $t'=0$ from the origin. Its world line is given by the equations

$$x' = u_{x'}t', \tag{α}$$

$$y' = u_{y'}t'. \tag{β}$$

One may expect that the same motion observed in S will be also a rectilinear motion with constant velocity components u_x, u_y. For the world line in S we have

$$x = u_x t, \tag{γ}$$

$$y = u_y t. \tag{δ}$$

In order to obtain a relation between the velocities in S and S' we introduce x, y, t into the equations (α) and (β) by means of the Lorentz transformation (70a). Considering first the equation (α) we get

$$(x-vt) = u_{x'}\left(t-\frac{v}{c^2}x\right) \quad \text{or} \quad \left(1+\frac{u_{x'}v}{c^2}\right)x = (u_{x'}+v)t.$$

By comparison with (γ) one has

$$u_x = \frac{u_{x'}+v}{1+\dfrac{u_{x'}v}{c^2}}. \tag{77a}$$

In the same way we get from (β)

$$u_y = \frac{u_{y'}\left(1-\dfrac{v}{c^2}u_x\right)}{\sqrt{1-\dfrac{v^2}{c^2}}},$$

and with (77*a*)

$$u_y = u_{y'} \frac{\sqrt{1-\frac{v^2}{c^2}}}{1+\frac{u_{x'}v}{c^2}}. \tag{77b}$$

Equations (77*a*, *b*) express Einstein's *addition theorem for velocities*, which takes the place of the simple formulae of the old kinematics

$$u_x = u_{x'}+v, \qquad u_y = u_{y'}.$$

If one wishes to express $u_{x'}$, $u_{y'}$ in terms of u_x, u_y one obtains formulae of the same structure, the only modification being that v is replaced by $-v$. This follows from the equivalence of the systems of reference and can also be checked by calculation.

If, in particular, we are dealing with a light ray which is travelling in the direction of motion of the system S' with respect to S, then $u_{x'}=c$, $u_{y'}=0$. And then formula (77) gives the expected result:

$$u_x = \frac{v+c}{1+\frac{v}{c}} = c, \qquad u_y = 0,$$

which expresses the theorem of the constancy of the velocity of light. Moreover, we see that for any body moving longitudinally $u_x < c$, as long as $u_{x'} < c$ and $v < c$. For if we divide formula (77*a*) by c, we can rearrange it to become

$$\frac{u_x}{c} = 1 - \frac{\left(1-\frac{u_{x'}}{c}\right)\left(1-\frac{v}{c}\right)}{1+\frac{u_{x'}v}{c^2}}.$$

From this formula we read off our statement, because under the conditions mentioned above, the second term of the right-hand side is always smaller than 1 (the denominator is larger than 1 and both factors in the numerator are smaller than 1). The corresponding result holds, of course, for transverse motion, and, indeed, for motion in any direction.

Hence the velocity of light is, kinematically, a limiting velocity which cannot be exceeded. This assertion of Einstein's theory has encountered much opposition. It seemed an unjustifiable limitation for future discoverers who were prepared to find motions occurring with velocities greater than that of light.

We know that the β-rays of radioactive substances are electrons moving nearly with the velocity of light. Why should it not be possible to accelerate them so that they move with a velocity exceeding that of light?

Einstein's theory, however, asserts that in principle this is not possible, because the inertial resistance or the mass of a body increases with the velocity approaching that of light. We thus arrive at a new system of dynamics which is built upon Einstein's kinematics.

7. Einstein's Dynamics

The mechanics of Galileo and Newton is intimately connected with the old kinematics. The classical principle of relativity, in particular, depends on the fact that changes of velocity—accelerations—are invariant with respect to Galileo transformations.

Now we cannot take one kinematics for one part of physical phenomena and another kinematics for another part, holding to invariance with respect to Galileo transformations for mechanics and invariance with respect to Lorentz transformations for electrodynamics.

We know, however, that the former transformations are a limiting case of the latter, namely, those for which the constant c is infinite. Accordingly, following Einstein, we shall assume that classical mechanics is not strictly valid but rather requires some kind of modification. The laws of the new mechanics must be invariant with respect to Lorentz transformations.

In order to find these laws we must decide which fundamental laws of classical mechanics are to be retained and which rejected or modified. The fundamental law of dynamics with which we started is the *law of momentum*. We defined momentum on p. 34 by

$$p = mv.$$

The change of momentum is mw, where w is the change of velocity. It is obvious that we cannot simply retain it in this form. For, whereas in classical mechanics the change of velocity w has always the same value for various inertial systems (see III, 5, p. 68), this is not the case here, on account of Einstein's addition theorem of velocities (77). Thus formula (7) has no meaning unless special instructions are given for the transformation of the momentum of one system of reference to another; hence it would not be expedient to start from (7) to reach the new fundamental law by generalization.

But we may certainly start from the *law of conservation of momentum* (II, 9, p. 35, formula (9)). This concerns the total momentum *carried along* by two bodies, and states that when the bodies collide this momentum remains preserved no matter how their velocities are changed in the process. Thus the assertion involves nothing but two bodies that act on each other, that suffer a *mutual* impact without external influences; hence there is no reference to a third body or coordinate system. Accordingly, we shall specify that this law of

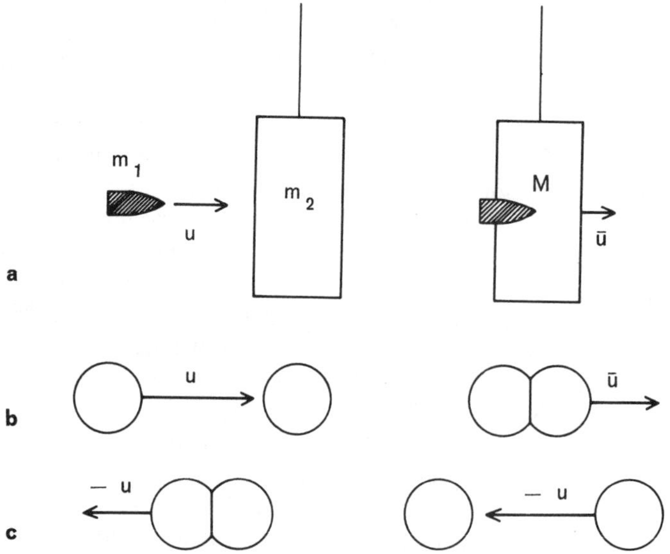

Fig. 126*a* *A block of wood (mass* m_2*) hangs in its equilibrium position from a long string like a pendulum. A pistol bullet (mass* m_1*) is shot with high velocity* u *against the block and sticks in it. Block and bullet together get a common velocity* ū *which is much smaller than* u *if* m_2 *is much larger than* m_1*, and can easily be measured by observing the swinging pendulum.*

b *Collision of two equal spheres, which stick together after the impact. The left sphere arrives with velocity* u*; the common velocity after the collision is* ū.

c *The same collision as in Fig. 126b, but observed from a system* S′ *which moves with the velocity* u *of the left sphere of Fig. 126b. Here the left sphere rests, the right moves with velocity* −u, and *the common velocity after collision is* −ū.

conservation of momentum remains valid in the new dynamics.

This is, of course, as we shall see presently, impossible if we retain the axiom of classical mechanics that mass is a constant quantity peculiar to each body. Hence we shall assume from the outset that *the mass of one and the same body is a relative quantity.* It is to have different values according to the systems of reference from which it is measured, or, if measured from a definite system of reference, according to the velocity of the moving body. It is clear that the mass with respect to a definite system of reference can depend only on the value of the velocity of the moving body with respect to this system, not on its direction.

To get the unknown dependence $m(u)$ of the mass m of a body on its velocity u, we choose a very special example, an "inelastic" collision between two moving bodies. Inelastic means that the two bodies stick together after the collision.* An example of such a collision is a pistol bullet m_1 shot against a block of wood m_2; after the collision the bullet and the block adhere and move with the same velocity (Fig. 126*a*).

Let us first treat the problem in Newton's mechanics by using the law of conservation of momentum. Before the collision the velocity of m_1 may be u, the velocity of m_2 zero, and the common velocity of the two bodies after the collision $\bar{u}$. Then the total momenta are

before the collision $m_1 u$,
after the collision $M\bar{u} = (m_1+m_2)\bar{u}$.

Since the law of conservation of momentum requires these two momenta to be equal, we have

$$m_1 u = M\bar{u}.$$

This equation enables one to calculate the velocity u of the bullet from the velocity $\bar{u}$ after the collision. This method was actually used to determine the velocities of bullets (before the development of high-speed photography and other modern methods), because when m_2 is much larger than m_1, the velocity $\bar{u}$ is much smaller than u and can be measured relatively easily.

*In contrast to the perfectly elastic collisions used in Chapter 2, p. 33, where the colliding bodies do not suffer permanent deformations and no mechanical energy is transformed into other forms of energy.

For sake of simplicity we take the colliding bodies now as equal, $m_1 = m_2 = m$ (for instance, two spheres of wax). Then one gets $\bar{u} = \frac{u}{2}$. We may mention that the mechanical energy is not conserved in this case. The kinetic energy is

before the collision $\frac{m}{2} u^2$,

after the collision $\frac{2m}{2} \bar{u}^2 = \frac{m}{4} u^2$.

The difference of these two energies, $\frac{m}{2} u^2 - \frac{m}{4} u^2 = \frac{m}{4} u^2$, is converted into heat during the impact.

That is the treatment according to classical mechanics.

Now we consider the two equal spheres in relativistic mechanics where a possible dependence of mass on velocity is taken into account. We shall indicate this by writing $m(u)$. For the same experiments (Fig. 126*b*) the law of conservation of momentum now reads

$$m(u)u = M(\bar{u})\bar{u}. \qquad (\alpha)$$

In equation (α) we have used the more general description $M(\bar{u})$ of the mass after the collision because it is not self-evident that $M(\bar{u})$ is just twice $m(\bar{u})$. As a matter of fact, we will see later that $M(\bar{u})$ is *not* equal to $2m(\bar{u})$.

Next we derive a relation between u and $\bar{u}$. The above equation refers to a system S (Fig. 126*b*) where the left sphere moves with the velocity u and the right sphere is at rest.

Now we observe the same collision from another system S' moving with respect to S with velocity $+u$. Then in S' the left sphere rests and the right one moves with $-u$. This may be easily seen from equation (77*a*): the velocity u is transformed into 0 and the velocity 0 is transformed into $-u$. The appearance of the collision in S is quite symmetrical to the appearance of the collision in S' as may be seen in Fig. 126*c*. Therefore, one can conclude that the common velocity after the push must be $-\bar{u}$. But we can express this velocity using (77*a*), setting $v = +u$, $u_x = \bar{u}$, $u_x' = -\bar{u}$. Then one has

$$\bar{u} = \frac{-\bar{u}+u}{1-\dfrac{\bar{u}u}{c^2}}$$

or, solved with respect to u,

$$u = \frac{2\bar{u}}{1+\dfrac{\bar{u}^2}{c^2}}. \tag{β}$$

Equation (β) shows that in the limit, the case of classical mechanics, $\left(\frac{\bar{u}}{c}\to 0\right)$ one has $\bar{u}=\frac{u}{2}$ as stated above.

Now we derive a further relation

$$m(u)+m(0) = M(\bar{u}) \tag{γ}$$

which may be called the law of conservation of mass. It can be easily proved by adding a small velocity v perpendicular to u or $\bar{u}$ and applying the law of conservation of momentum for the y-component v (Fig. 127). For this purpose we introduce a system of

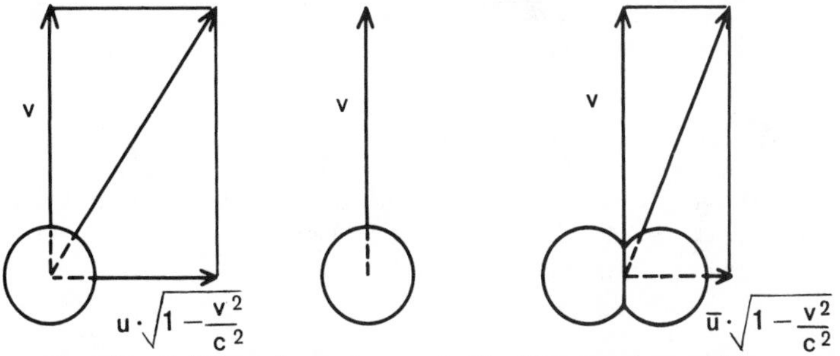

Fig. 127 *Collision of two spheres as in Fig. 126*b, *but observed from a system in which all spheres have a velocity* v *perpendicular to* u *and* ū *of Fig. 126*b.

reference S' which moves in the y-direction relative to the original system S with the velocity v. One can then apply the formulae (77*a*) and (77*b*) with the modification that the x- and y-directions are interchanged:

$$u_x' = u_x \frac{\sqrt{1-\dfrac{v^2}{c^2}}}{1+\dfrac{u_y v}{c^2}}, \qquad u_y' = \frac{u_y+v}{1+\dfrac{u_y v}{c^2}}.$$

Since in S the velocity is in the x-direction for the colliding spheres and for the combined body, one has for each of them $u_y=0$, and the last equations reduce to

$$u_x' = u_x\sqrt{1-\frac{v^2}{c^2}}, \qquad u_y' = v.$$

When the values of the velocity components in the system S are

	Left Sphere	*Right Sphere*	*Combined Body*
u_x	u	0	$\bar{u}$
u_y	0	0	0

for the three bodies, respectively, their values in S' are

	Left Sphere	*Right Sphere*	*Combined Body*
u_x'	$u\sqrt{1-\frac{v^2}{c^2}}$	0	$\bar{u}\sqrt{1-\frac{v^2}{c^2}}$
u_y'	v	v	v

respectively. Now the masses depend only on the absolute values of the velocities, i.e., on $\sqrt{u_{x'}^2+u_{y'}^2}$. Hence the y-component of the conservation law of momentum in S' reads

$$m(u')\times v+m(v)\times v = M(\bar{u}')\times v,$$

where we write briefly for $\sqrt{u_x'^2+u_y'^2}$ in the case of the left sphere u' and in the case of the combined body $\bar{u}'$; dividing by v we obtain

$$m(u')+m(v) = M(\bar{u}'). \qquad (\delta)$$

This equation must hold for every value of v. In particular, for $v=0$ we get (γ). Equation (δ) states the general form of the law of conservation of mass for arbitrary velocities, whereas (γ) is a special case which we use now to derive the dependence of mass on velocity.

If we replace $M(\bar{u})$ in equation (α) by $m(u)+m(0)$ according to (γ), the result is

$$m(u)u = [m(u)+m(0)]\bar{u}$$

or

$$m(u) = m(0)\frac{\bar{u}}{u-\bar{u}}.$$

Using (β) one gets* finally

$$m(u) = \frac{m(0)}{\sqrt{1-\frac{u^2}{c^2}}}. \tag{78}$$

In this way we have found how the mass depends on the velocity. The mass $m(0)=m_0$ is called the *rest mass* of the body, i.e., the mass measured in a system where the body is at rest. In classical mechanics only this limiting case of the mass is used.

For the momentum of a body moving with velocity v we get

$$p = mv = \frac{m_0}{\sqrt{1-\frac{v^2}{c^2}}}v, \tag{79}$$

as a function of the velocity of the body where m denotes the mass.

* The calculation runs like this: Substituting u from (β), one has

$$\frac{\bar{u}}{u-\bar{u}} = \frac{\bar{u}}{\frac{2\bar{u}}{1+\frac{\bar{u}^2}{c^2}}-\bar{u}} = \frac{1}{\frac{2}{1+\frac{\bar{u}^2}{c^2}}-1} = \frac{1+\frac{\bar{u}^2}{c^2}}{2-\left(1+\frac{\bar{u}^2}{c^2}\right)} = \frac{1+\frac{\bar{u}^2}{c^2}}{1-\frac{\bar{u}^2}{c^2}}.$$

On the other hand,

$$\left(1-\frac{\bar{u}^2}{c^2}\right)^2 = \left(1+\frac{\bar{u}^2}{c^2}\right)^2 - 4\frac{\bar{u}^2}{c^2},$$

hence, using (β) again,

$$\left(\frac{1-\frac{\bar{u}^2}{c^2}}{1+\frac{\bar{u}^2}{c^2}}\right)^2 = 1-\frac{4\frac{\bar{u}^2}{c^2}}{\left(1+\frac{\bar{u}^2}{c^2}\right)^2} = 1-\frac{u^2}{c^2},$$

and combining both results,

$$\frac{\bar{u}}{u-\bar{u}} = \frac{1}{\sqrt{1-\frac{u^2}{c^2}}}.$$

Now we can proceed to the law of motion for forces that act continuously. In doing so we must use the formulation of classical mechanics (II, 10, p. 36), which is based on the momentum carried along by the moving body. It may at once be applied to the new dynamics, but the law for the longitudinal and the transverse component must be formulated separately, thus:

A force K *produces a change in the momentum in such a way that the change of the longitudinal or, respectively, transverse component of momentum per unit of time is equal to the corresponding component of the force.*

The equations of motion may then easily be set up. The momentum p of a body at the instant $t=0$ may have the components $p_x(0)$ and $p_y(0)$; its velocity in the x-direction at the time $t=0$ may be v. Now a force with the components K_x and K_y may act for a short time τ; it will change the components of momentum into $p_x(\tau)$ and $p_y(\tau)$. The mathematical expression for this is

$$p_x(\tau)-p_x(0) = K_x\tau,$$
$$p_y(\tau)-p_y(0) = K_y\tau.$$

The force produces small additions w_x and w_y to the velocity components (Fig. 128); the resultant velocity may be $\bar{v}$. Then we have in the

$$x\text{-direction} \quad m(\bar{v})(v+w_x)-m(v)v = K_x\tau,$$
$$y\text{-direction} \quad m(\bar{v})w_y = K_y\tau.$$

Since w_x and w_y are small we can approximate $\bar{v}$ by neglecting squares of these quantities:

$$\bar{v} = \sqrt{(v+w_x)^2+w_y{}^2} \cong v\sqrt{1+\frac{2w_x}{v}} \cong v\left(1+\frac{w_x}{v}\right) = v+w_x;$$

(we have used the approximation $\sqrt{1+2x}=1+x$ when x is small). Thus we get

$$m(\bar{v}) = m(v+w_x) = \frac{m_0}{\sqrt{1-\dfrac{(v+w_x)^2}{c^2}}}.$$

Now the square root in the denominator can be written

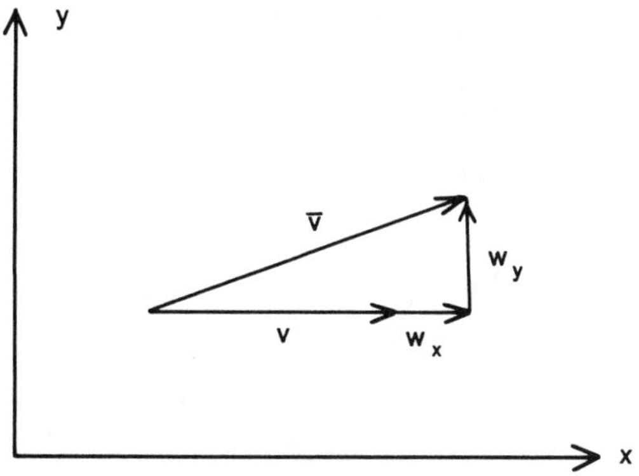

Fig. 128 *A velocity* v *in the* x*-direction is changed by small velocity components* w_x *and* w_y. *The resulting velocity is* $\bar{v}$.

$$\sqrt{1-\frac{v^2}{c^2}\left(1+\frac{w_x}{v}\right)^2} = \sqrt{1-\frac{v^2}{c^2}\left(1+\frac{2w_x}{v}+\frac{w_x^{\,2}}{v^2}\right)},$$

and if $\frac{w_x^{\,2}}{c^2}$ is neglected and the abbreviation $\alpha^2=1-\frac{v^2}{c^2}$ is used, this is equal to

$$\sqrt{\alpha^2-\frac{2vw_x}{c^2}} = \alpha\sqrt{1-\frac{2vw_x}{c^2\alpha^2}} = \alpha\left(1-\frac{vw_x}{\alpha^2c^2}\right),$$

according to the approximation formula used above. Considering that in the same approximation $\frac{1}{(1-x)}=1+x$, we obtain

$$m(\bar{v}) = \frac{m_0}{\alpha\left(1-\frac{vw_x}{\alpha^2c^2}\right)} = \frac{m_0}{\sqrt{1-\frac{v^2}{c^2}}}\left(1+\frac{vw_x}{\alpha^2c^2}\right) = m(v)\left(1+\frac{vw_x}{\alpha^2c^2}\right). \qquad (\epsilon)$$

This has to be substituted in the conservation laws of momentum components. The left side of the equation for the x-direction is now (with the usual approximation)

$$m(v)\left(1+\frac{vw_x}{\alpha^2c^2}\right)(v+w_x)-m(v)v = m(v)w_x\left(\frac{v^2}{\alpha^2c^2}-1\right) = \frac{m(v)}{1-\frac{v^2}{c^2}}\,w_x,$$

and correspondingly for the y-direction

$$m(v)\left(1+\frac{vw_x}{\alpha^2c^2}\right)w_y = m(v)w_y,$$

where we have neglected the terms $w_x{}^2$ and w_xw_y as small. Therefore our result is

$$\frac{m(v)}{1-\dfrac{v^2}{c^2}}w_x = \frac{m_0}{\left(\sqrt{1-\dfrac{v^2}{c^2}}\right)^3}w_x = K_x\tau,$$

$$m(v)w_y = \frac{m_0}{\sqrt{1-\dfrac{v^2}{c^2}}}w_y = K_y\tau.$$

Now we introduce the components of the acceleration

$$b_x = \frac{w_x}{\tau} \quad \text{and} \quad b_y = \frac{w_y}{\tau}$$

and get for the components of force the expressions:

$$K_x = \frac{m_0b_x}{\left(\sqrt{1-\dfrac{v^2}{c^2}}\right)^3}, \qquad K_y = \frac{m_0b_y}{\sqrt{1-\dfrac{v^2}{c^2}}}. \tag{80}$$

The relationship between the force and the acceleration produced by it is thus different according to whether the force acts in the direction of the velocity that is already present or in a direction perpendicular to this.

In the first years of relativistic theory it was usual to bring these formulae into a form in which they resemble the fundamental law of classical dynamics (II, 10, formula (10), p. 35) as much as possible. For this purpose one introduced the symbols

$$m_x = \frac{m_0}{\left(\sqrt{1-\dfrac{v^2}{c^2}}\right)^3}, \qquad m_y = \frac{m_0}{\sqrt{1-\dfrac{v^2}{c^2}}} \tag{81}$$

and called these quantities *longitudinal* and *transverse mass*. The latter is identical with the quantity m, simply called *relativistic mass*, in formula (78).

Then one can write in place of (80)

$$K_x = m_x b_x, \qquad K_y = m_y b_y \tag{82}$$

which agrees in form with the fundamental classical law.

To avoid confusion we will use in the following text only the relativistic mass $m = \dfrac{m_0}{\sqrt{1-\dfrac{v^2}{c^2}}}$. Nevertheless we see how necessary it is from the very beginning to define the concept of mass exclusively by the inertial resistance. Otherwise it would not be possible to apply it in relativistic mechanics, since a different expression "mass" comes into account for the total momentum carried along, for the longitudinal and the transverse force; moreover, these masses are not characteristic constants of the body, but depend on its velocity. Thus the concept of mass in Einstein's dynamics differs greatly from that to which we are accustomed, in which mass denotes quantity of matter in some way. In a certain sense the rest mass m_0 is a measure of the Einsteinian mass, but again, unlike the mass of ordinary mechanics, it is not, in an arbitrary system of reference, equal to the ratio of momentum to velocity or of force to acceleration.

A glance at formula (78) for the mass tells us that the values of the relativistic mass m become greater as the velocity v of the moving body approaches the velocity of light. For $v=c$ the mass becomes infinitely great.

From this it follows that it is impossible to make a body move with a velocity greater than that of light by applying forces: Its inertial resistance grows to an infinite extent and prevents the velocity of light from being reached.

Here we see how Einstein's theory rounds off to a harmonious whole. The assumption that there is a limiting velocity that cannot be exceeded, which seems almost paradoxical, is itself required by the physical laws in their new form.

Formula (78), giving the dependence of the mass on the velocity, is the same as that already found by Lorentz from electrodynamic calculations for his flattened electron. In it m_0 was expressed in terms of the electrostatic energy S of the stationary electron just as in Abraham's theory (V, 13, p. 211, formula (69)), namely,

$$m_0 = \frac{4}{3}\frac{S}{c^2}.$$

We now see that Lorentz's formula for the dependency of mass on velocity has a much more general significance than is at first apparent. It must hold for every kind of mass, no matter whether it is of electrodynamic origin or not.

Experiments by Kaufmann (1901) and others who have deflected cathode rays by electric and magnetic fields have shown very accurately that the mass of electrons grows with velocity according to Lorentz's formula (78). On the other hand, these measurements can no longer be regarded as a confirmation of the assumption that all mass is of electromagnetic origin. For Einstein's theory of relativity shows that mass as such, regardless of its origin, must depend on velocity in the way described by Lorentz's formula.

Another confirmation of (78) has been provided by spectroscopy. An atom consists of a heavy, positively charged nucleus which is surrounded by a number of electrons in such a way that the whole atom is electrically neutral. Spectroscopy deals with the interaction of these electrons with light. The motion of the electrons is governed by the laws of mechanics. Because the spectroscopic measurements are extremely accurate, one can easily detect deviations from classical dynamics of electronic motion. The results have fully proved the validity of Einstein's dynamics.

8. The Inertia of Energy

There is a point concerning inelastic collision which we have not considered in the preceding section. We have not discussed the relation between the masses m of the single spheres and the mass M of the two united spheres after the collision. This can be investigated with the help of the equation (γ)

$$m(u)+m(0) = M(\bar{u}).$$

To begin with, let us consider the rest mass $M(0)=M_0$ of the joined spheres. We get it by going over to a system S'' in which the velocity of M equals zero (Fig. 129). Obviously S'' moves with the velocity $\bar{u}$ relative to S. Since the two spheres are equal, it follows from symmetry that before the collision they have opposite velocities $\pm\bar{u}$ in S''. (By the transformation $\bar{u}$ goes over into 0, u into $\bar{u}$, and 0 into $-\bar{u}$.)

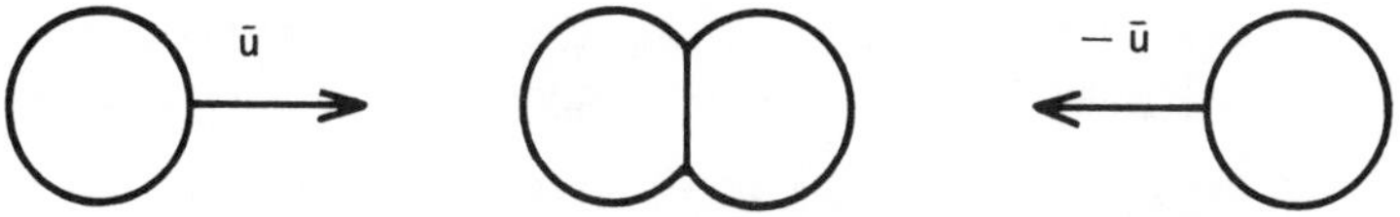

Fig. 129 *The collision of two equal spheres (Fig. 126b) observed from a system S″ in which the common velocity after the collision is zero. The velocities before they collide were* ū *and* −ū.

Then the conservation of mass reads

$$m(\bar{u})+m(-\bar{u}) = 2m(\bar{u}) = M(0) = M_0 \qquad (\alpha)$$

or

$$M_0 = 2\frac{m_0}{\sqrt{1-\frac{\bar{u}^2}{c^2}}}. \qquad (\alpha')$$

In the case of the velocity being small, $\bar{u} \ll c$, we have

$$M_0 = 2m_0\left(1+\tfrac{1}{2}\frac{\bar{u}^2}{c^2}\right) = 2m_0+2\times\frac{1}{c^2}\times\tfrac{1}{2}m_0\bar{u}^2. \qquad (\beta)$$

The rest mass M_0 does not equal the rest mass $2m_0$ of the two colliding spheres as one might have expected. There is another contribution of second order: a term which is the kinetic energy of the two spheres before the collision divided by c^2. The kinetic energy of one sphere is $T=\frac{m_0}{2}\bar{u}^2$. When the two spheres come to rest during the collision, their kinetic energy $2T$ is converted into heat of amount $Q=2T$. One sees that in our case one has

$$M_0 = 2m_0+\frac{Q}{c^2}. \qquad (\gamma)$$

This can be interpreted by saying that an addition of heat energy Q increases the mass by $\frac{Q}{c^2}$.

Another example of how mass is changed is in the addition of kinetic energy. Consider the normal dependence of mass $m(v)$ on velocity v for $v \ll c$

$$m(v) = m_0+\frac{1}{c^2}\times\tfrac{1}{2}m_0v^2 = m_0+\frac{T}{c^2}. \qquad (\delta)$$

Again the mass is increased by a term energy divided by c^2, where the energy concerned is the kinetic energy T of the moving body.

We may generalize these results by saying that addition of energy e to a body changes its mass by $\frac{e}{c^2}$. We have shown this only for kinetic energy and heat, but we expect that the same will hold for other forms of energy (electrical, chemical energy, etc.).

When we multiply equation (β) by c^2 and use (γ), it reads

$$2m_0c^2+2T = M_0c^2 = 2m_0c^2+Q \tag{β'}$$

or

$$2T = Q. \tag{β''}$$

This equation represents the law of conservation of energy: *The energy before the collision* (*kinetic energy* 2T) *equals the energy after* (*heat energy* Q). It is a special case of equation (17) where only kinetic and heat energy are involved.

We see that from the relativistic standpoint the law of conservation of mass is nothing less than the law of conservation of energy. Therefore one is justified in asserting that the energy content E of a body equals its mass times c^2:

$$E = mc^2. \tag{83}$$

Generally the kinetic energy T is defined by the difference of energy $E=m(v)c^2$ of the moving body and the energy $E_0=m_0c^2$ of the body at rest. Thus the general definition of T is

$$T = (m(v)-m_0)c^2. \tag{84}$$

This definition reduces to the classical definition $T=\frac{m_0}{2}v^2$ for small values of v. With the new definition of T the equations (β') and (β'') hold also for velocities which are not small as compared with c.

Here an essential difference between classical and relativistic mechanics comes to light. In classical mechanics we have to distinguish between processes which conserve mechanical energy and those which do not but where it is changed into heat or other energy forms. Returning, for instance, to our inelastic collision, we see that one-half of the kinetic energy (as observed in the system S) is changed into heat. Therefore mechanical energy is not conserved.

In relativistic mechanics, however, we have the law of conservation of energy ((γ) of the preceding section), which includes all energy forms:

$$m(u)c^2+m_0c^2 = M(\bar{u})c^2.$$

This equation reduces in the classical (limit) case to the usual treatment of the problem: we get

$$m_0c^2+\frac{m_0}{2}u^2+m_0c^2 = M_0c^2+\tfrac{1}{2}M_0\bar{u}^2.$$

We know that $M_0=2m_0+\frac{Q}{c^2}$, where $Q=2\times\frac{1}{2}m_0\bar{u}^2$. Therefore we get

$$\frac{m_0}{2}u^2 = Q+\tfrac{1}{2}\,2m_0\bar{u}^2+\tfrac{1}{2}\,\frac{Q}{c^2}\,\bar{u}^2.$$

The last term is to be neglected in the classical approximation ($\bar{u}\ll c$), for it is a part of the relativistic correction of the kinetic energy; it is smaller by the factor $\frac{\bar{u}^2}{c^2}$ than the two other terms and negligible for small velocities u and $\bar{u}$. Hence the equation reduces to

$$\frac{m_0}{2}u^2 = Q+\tfrac{1}{2}\,(2m_0)\bar{u}^2. \tag{85}$$

On the left-hand side we have the kinetic energy T_1 of the system before the collision. On the right-hand side Q represents that part of T_1 which is changed into heat by the collision. The second term is the remaining kinetic energy T_2 of the two spheres adhering. Since we know that in classical mechanics $\bar{u}=\frac{u}{2}$, we see that $T_2 = \frac{m_0}{4}u^2=\frac{T_1}{2}$ and $Q=\frac{T_1}{2}$. These are the classical formulae.

Now we derive the change of T or E by a force which operates for a short time τ. Let v, $T(0)$, and $E(0)$ be the velocity in the x-direction, the kinetic and the total energy of our body before the force has acted, while $\bar{v}$, $T(\tau)$ and $E(\tau)$ are the velocity, the kinetic and the total energy of the same body after the force has acted. Then with (83) we have

$$E(\tau)-E(0) = T(\tau)-T(0) = [m(\bar{v})-m(v)]c^2.$$

From the preceding section (equation (ϵ)) we know that

$$m(\bar{v}) = m(v)\left[1+\frac{vw_x}{c^2\left(1-\frac{v^2}{c^2}\right)}\right],$$

where w_x is a small additive velocity in the x-direction produced by the force (Fig. 128). Thus we get

$$E(\tau)-E(0) = \frac{w_x v}{1-\frac{v^2}{c^2}}\, m(v) = K_x v\tau.$$

The last part of this equation follows from (80). Now

$$\frac{E(\tau)-E(0)}{\tau} = \frac{T(\tau)-T(0)}{\tau}$$

is the rate of change of energy, and by introducing the component of acceleration $b_x = \frac{w_x}{\tau}$ we get

$$\frac{T(\tau)-T(0)}{\tau} = \frac{m(v)}{1-\frac{v^2}{c^2}}\, b_x v = K_x v. \tag{86}$$

The corresponding formula in classical mechanics would be

$$\frac{T(\tau)-T(0)}{\tau} = m_0 b_x v = K_x v.$$

The quantity on the right-hand side in (86) is the negative time change of the potential energy U. For during a sufficiently small interval of time τ the force may be regarded as approximately constant, and we may proceed as if we were dealing with a gravitational force. We have shown (II, 14, formula (15), p. 48) that its potential energy would then be Gx; there, however, we took the direction of x opposite to that of gravity so that we had $G = -K_x$. If $x(0)$ and $x(\tau)$ are the positions before and after the force has acted, the time change of the potential energy becomes

$$\frac{U(\tau)-U(0)}{\tau} = G\,\frac{x(\tau)-x(0)}{\tau} = Gv = -K_x v,$$

since $\frac{x(\tau)-x(0)}{\tau}=v$ according to the definition of velocity. Introducing this in equation (86), one obtains the result

$$T(\tau)+U(\tau) = T(0)+U(0),$$

that is, $T+U$ is constant in time as in classical mechanics.

Einstein's equation (83)

$$E = mc^2,$$

which states the *proportionality of energy and inertial mass* and is often called the *law of inertia of energy*, is perhaps the most important result of the theory of relativity. We shall give another simple proof of it due to Einstein himself, a proof which does not make use of the mathematical formalism of the theory of relativity.

It is based on the fact that radiation exerts a pressure. From Maxwell's field equations, supplemented by a theorem first deduced by Poynting (1884), it follows that a light wave which falls on an absorbing body exerts a pressure on it. It is found that the momentum transferred to an absorbing surface by a short flash of light is equal to $\frac{E}{c}$, where E is the energy of the light flash. This fact, which we will prove in the next section (9), was confirmed experimentally by Lebedew (1890) and again later with greater accuracy by Nichols

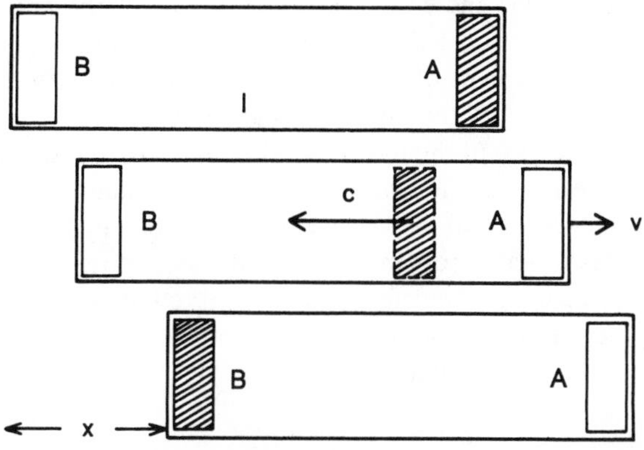

Fig. 130 *A tube with two equal bodies,* A *and* B, *at its ends.* A *carries an energy* E *which is sent from* A *to* B *in the form of a light flash with velocity* c; *the recoil produces a velocity* v *of the tube. When* E *is absorbed by* B, *the tube is at rest again, but displaced by a distance* x.

and Hull (1901) and others. Exactly the same pressure is experienced by a body which emits light, just as a gun experiences a recoil when a shot is fired.

We next imagine a long tube at whose ends are two bodies A and B which are exactly equal and are composed of the same material and which, according to ordinary ideas, have the same mass (Fig. 130). But the body A is to have an excess of energy E over that of B, say in the form of heat, and there is to be an arrangement (concave mirror or something similar) by which this energy E can be sent in the form of radiation to B. Let the spatial extent of this flash of light be small compared with the length l of the tube (Fig. 130).

Then A experiences the recoil E/c. If this were transferred to the whole tube of mass M this would acquire a velocity v given by the momentum equation

$$Mv = E/c.$$

Now the transfer of momentum to the tube does not happen instantaneously; for if the tube were rigid the forces would propagate with a velocity larger than that of light. In fact, the propagation of the recoil through the tube from A to B is due to the elastic forces in the wall of the tube which are much slower than light. One has therefore to regard the process as consisting of two separate parts: (1) the emission from A, and (2) the absorption at B, and then to consider their effect on the tube, independent of one another, at a moment of time so late that not only has the elastic movement excited by the impacts expanded over the whole tube but also all elastic vibrations have died out and only the displacements of the whole tube are left over. In order to obtain the total effect one has to add the two displacements due to the impacts at A and B because elastic waves (of small amplitude) superpose undistortedly.

(1) The recoil at A transfers a movement to the tube in such a way that at the late instant t_1 when all vibrations have disappeared its velocity is v and its displacement

$$x_1 = vt_1.$$

(2) When the light is absorbed at B the tube receives a movement of which at the instant t_1 only a resultant velocity in the opposite

direction, $-v$, is left over; the corresponding displacement is

$$x_2 = -v(t_1 - t),$$

if t is the time the light needs to travel from A to B; for the impact on B happens the time interval t later, The sum of the two displacements is

$$x = x_1 + x_2 = vt,$$

the same as if the tube were rigid.* If we substitute here $v = E/Mc$ and $t = l/c$ we obtain for the displacement of the tube

$$x = \frac{El}{Mc^2}.$$

Now the bodies A and B may be exchanged (this may be done without using external influences). Let us suppose that two men are situated in the tube, who put A in the place of B, and B in the place of A, and then themselves return to their original positions. According to ordinary mechanics the tube as a whole must suffer no displacement, for changes of position can be effected only by external forces.

If this exchange were to be carried out, everything in the interior of the tube would be as at the beginning, the energy E would again be at the same place as before, and the distribution of mass would be exactly the same. But the whole tube would be displaced a distance x with respect to its initial position by the light impulse. This, of course, contradicts all the fundamental canons of mechanics. We could repeat the process and thus impart any arbitrary change of position to the system without applying external forces. This is, however, an impossibility. The only escape from the difficulty is to assume that when the bodies A and B are exchanged, these two bodies are not mechanically equivalent but that B has a mass greater by m than A in consequence of its excess of energy E. Then the symmetry during the exchange is not maintained, and the mass m is displaced from right to left by a distance l. At the same time the whole tube is displaced a distance x in the reverse direction. This distance is determined by the circumstance that the process occurs

* Einstein's first derivation (1905) supposed the tube to be rigid. Later (1907) he himself criticized the concept of a rigid body in the theory of relativity. Our derivation is a simplified version of a consideration by E. Feenberg.

without the intervention of external influences. The total momentum, consisting of that of the tube $M\frac{x}{t}$ and that of the transported mass $-m\frac{l}{t}$, is thus zero. Then

$$Mx - ml = 0,$$

from which it follows that

$$x = \frac{ml}{M}.$$

Now this displacement must exactly counterbalance that produced by the light impulse, hence we must have

$$x = \frac{ml}{M} = \frac{El}{Mc^2}.$$

This allows us to calculate m and we get

$$m = \frac{E}{c^2}.$$

This is the amount of inertial mass that must be ascribed to the energy E in order that the principle of mechanics which states that no changes of position can occur without the action of external forces remain valid.

Since every form of energy is finally transformable into radiation by some process or other, this law must be universally valid. Thus we have a great unification in our knowledge of the material world: *Matter in the widest meaning of the word* (including light and other forms of pure energy, in the language of classical physics) *has two fundamental qualities: inertia, measured by its mass, and the capability of performing work, measured by its energy.* These two are strictly proportional to one another. Wherever electric and magnetic fields or other effects lead to intense accumulations of energy, they are accompanied by inertia. Electrons and atoms are examples of enormous concentrations of energy.

We can touch on only a few of the numerous important consequences of this theorem.

Concerning the mass of the electron, formula (69) (p. 211) shows that for the statical mass $m_0=\frac{4}{3}\frac{S}{c^2}$ the electrostatic energy S cannot be the total energy E of the electron at rest. There must be present another amount of energy V, $E=S+V$, such that

$$m_0 = \frac{4}{3}\frac{S}{c^2} = \frac{E}{c^2} = \frac{S+V}{c^2}.$$

Hence $V=\frac{4}{3}S-S=\frac{1}{3}S=\frac{1}{4}E$. Thus the total energy is three-quarters electrostatic and one-quarter of a different kind. This part must be due to cohesive forces which hold the electron together by counterbalancing the electrostatic repulsion; such a force must be assumed to stabilize the electron.

Now we turn to other examples taken from recent researches. These have shown that there exist three kinds of π-mesons (see page 260); the first kind positively charged, the second kind negatively charged (with the same amount of charge as the electron), the third kind electrically neutral (π^+, π^-, and π^0-mesons). Exact measurements of the masses of these particles have revealed that the masses of π^+ and π^- are equal and are 273 times the mass of the electron, while the mass of π^0 is only 264 times that of the electron. As in the case of the electron, we can interpret this mass difference as due to the electric charge. The charge causes an electrical field with a certain energy content S which is the same for positively and negatively charged mesons. Therefore the charged particles $\pi^\pm$ must be heavier by $\frac{S}{c^2}$ than the particles π^0. The mass difference $m_\pm - m_0 = 9m_{el}$ (where m_{el} is the mass of the electron) shows that in the charged mesons there is accumulated much more electric energy than in the electron. In Chapter V, 13 (Electromagnetic Mass) we have established a formula which connects the energy content S of a charge e distributed on the surface of a sphere with its radius a, namely

$$S = \frac{1}{2}\frac{e^2}{a}.$$

If we apply the same model to the charged meson, then $c^2(m_\pm - m_0) = \frac{1}{2}\frac{e^2}{a}$, where a is the "radius" of the charge distribution on the meson.

This radius then can be computed; one finds $a = 1.5 \times 10^{-14}$ cm., a value much smaller than the radius of the electron, considering the larger mass difference.

As we have explained above, an atom consists of a small, positively charged nucleus (diameter of the order of 10^{-13} cm.) and a surrounding cloud of electrons which just neutralize the charge of the nucleus. A nucleus is built up from protons and neutrons. Protons have a positive charge of the same amount as the electron charge; their mass is about 2000 times m_{el}. The proton is the nucleus of the lightest atom (hydrogen) consisting of one proton and one surrounding electron. The neutron has about the same mass as the proton but—as its name indicates—no charge. An atom can be characterized by two quantities: its mass, which is essentially the mass of the nucleus (the contribution of the electrons can be disregarded), and its charge, which is determined by the number of protons (or of the surrounding electrons). The chemical behavior of an atom is determined by the surrounding cloud of electrons, which extends to distances of about 10^{-8} cm. from the nucleus. Therefore, the number of protons or electrons alone decides the chemical properties of the atom. One finds nuclei with the same number of protons but different numbers of neutrons. The corresponding atoms display the same chemical behavior, but have different masses. Such atoms are called *isotopes*. A chemical element as it occurs in nature is generally a mixture of several isotopes.

The simplest example is hydrogen, which has an isotope, the so-called heavy hydrogen or deuterium, whose nucleus, the deuteron, consists of a proton and a neutron. When we add their masses

$$\begin{aligned} &\text{mass of proton} & m_P &= 1.6724 \times 10^{-24} \text{ gm.} \\ &\text{mass of neutron} & m_N &= 1.6747 \times 10^{-24} \text{ gm.} \\ &\text{we get} & m_P + m_N &= 3.3471 \times 10^{-24} \text{ gm.} \end{aligned}$$

But the measured mass of the deuteron is only

$$m_D = 3.3433 \times 10^{-24} \text{ gm.}$$

According to equation (83), the mass difference

$$m_P + m_N - m_D = 0.0038 \times 10^{-24} \text{ gm.}$$

(about four electron masses) tells us how much energy must be added to a deuteron to split it into a proton and a neutron. The experiment exactly confirms this consideration. The same energy will be set free if a neutron and a proton combine to become a deuteron (nuclear fusion).

Another example of this kind is of great technical importance and is used in atomic reactors. In these heat is produced by the following process: A nucleus of the uranium isotope U_{235} (whose nucleus consists of 235 particles—92 protons and 143 neutrons) absorbs a neutron, becomes unstable, and splits into two smaller nuclei and some neutrons. The neutrons are absorbed by other uranium nuclei and induce them to split. Thus a chain reaction develops which sustains itself. The products of each fission fly apart with large kinetic energies, are stopped by the surrounding material, and heat it up. If we add the masses of the fission products, we find a sum which is smaller than the mass of the U_{235} nucleus. The mass difference is about 400 times the electron mass. The energy of this mass difference is the kinetic energy of the fragments of the uranium nucleus. This energy is transformed into heat.

To illustrate the magnitude of this effect, one can compare the heat produced by fission from U_{235} with the heat produced by burning coal. One gets the same energy by fission from 1 gm. U_{235} as by burning 20 tons of coal.

These two examples show that heat may be obtained by two processes: by fission of large nuclei and by building up small nuclei from their constituents (nuclear fusion). The latter process is the heat source of the stars.

It is well known that the process of fission is used in the atomic bomb (A-bomb) and that of fusion in the hydrogen bomb (H-bomb). But we shall not discuss these bleak aspects of technical developments connected with Einstein's formula here.

9. Energy and Momentum

In section 7 of this chapter (p. 271) we derived the law of conservation of mass or energy (γ) by using only the law of conservation of momentum. This reveals a very close connection between momentum and energy which constitutes a new characteristic relationship in the theory of relativity analogous to that connecting space and time.

Earlier, in section 3, we derived from the Lorentz transformation the invariant $F = x^2 - c^2t^2$, where x and t are the coordinates of any world point P. Let us recall what the invariance of F means: If in two systems S and S' with the same origin we represent P by its coordinates (x, t) and (x', t'), then

$$x^2 - c^2t^2 = x'^2 - c^2t'^2 = F$$

is independent of the chosen frame of reference.

From $p = m(u)u$ and $E = m(u)c^2$ we form a quadratic expression which, after simplifying with the help of (78), becomes

$$p^2 - \frac{E^2}{c^2} = m^2(u)(u^2 - c^2) = m_0^2 \frac{u^2 - c^2}{1 - \frac{u^2}{c^2}} = -m_0^2c^2. \tag{87}$$

Hence the expression on the left-hand side is an invariant, i.e., independent of the special system in which we have measured momentum and energy. The following is then suggested: *Momentum* p *and energy* E *divided by* c² *(or in other words momentum* p *and mass* m) *transform from one system* S *into another system* S′ *in the same way as* x *and* t, *namely by the Lorentz transformation.*

We can prove this without difficulty. The system S' may have the velocity v, as observed from S. Then (for simplicity we assume all velocities to be in the v-direction) we have the following table:

in S	momentum:	$p = m(u)u,$	energy:	$E = m(u)c^2;$
in S'	momentum:	$p' = m(u')u',$	energy:	$E' = m(u')c^2,$

where the velocities u, u' are connected by the addition theorem (77*a*) $u = \dfrac{u'+v}{1+\dfrac{u'v}{c^2}}$. Therefore we get

$$p = m(u)u = m_0 \frac{u}{\sqrt{1-\frac{u^2}{c^2}}} = m_0 \frac{u'+v}{\sqrt{\left(1+\frac{u'v}{c^2}\right)^2 - \frac{(u'+v)^2}{c^2}}}$$

or

$$p = m_0 \frac{u'+v}{\sqrt{1-\frac{v^2}{c^2}}\sqrt{1-\frac{u'^2}{c^2}}} = \frac{p' + v\frac{E'}{c^2}}{\sqrt{1-\frac{v^2}{c^2}}} \tag{88a}$$

and

$$\frac{E}{c^2} = m(u) = m_0 \frac{1+\dfrac{v}{c^2}u'}{\sqrt{1-\dfrac{v^2}{c^2}}\sqrt{1-\dfrac{u'^2}{c^2}}} = \frac{\dfrac{E'}{c^2}+\dfrac{v}{c^2}p'}{\sqrt{1-\dfrac{v^2}{c^2}}}. \qquad (88b)$$

These transformation formulae are completely analogous to the first and the fourth formula of (70*b*).

If the momentum is not directed parallel to the x-axis, but has components p_x, p_y, p_z in S and p_x', p_y', p_z' in S', we must replace p by p_x and p' by p_x' in (88*a*, *b*) and complete the equations by $p_y = p_y'$ and $p_z = p_z'$.

If we invert these formulae, we get

$$p_x' = \frac{p_x - v\dfrac{E}{c^2}}{\sqrt{1-\dfrac{v^2}{c^2}}}, \quad p_y' = p_y, \quad p_z' = p_z, \quad \frac{E'}{c^2} = \frac{\dfrac{E}{c^2}-\dfrac{v}{c^2}p_x}{\sqrt{1-\dfrac{v^2}{c^2}}}. \qquad (88c)$$

in accordance with (70*a*).

Formula (87) is of great importance. It allows us to compute p if E is known, and vice versa:

$$p = \sqrt{\frac{E^2}{c^2} - m_0^2c^2} = \frac{\sqrt{E^2-E_0^2}}{c}, \qquad (87a)$$

$$E = c\sqrt{p^2+m_0^2c^2}. \qquad (87b)$$

In Einstein's proof of the fundamental law of inertia of energy a relation between the momentum and the energy of a flash of light was used: $p = \frac{E}{c}$. Now light is a flow of energy with velocity c. Hence, according to the theory of relativity, its rest mass must be zero: $m_0 = 0$. For then (87*a*) reads $p = \frac{E}{c}$ as required.

An interesting application of the transformation formula of p and E is made in quantum theory, of which we shall speak briefly. It is the theory on which the explanation of atomic phenomena is based and was founded by Planck (1900). One of its main results is the "quantization" of the energy of light, a phrase which means that the energy in a beam of light of frequency ν cannot have an arbitrary value, but must consist of quanta of finite size

$$\epsilon = h\nu, \tag{89}$$

where $h = 6.6 \times 10^{-27}$ gm. $\frac{\text{cm}^2.}{\text{sec.}}$ is the fundamental Planck's constant which governs all processes in atoms. From (89) one can derive the mass and momentum of a quantum of light:

$$m = \frac{\epsilon}{c^2} = \frac{h\nu}{c^2}; \qquad p = \frac{\epsilon}{c} = h\frac{\nu}{c} = \frac{h}{\lambda} \tag{90}$$

$\left[\lambda = \frac{c}{\nu}\right.$ is the wave length of the light: see equation (35)$\left.\right]$.

Hence one can interpret a light wave as consisting of particles with rest mass zero, momentum $p = \frac{\epsilon}{c}$, and energy $\epsilon = h\nu$.

These "light quanta," or "photons," can be converted into other particles, provided the law of conservation of energy and momentum can be fulfilled. We will discuss a special example where one has to take into account another law of conservation, namely, the conservation of charge. According to this law the total amount of charge of colliding and reacting particles is the same before and after the collision. Now it was found by Anderson (1932) that light quanta colliding with other particles (atomic nuclei) are converted into a pair of particles, one of which is an electron, the other called a positron, its positive counterpart. The latter has exactly the same properties as the electron, apart from its charge, which is the same in magnitude but with the opposite sign. The condition of charge conservation is fulfilled by the production of a pair. The conservation of energy requires the energy of the light quantum ϵ to be larger than $2m_{el}c^2$, the rest energy of the pair. But this is not enough. One has also to postulate the conservation of momentum.

Now it can easily be seen that both conservation laws cannot hold simultaneously for pair production by light quanta in free space. Without going into details, one can argue thus: If such a transmutation were possible, it could be described in any frame of reference. Now for an observer in a reference system S' moving with the velocity v in the direction of the light quantum, its energy ϵ' is, according to (88*b*) (putting $E' = \epsilon' = p'c$, $E = \epsilon$ and solving for ϵ'):

$$\epsilon' = \epsilon \frac{\sqrt{1-\frac{v^2}{c^2}}}{1+\frac{v}{c}} = \epsilon \sqrt{\frac{1-\frac{v}{c}}{1+\frac{v}{c}}}. \tag{91}$$

Hence by choosing the velocity v large enough one can make ϵ' arbitrarily small. So one sees immediately that in S' the energy cannot be conserved because, as we have seen, it must be larger than $2m_{el}c^2$ in any system. (In the extreme case of $v = c$ the energy of the light quantum is transformed into zero: it does not exist any more.)

For this process of pair production to occur, another particle must be present to carry off some energy and momentum in such a way that the conservation laws are satisfied. Therefore this transmutation can be observed only when an atomic nucleus is present. The nucleus is not changed during the process; it only takes care of the conservation conditions. The previous argument, that in a fast-moving system S' the energy of the photon is very small so that the condition $\epsilon' > 2m_{el}c^2$ cannot be satisfied, does not apply any more, for the atomic nucleus has a very high energy in S' and can supply the difference.

There exists a relation between current and charge similar to that between space and time coordinates or that between momentum and energy which we have just discussed. We shall consider this now and get a glimpse into the relativistic theory of electricity.

Assume N electrons with charge $N \times e$ in a cube whose edge has the length l_0. Then the charge density ρ_0 in this volume is

$$\rho_0 = \frac{Ne}{l_0^3}.$$

If the charges are at rest the current density j_0 is zero.

An observer relative to whom the charges move with the velocity v in the direction of one edge of the cube will measure a contracted volume $l_0^3\sqrt{1-\frac{v^2}{c^2}}$, for the edge of the cube parallel to the direction of the velocity is contracted by $\sqrt{1-\frac{v^2}{c^2}}$. The number of electrons, however, cannot change with the frame of reference. Therefore, the observer will measure a charge density

$$\rho = \frac{Ne}{l_0^3\sqrt{1-\dfrac{v^2}{c^2}}} = \frac{\rho_0}{\sqrt{1-\dfrac{v^2}{c^2}}}. \tag{92}$$

The moving charges represent a current whose density j inside the cube is obviously given by

$$j = \frac{Nev}{l_0^3\sqrt{1-\dfrac{v^2}{c^2}}} = \frac{\rho_0 v}{\sqrt{1-\dfrac{v^2}{c^2}}} = \rho v. \tag{93}$$

With the help of (92) and (93) one can derive an invariant which is formed from j and ρ in the same manner as those previously constructed from x and t or from p and m:

$$j^2-c^2\rho^2 = \rho^2(v^2-c^2) = -\rho_0^2c^2. \tag{94}$$

Equations (92) and (93) can also be written in the form

$$\rho = \frac{\rho_0}{m_0}m, \qquad j = \frac{\rho_0}{m_0}p,$$

where m_0 is the rest mass of the electrons: $m_0 = N\times m_{el}$.
Therefore the quantities j and ρ have the same transformation law as p and m, the Lorentz transformation, namely

$$j' = \frac{j-v\rho}{\sqrt{1-\dfrac{v^2}{c^2}}}, \qquad \rho' = \frac{\rho-\dfrac{v}{c^2}j}{\sqrt{1-\dfrac{v^2}{c^2}}},$$

and inverted

$$j = \frac{j'+v\rho'}{\sqrt{1-\dfrac{v^2}{c^2}}}, \qquad \rho = \frac{\rho+\dfrac{v}{c^2}j'}{\sqrt{1-\dfrac{v^2}{c^2}}}$$

in agreement with (70*a*, *b*) and (88*a*, *b*, *c*).

Thus we are led to a remarkable effect. Consider a long straight wire at rest and carrying a current. The wire is electrically neutral, for there are as many positive ions at rest as moving negative electrons (n particles per volume). These may have the velocity u.

We can now write the densities of charge and current in this way:

$$\text{electrons} \quad \rho_- = -ne \qquad j_- = -neu = \rho_- u$$

$$\rho_+ + \rho_- = 0;$$

$$\text{ions} \quad \rho_+ = +ne \qquad j_+ = 0.$$

An observer moving with the velocity v in the direction of the wire now finds it charged. For the charge densities of electrons and ions which he measures are the transformed expressions

$$\rho_-' = \frac{\rho_- - \frac{v}{c^2} j_-}{\sqrt{1-\frac{v^2}{c^2}}} = \frac{\rho_-}{\sqrt{1-\frac{v^2}{c^2}}}\left(1-\frac{vu}{c^2}\right),$$

$$\rho_+' = \frac{\rho_+}{\sqrt{1-\frac{v^2}{c^2}}};$$

adding these one obtains the total density

$$\rho_+' + \rho_-' = -\rho_- \frac{\frac{vu}{c^2}}{\sqrt{1-\frac{v^2}{c^2}}} = \frac{nevu}{c^2\sqrt{1-\frac{v^2}{c^2}}}.$$

We see that a positive charge results.

This can be illustrated in the x, ct-diagram (Fig. 131*b*). For simplicity we consider a linear model of the wire and assume the positive ions and the negative electrons equidistant (distance a) on it (Fig. 131*a*). At $t=0$ electrons and ions may be at the same points on the wire; the world lines of the ions at rest are lines parallel to the ct-axis and the world lines of the electrons are also parallel lines but inclined corresponding to the velocity u. When we consider the wire from a system x', ct', which moves with velocity v relative to x, ct, we see that the distance $\overline{BO}$ of the ions on the x'-axis is different from the distance $\overline{DO}$ of the electrons. In Fig. 131*b* $\overline{BO}$ is smaller than $\overline{DO}$. Hence the density of the ions is larger than that of the electrons, and the wire is positively charged.

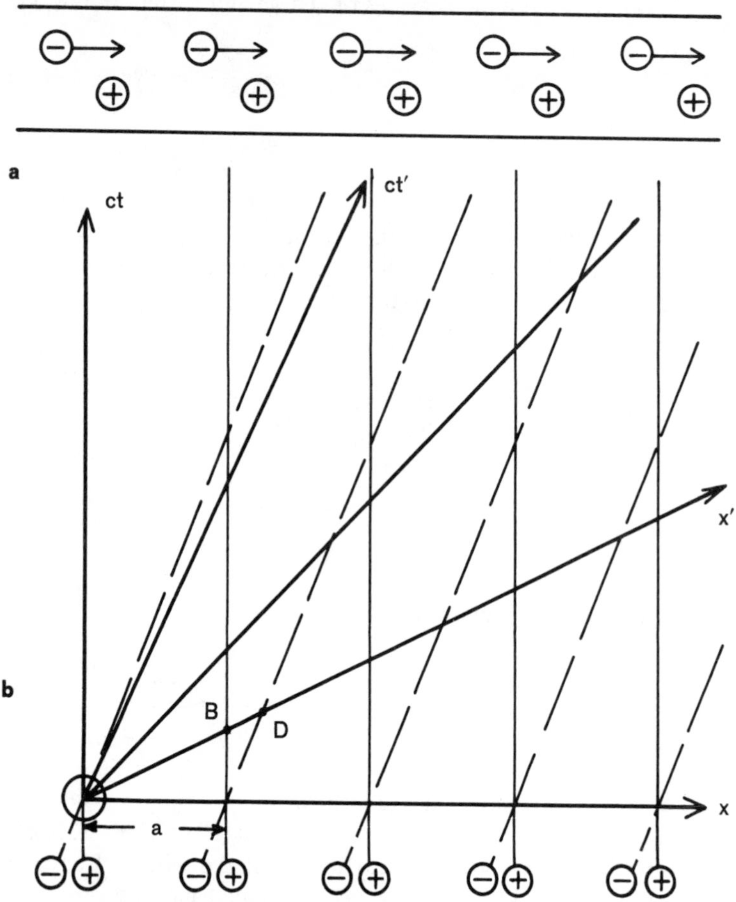

Fig. 131*a* *Resting ions* ⊕ *and moving electrons* ⊖ *in a wire.*
b *The world lines of the ions parallel to the* ct-*axis and the world lines of the moving electrons parallel, but inclined, at the same distance* a *from one another. If observed from an* x′, ct′-*system which moves with velocity* v *relative to the* x, ct-*system, the distances of ions* $\overline{OB}$ *are different from those of the electrons* $\overline{OD}$.

Let us consider the fields produced by the currents and charges in the two systems. We know that the electrically neutral wire with its current in the x, ct-system is surrounded only by a magnetic field H, while the positively charged wire in the x', ct'-system has in addition an electric field E. Therefore in the theory of relativity the concepts of electric and magnetic fields have no separate meaning; they must be unified into the concept of an electromagnetic field (E and H) whose components depend on the frame of reference. That means, for example, that if in one frame of reference S there is only a magnetic field, an observer in S' will find also an electric field or vice versa. Thus one arrives at an extremely simple qualitative explanation of the electromagnetic phenomena in moving matter discussed earlier (V, 11): In Figs. 103 and 105 a piece of matter moves in a magnetic field and therefore an observer moving with this piece observes an electric field. This electric field gives rise to an electric current in a conductor, whereas on an insulator surface charges are produced.

We now proceed to discuss the optical phenomena in moving matter.

10. Optics of Moving Bodies

Now that we have drawn the most important inferences from the modified mechanics, it is time to return to those problems which were the main source of Einstein's theory of relativity, namely, the optics of moving bodies. The fundamental laws of these phenomena are condensed into Maxwell's field equations. Lorentz had already recognized that these are invariant for empty space ($\epsilon=1$, $\mu=1$, $\sigma=0$) with respect to Lorentz transformations. The exact invariant field equations for moving bodies have been formulated by Minkowski (1907). They differ from the Lorentz formulae of the theory of electrons only in minor terms which cannot be tested by observation, but have in common with these the partial convection of the dielectric polarization, and hence account satisfactorily for all electromagnetic and optical phenomena involving moving bodies. We recall, in particular, the experiments of Röntgen, Eichenwald, and Wilson (V, 11, p. 194) (although we shall not discuss them in more detail since this would require elaborate mathematical calculation). But the optics of moving bodies may be treated in quite an elementary way, and we shall describe them here because they constitute one of the most illuminating applications of Einstein's theory.

According to the theory of relativity, there is no ether but only bodies moving relative to each other; hence it is self-evident that all optical phenomena which occur when the source of light, the substances traversed by radiation, and the observer are at rest in one inertial system, are the same for all inertial systems. Thus the Michelson-Morley experiment is also explained. The question now is whether the phenomena which occur when the source of light, the medium traversed by radiation, and the observer are in relative motion are correctly represented by the theory.

Let us imagine a light wave in a material body which is at rest in the system of reference S. Let the velocity of the light be $c_1 = \frac{c}{n}$ (n being the index of refraction), its vibration number be ν, and its direction relative to the system S be the x-direction. We inquire how these three characteristics of the wave are judged by an observer who is at rest in a system of reference S' moving with the velocity v parallel to the x-direction of the system S.

To answer this question we use the same method we applied earlier (IV, 7, p. 120), except that now we use Lorentz transformations as our basis for reasoning in place of Galileo transformations. We showed then that the wave number

$$\nu\left(t_1 - t_0 - \frac{x_1 - x_0}{c}\right)$$

is an invariant, for it denotes the number of waves which have reached the point x_0 up to the moment t_0 and have left the point x_1 after the moment t_1 (Fig. 69, p. 120). This invariance now holds, of course, for Lorentz transformations. Thus we have

$$\nu\left(t_1 - t_0 - \frac{x_1 - x_0}{c_1}\right) = \nu'\left(t_1' - t_0' - \frac{x_1' - x_0'}{c_1'}\right),$$

where ν, ν' and c_1, c_1' are the frequencies and velocities of the wave relative to the systems S and S'. If on the right we insert the expressions for x' and t' given by the Lorentz transformation (70*a*) (p. 236), we get

$$\nu\left(t_1-t_0-\frac{x_1-x_0}{c_1}\right) = \frac{\nu'}{\alpha}\left[t_1-t_0-\frac{v}{c^2}(x_1-x_0)-\frac{x_1-x_0-v(t_1-t_0)}{c_1'}\right],$$

where $\alpha=\sqrt{1-\beta^2}=\sqrt{1-\frac{v^2}{c^2}}$. We now first observe the whole wave train at the same moment, that is, we set $t_1-t_0=0$. Dividing by (x_1-x_0), we obtain

$$\frac{\nu}{c_1} = \frac{\nu'}{\alpha}\left(\frac{v}{c^2}+\frac{1}{c_1'}\right). \tag{95a}$$

Second, we observe the waves passing the same space point $x_1=x_0$. Dividing by (t_1-t_0), we get

$$\nu = \frac{\nu'}{\alpha}\left(1+\frac{v}{c_1'}\right). \tag{95b}$$

If we divide the second equation by the first, we find

$$c_1 = \frac{1+\frac{v}{c_1'}}{\frac{v}{c^2}+\frac{1}{c_1'}} = \frac{c_1'+v}{1+\frac{vc_1'}{c^2}}.$$

This agrees exactly with Einstein's addition theorem of velocities for longitudinal motion (formula (77*a*), p. 265) if we replace u_x by c_1 and u_x' by c_1'. The same rule which holds for calculating the velocities of material bodies relative to various systems of reference may also be applied to the velocity of light.

If, conversely, we solve for c_1' we get the *convection formula*

$$c_1' = \frac{c_1-v}{1-\frac{vc_1}{c^2}}.$$

If terms of order higher than the second in $\beta=\frac{v}{c}$ be neglected, this law becomes identical with Fresnel's convection formula (44) (p. 136). For, with this approximation, we may write

$$\frac{1}{1-\frac{vc_1}{c^2}} = \frac{1}{1-\frac{\beta}{n}} = 1+\frac{\beta}{n} = 1+\frac{v}{nc}.$$

Thus

$$c_1' = (c_1 - v)\left(1 + \frac{v}{nc}\right)$$
$$= c_1 - v + \frac{vc_1}{nc} - \frac{v^2}{nc}$$
$$= c_1\left(1 - \frac{v}{c_1} + \frac{v}{nc} - \frac{c}{nc_1}\frac{v^2}{c^2}\right),$$

and if we omit the last second-order term and set $\frac{c_1}{c} = \frac{1}{n}$, we get

$$c_1' = c_1 - v\left(1 - \frac{1}{n^2}\right).$$

This is precisely *Fresnel's convection formula.*

The formula (95*b*) represents Doppler's principle. This is usually applied to a vacuum, so that $c_1 = c$; then, as we know, it follows from the addition theorem of velocities (p. 266) that $c_1' = c$. And formula (95*b*) gives

$$\nu' = \nu \frac{\alpha}{1 + \frac{v}{c}} = \nu \frac{\sqrt{1-\beta^2}}{1+\beta}. \tag{95c}$$

But $1 - \beta^2 = (1-\beta)(1+\beta)$; hence we may write

$$\nu' = \nu \frac{\sqrt{(1-\beta)(1+\beta)}}{1+\beta} = \nu \sqrt{\frac{1-\beta}{1+\beta}}.$$

Thus the *rigorous formula for the Doppler effect* assumes the symmetrical form

$$\nu' \sqrt{1 + \frac{v}{c}} = \nu \sqrt{1 - \frac{v}{c}}, \tag{96}$$

which illustrates the equivalence of the systems of reference S and S'.

For small $\beta = \frac{v}{c}$ one obtains the ordinary formula (41) (p. 124) for the Doppler effect by neglecting β^2 in (95*c*):

$$\nu' \simeq \frac{\nu}{1+\beta} \simeq \nu(1-\beta).$$

One can derive formula (96) also by using the concept of light quanta with the energy $\epsilon=\frac{h\nu}{c}$ and the momentum $p=\frac{\epsilon}{c}\cdot$ If we substitute $\epsilon=h\nu$ and $\epsilon'=h\nu'$ in (91) the resulting formula is identical with (96).

For large velocities the relativistic formula (96) differs from the classical one (41). This becomes more apparent if one considers the case in which the direction of the propagation of light and of the relative velocity v do not coincide, particularly when they are perpendicular. According to the classical theory there should then be no Doppler effect at all, while in the relativistic theory there is one. Therefore one can speak of a new relativistic effect, which is often called *transversal Doppler effect*. It can be treated in the same way as the ordinary longitudinal effect.

The relative velocity of the system S and S' is assumed, as before, to be in the direction of the coinciding x- or x'-axes, but the propagation of the light is perpendicular to this direction, say parallel to the y'-axis. It is, however, not permissible to assume that the normal of the light waves as seen from S is parallel to the y-axis.

While the distance from the beginning of the train of light at t_0' up to the end at t_1' as observed in S' is $y_1'-y_0'$, in S it will not simply be y_1-y_0 but will also depend on x_1-x_0, say $a(x_1-x_0)+b(y_1-y_0)$. The invariance of phase then reads

$$\nu\left(t_1-t_0-\frac{a(x_1-x_0)+b(y_1-y_0)}{c_1}\right) = \nu'\left(t_1'-t_0'-\frac{y_1'-y_0'}{c_1'}\right).$$

We have now to apply the Lorentz transformation (70*a*) or (70*b*). If we take (70*b*) and put therein $x_1'=x_0'$, $y_1'=y_0'$, expressing the fact that the experiment is performed at a place fixed in S', the calculation becomes involved, because we need then to know the quantity a. However, if we use (70*a*) and put $x_1=x_0$, $y_1=y_0$, we obtain easily

$$\nu = \frac{\nu'}{\alpha}.$$

An observer moving relative to S with the velocity v, looking perpendicularly at a light source in S which emits the frequency ν, measures the modified frequency

$$\nu' = \alpha\nu = \sqrt{1-\frac{v^2}{c^2}}\,\nu$$

which is smaller than ν. This is the transversal Doppler effect. As the derivation shows, it is closely related to the time dilation $t_1' - t_0' = \frac{1}{\alpha}(t_1 - t_0)$; it means simply that the number of beats of a clock is the same for all observers.

Now the transversal Doppler effect has been observed in the laboratory by using canal rays (IV, 8, p. 127) of a well-defined velocity and direction (Ives and Stilwell, 1938; Otting, 1939). The difficulty of these measurements consists of the contribution which the ordinary longitudinal Doppler effect would give if the direction of observation should deviate a little from the perpendicular. This is overcome by the remark that the longitudinal Doppler displacement is opposite and equal for two light rays emitted by the canal ray in opposite directions; if one observes both light rays simultaneously and takes the mean of the results, the longitudinal effect is eliminated. In this way the transversal effect was confirmed and thus also the time dilation proved in a rather direct way.*

We proceed now to discuss the aberration of light. This can be done by the method just used: by determining the effect of the Lorentz transformation on the direction of the ray (i.e., on the coefficients a, b above). But as this method leads to somewhat complicated calculations, we prefer another one, namely, the application to light quanta of the addition theorem of velocities.

A beam of light is propagated in the y-direction of the system S' so that $u_x' = 0$, $u_y' = c$. The transformation formulae (77a, b) yield as velocities in S

$$u_x = v \qquad \text{and} \qquad u_y = c\sqrt{1 - \frac{v^2}{c^2}}.$$

Therefore in S the direction of the light is not perpendicular to the x-direction, but inclined. First we can confirm that the velocity of light in S is c again, for one has $\sqrt{u_x^2 + u_y^2} = c$. The ratio of the components $\frac{u_x}{u_y}$ corresponds to the elementary definition of the

*The transversal Doppler effect has recently been accurately measured with the help of the Mössbauer effect (see p. 360). On a circular disk the emitter S of Mössbauer radiation was fixed in the center, the receiver R at the distance r from the center. The disk was brought into rapid rotation (period T), so that R had a velocity $u = 2\pi r/\tau$ relative to S in the direction normal to $SR = r$. A Doppler displacement of the resonance frequency of R relative to that of S was found in exact agreement with the formula of the text.

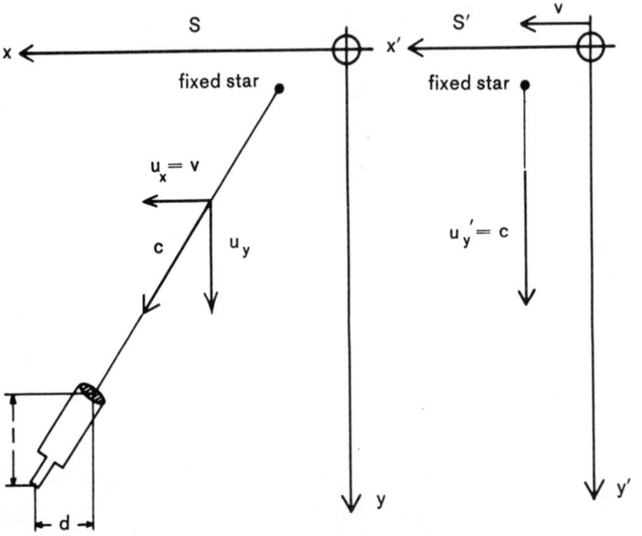

Fig. 132 *The aberration in the theory of relativity. To have positive components of velocity* (u_x, u_y, *etc.*) *we use a turned coordinate system. In* S′, *the system which moves with velocity* v *relative to the earth and in which the fixed star rests, the light travels in the* y-*direction. In* S, *in which the earth rests, we have two components of velocity,* u_x *and* u_y.

aberration constant $\frac{d}{l}$ (Fig. 132). Here l is the length of the telescope and d is the displacement of the telescope during the time light needs to traverse the tube (IV, 3). Thus we get

$$\frac{d}{l} = \frac{u_x}{u_y} = \frac{\frac{v}{c}}{\sqrt{1-\frac{v^2}{c^2}}}.$$

Another way to derive the aberration formula is to use the concept of light quanta and the transformation formulae (88*a*) and (88*b*) for the components of momentum and energy. We write these, with the usual abbreviations, $\alpha = \sqrt{1-\beta^2}$, $\beta = \frac{v}{c}$:

$$\alpha p_x = p_x' + \beta\frac{E'}{c}, \qquad p_y = p_y', \qquad \alpha E = E' + v p_x'.$$

Assuming the light quantum moving in the y-direction of S', its components in S' are $p_x' = 0$, $p_y' = p' = \frac{E'}{c}$, hence in S:

$$\alpha p_x = \beta \frac{E'}{c} = \beta p_y' = \beta p_y \qquad \text{and} \qquad \alpha E = E'.$$

Therefore, $\frac{p_x}{p_y} = \frac{\beta}{\alpha}$, which agrees with the expression $\frac{d}{l}$ given above. According to Planck the energy of a light quantum is $E = h\nu$ in S, $E' = h\nu'$ in S'. If S' is taken to be the system S_0 in which the light-emitting atoms are at rest, one has $\nu' = \nu_0$, and $\alpha E = E'$ is equivalent to $\alpha\nu = \nu_0$. Thus we have another derivation of the transversal Doppler effect.

The elementary aberration formula is obtained from the exact one by neglecting β^2:

$$\frac{d}{l} = \beta = \frac{v}{c}.$$

This result is particularly remarkable because all the ether theories have considerable difficulty in explaining aberration. From the Galileo transformation one obtains no deflection at all of the wave plane and the wave direction (IV, 10, p. 141), and to explain aberration one has to introduce the concept "ray," which in moving systems need not coincide with the direction of propagation. In Einstein's theory this difficulty disappears. In every inertial system S the direction of the ray (that is, the direction along which the energy is transported) coincides with the perpendicular on the wave planes, and the aberration results in the same way as the Doppler

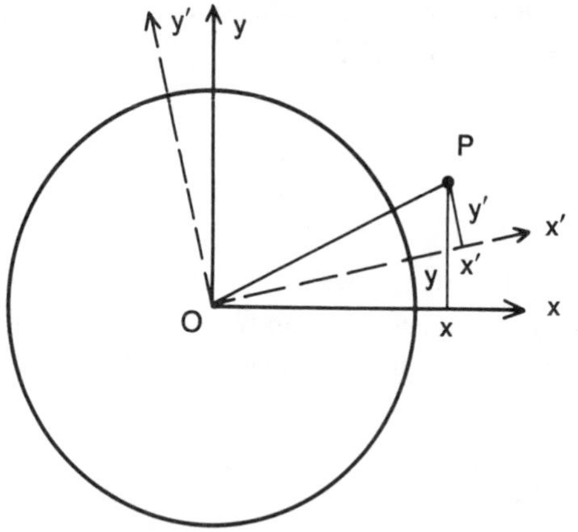

Fig. 133 *The fundamental invariant in Euclidean geometry.*

effect and Fresnel's convection coefficient, from the concept of a wave with the help of the Lorentz transformation.

This method of deriving the fundamental laws of the optics of moving bodies shows very strikingly that Einstein's theory of relativity is superior to all other theories.

11. Minkowski's Absolute World

The essence of the new kinematics consists of the inseparability of space and time. The world is a four-dimensional manifold, its element the world point. Space and time are forms of arrangement of world points, and this arrangement is, to a certain extent, arbitrary. Minkowski has expressed this idea thus: "From now onwards space and time are to degenerate to mere shadows and only a sort of union of both retain independent existence." He has worked out this suggestion by developing kinematics as four-dimensional geometry. We have made use of his method of description throughout, omitting the *y*- and *z*-axis only for the sake of simplicity and working in the *xt*-plane. If we cast a glance at the geometry in the *xt*-plane from the mathematical point of view, we see that we are not dealing with ordinary Euclidean geometry. For in this all straight lines that radiate from the origin are equivalent, the unit of length on them is the same, and the calibration curve is thus a circle (Fig. 133). But in the *xt*-plane the spacelike and timelike straight lines are not equivalent. There is a different unit of length on each, and the calibration curve consists of the hyperbolae

$$F = x^2 - c^2t^2 = \pm 1.$$

In Euclidean geometry we can construct an infinite number of rectangular coordinate systems with the same origin O, each of which can be transformed into another by a rotation. In the *xt*-plane there are likewise an infinite number of equivalent coordinate systems, for which the one axis can be chosen at will within a certain angular region.

In Euclidean geometry the distance s of a point P with the coordinates x, y from the origin is an invariant with respect to rotations of the coordinate system (see III, 7, formula (28), p. 73). By Pythagoras' theorem we have (Fig. 125) in the *xy*-system

$$s^2 = x^2 + y^2,$$

and in any other system x', y' we likewise have $s^2 = x'^2 + y'^2$. The calibration curve, the circle of radius unity, is represented by $s=1$. Hence s (or s^2) can be regarded as the fundamental invariant of Euclidean geometry.

In the xt-plane the fundamental invariant is

$$F = x^2 - c^2t^2,$$

and the calibration curve is

$$F = \pm 1.$$

Now Minkowski observed that a parallel presents itself here which throws light on the mathematical structure of the four-dimensional world (or, in our case, the xt-plane). For if we set $-c^2t^2 = u^2$, we get

$$F = x^2 + u^2 = s^2,$$

and this may be regarded as the fundamental invariant s^2 of a Euclidean geometry having the rectangular coordinates x, u.

It is true that in the domain of ordinary number one cannot extract the square root of the negative quantity $-c^2t^2$; hence u has no elementary meaning. But mathematicians have long been accustomed to overcoming such difficulties. The "imaginary" quantity $\sqrt{-1} \equiv i$ has been firmly established in mathematics since the time of Gauss. We cannot here enter into the question of how the doctrines of imaginary numbers can be rigorously established. These numbers are essentially no more "imaginary" than a fraction such as $\frac{2}{3}$, for numbers with which we number things or count properly comprise only the natural integers 1, 2, 3, 4. . . . The number 2 is not divisible by 3, so that $\frac{2}{3}$ is an operation that can be carried out just as little as $\sqrt{-1}$. Fractions such as $\frac{2}{3}$ signify an extension of the natural concept of numbers; however, they have become familiar through education and custom, and excite no feeling of strangeness. The introduction of imaginary numbers is a similar extension: all formulae that contain imaginary numbers have just as definite a meaning as those formed from ordinary "real" numbers, and the inferences drawn from them are just as convincing.

If we here use the symbol $\sqrt{-1} \equiv i$, we may write

$$u = ict.$$

The non-Euclidean geometry of the xt-plane is thus formally identical with Euclidean geometry in the xu-plane if imaginary values of u are made to correspond to real times t.

This theorem is of considerable advantage for the mathematical treatment of the theory of relativity. For in numerous operations and calculations the problem has nothing to do with the reality of the quantities considered, but with the algebraic relations that exist between them, and these hold just as well for imaginary numbers as for real numbers. Hence, we can apply the laws known from Euclidean geometry to the four-dimensional world. Minkowski replaces

$$x, y, z, ict$$

by

$$x, y, z, u$$

and then operates with these four coordinates in a fully symmetrical way. The fundamental invariant then clearly becomes

$$F = s^2 = x^2+y^2+z^2+u^2.$$

The particular feature of the time variable (that its square appears with the negative sign in F) thus vanishes from all formulae, and this, to a considerable degree, facilitates the calculations and allows us to survey them readily as a whole. In the final result we again replace u by ict, and then only those equations retain a physical meaning which are formed exclusively from real numbers.

In the xt-plane t is clearly not interchangeable with x. The light lines X and Y are the insuperable barriers between the timelike and the spacelike world lines. Thus Minkowski's transformation $u=ict$ is to be valued only as a mathematical artifice which illuminates certain formal analogies between space coordinates and time coordinates without, however, allowing them to be interchanged. We may call

$$s = \sqrt{F} = \sqrt{x^2+y^2+z^2+u^2} = \sqrt{x^2+y^2+z^2-c^2t^2}$$

the "four-dimensional distance," but we must remember that the expression is used in a symbolic way. According to our previous discussion of the invariant F, the real meaning of the quantity s is easy to understand. Let us confine our attention to the xt-plane (or xu-plane); then

$$s = \sqrt{F} = \sqrt{x^2+u^2} = \sqrt{x^2-c^2t^2}.$$

Now for every spacelike world line, F is positive and thus s, as the square root of a positive number, is a real quantity. We can then make the world point x, t simultaneous with the origin by choosing a suitable system of reference S. For $t=0$, we have $s=\sqrt{x^2}=x$ as the spatial distance of the world point from the origin.

For every timelike world line, F is negative and hence s is imaginary. There exists then a coordinate system in which $x=0$ and therefore $s=\sqrt{-c^2t^2}=ict$. Thus, in any case, s has a simple meaning and is to be regarded as a measurable quantity.

Our account of Einstein's special theory of relativity may be condensed into the following statements:

Not only the laws of mechanics but those of all physical events—in particular, of electromagnetic phenomena—are completely identical in an infinite number of systems of reference which are moving with constant velocity relative to each other and which are called inertial systems. In any of these systems lengths and times measured with the same physical rods and clocks appear different in any other system, but the results of measurements are connected with each other by Lorentz transformations.

Systems of reference which move with acceleration relative to each other are, no more than in ordinary mechanics, identical with inertial systems. If we refer physical laws to such accelerated systems, they become different. In mechanics there appear centrifugal forces and in electrodynamics there are analogous effects, the study of which would take us too far. Thus Einstein's special theory of relativity does not do away with Newton's absolute space in the restricted sense in which we used this term (III, 6, p. 70). In a certain sense it puts the whole of physics, including electrodynamics, into the same state as that in which mechanics has been since the time of Newton. The fundamental questions of absolute space which troubled us then are not yet solved.

We shall now see how Einstein has overcome these difficulties.

CHAPTER VII

EINSTEIN'S GENERAL THEORY OF RELATIVITY

1. Relativity in the Case of Arbitrary Motions

In dealing with classical mechanics we discussed in detail the reasons that led Newton to conceive the idea of absolute space and absolute time. At the same time we emphasized the objections that can be raised against these abstractions from the point of view of the theory of knowledge.

Newton based his assumption of absolute space on the existence of inertial resistances and centrifugal forces. It is clear that these cannot depend on interactions between bodies, since they occur in the same way, independent of the local distribution of masses, in all the universe as far as observation can reach. Hence Newton concludes that they depend on absolute accelerations. In this way absolute space is introduced as the fictitious cause of physical phenomena.

The unsatisfactory features of this theory may be recognized from the following example:

Suppose two bodies S_1 and S_2 of the same deformable (fluid) material and the same size to be present in astronomic space and at such a distance from each other that ordinary gravitational effects of the one on the other are inappreciably small (Fig. 134). Each of these bodies is to be in equilibrium under the action of the gravitation of its parts on each other and the remaining physical forces so that no relative motions of its parts with respect to each other occur. But the two bodies are to execute a relative motion of rotation with constant velocity about the line connecting their centers. This signifies that an observer on the one body S_1 notes a uniform rotation of the other body S_2 with regard to his own system of reference, and vice versa. Now suppose each of these observers determines the

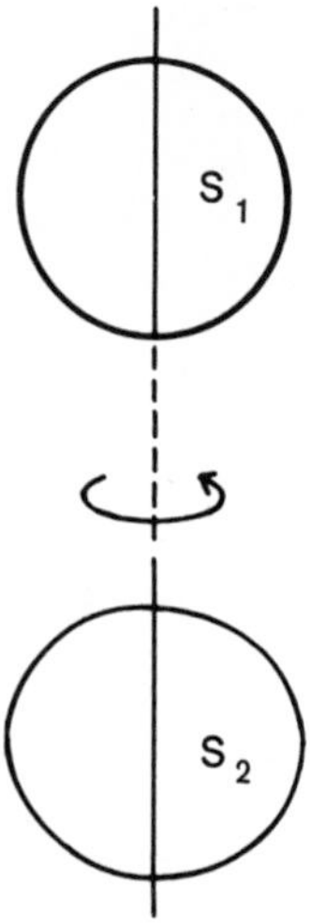

Fig. 134 *Two originally spherical liquid bodies* S_1 *and* S_2 *in relative rotation about a common axis.*

shape of the body on which he stands and that it is found S_1 is a sphere and S_2 is a flattened ellipsoid of rotation.

Newtonian mechanics would infer from the different shapes of the two bodies that S_1 is at rest in absolute space but that S_2 executes an absolute rotation. The flattening of S_2 is then explicable by centrifugal forces.

This example illustrates clearly how absolute space is introduced as the (fictitious) cause. For S_1 cannot be responsible for the flattening of S_2, since the two bodies are in exactly the same condition relative to each other and therefore cannot deform each other differently.

To take space as a cause does not satisfy the requirements of logic with regard to causality. For as we have no other indication of its existence than centrifugal forces, we are supporting the hypothesis of absolute space only by the fact it was introduced to explain. Sound epistemological criticism refuses to accept such made-to-order hypotheses. They are too facile and are at odds with the aim of scientific research, which is to determine criteria for distinguishing its results from dreams of fancy. If the sheet of paper on which I have just written suddenly flies up from the table, I should be free to make the hypothesis that a ghost, say the specter of Newton, had spirited it away. But common sense leads me instead to think of a draft coming from the open window because someone is entering by the door. Even if I do not feel the draft myself, this hypothesis is reasonable because it brings the phenomenon which is to be explained into relationship with other observable events. This critical choice of admissible causes distinguishes the attitude towards the world based on reason which physical research aspires to, from mysticism, spiritualism and similar manifestions of unbridled fancy.

Indeed, the concept of absolute space is almost spiritualistic in character. If we ask, "What is the cause of centrifugal forces?" the answer is: "Absolute space." If, however, we ask what absolute space is and in what other way it expresses itself, no one can furnish an answer other than that absolute space is the cause of centrifugal forces but has no further properties. This consideration shows that space as the cause of physical occurrences must be eliminated from the world picture.

It is, perhaps, not superfluous to mention that this opinion of absolute space is in no way affected by the existence of electromagnetic phenomena. In rotating coordinate systems effects are observed which are analogous to the centrifugal forces of mechanics. But, of course, this does not give new and independent proofs of the existence of absolute space, for, as we know, the theorem of the inertia of energy amalgamates mechanics and electrodynamics to a unit. It is merely more convenient for us to operate with the concepts of mechanics alone.

Let us now again consider the two bodies S_1 and S_2. If space is not accepted as the cause of their different shapes we must look for other and more convincing causes.

Let it be supposed that there are no other material bodies outside the two bodies S_1 and S_2. The different shapes of S_1 and S_2 would then be really inexplicable. But is this difference in shapes, then, an empirical fact? There is no doubt that it is *not*. We have never had the experience of observing two bodies that are poised alone in the universe. The assumption that two real bodies S_1 and S_2 would behave differently under these circumstances is supported by *no evidence* at all. Rather, we must demand of a satisfactory mechanics that it exclude this assumption. But if we observe in the case of two real bodies S_1 and S_2 the different behavior above described (we know that planets are more or less flattened), we can take as the cause of this only *distant masses*. In the real world such masses are actually present, namely, the countless legion of stars. Whatever stellar body we select, it is surrounded by innumerable others which are enormously distant from it and which move so slowly relative to one another that, as a whole, they exert the effect of a solid mass containing a cavity in which the body under consideration is situated.

The idea that the totality of distant masses must be the cause of the centrifugal forces was first expressed by the philosopher-physicist Ernst Mach, whose writing had a profound influence on Einstein. There is no experience to contradict it, for the system of reference of astronomy with respect to which the rotations of heavenly bodies are determined has been chosen so that it is at rest relative to the stellar system as a whole, or, more accurately, so that the apparent motions of the fixed stars relative to the system of reference are quite irregular and have no favored direction. The flattening of a planet is the greater, the greater its velocity of rotation with respect to this system of reference attached to the distant masses.

Accordingly we shall demand that the laws of mechanics—and, indeed, of physics in general—involve only the relative positions and motions of bodies. No system of reference may be favored a priori as with the inertial systems of Newtonian mechanics and Einstein's special theory of relativity; otherwise absolute accelerations with respect to these favored systems of reference, and not only relative motions of bodies, would enter into physical laws.

We thus arrive at the postulate that the true laws of physics must hold in exactly the same way in systems of reference that are moving arbitrarily. This denotes a considerable extension of the principle of relativity.

2. The Principle of Equivalence

Fulfillment of this postulate requires an entirely new formulation of the law of inertia, since this is what gives inertial systems their favored position. The inertia of a body is to be regarded no longer as an effect of absolute space but rather as one due to other bodies.

Now we know of only one kind of interaction between all material bodies, namely, gravitation. Further, we know that experiment has exhibited a remarkable relationship between gravitation and inertia, which is expressed in the law of the equality of gravitational and inertial mass (II, 12, p. 41–2). Thus the two phenomena of inertia and attraction which are so different in Newton's formulation must have a common root.

This is the great discovery of Einstein which has transformed the general principle of relativity from an epistemological postulate into a law of exact science.

We may characterize the object of the following investigation in this way: In ordinary mechanics the motion of a heavy body (on which no electromagnetic or other forces act) is determined by two causes: (1) its inertia, tending to prevent accelerations with respect to absolute space; (2) the gravitation of the remaining masses. Now a formulation of the law of motion is to be found in which inertia and gravitation amalgamate to a concept of higher order in such a way that motion is determined only by the distribution of the remaining masses in the universe. But before we establish the new law we must follow a somewhat lengthy road in order to overcome certain conceptual difficulties.

Earlier we discussed the law of the equality of gravitational and inertial mass in detail. For events on the earth it states that all bodies fall equally quickly; for motions of the heavenly bodies it says that acceleration is independent of the mass of the moving body. We have already mentioned that according to measurements of Eötvös this law is valid to an extraordinary degree of accuracy, but that, in spite of this, it is not reckoned among the fundamental laws in classical mechanics but rather is accepted, so to speak, as an accidental gift of nature.

This is now to become different. This law plays a fundamental part not only in mechanics but, indeed, in the whole of physics. We must therefore illustrate it so that its essential content emerges quite clearly. We advise the reader to make the following simple experiment: Let him take two light objects of different heaviness, say a coin and a piece of india rubber, and place them on the palm of his hand. He then experiences the weights of the two bodies as pressures on his hand and finds them different. Now if he moves his hand rapidly downwards, he experiences a diminution in the pressure of both bodies. If this motion is continued more and more rapidly, a moment will finally come when the bodies will release themselves from his palm and lag behind in the motion. This will clearly occur as soon as the hand is drawn down more rapidly than the bodies can fall freely. Now, since they fall equally quickly in spite of their different weights, they always remain together at the same height even when they are no longer in contact with the hand.

Let us imagine, if you will, little imps living on the surface of the hand who know nothing of the outer world. How would they judge this whole process? It is easy to imagine oneself in the position of such little observers moving with the hand while we make the experiment and pay attention to the changing pressures and motions of the bodies with respect to the hand. When the hand is at rest the imps will establish that the two bodies have different weights. When the hand sinks they will note a decrease of weight of the bodies. They will look for a cause and will observe that their point of vantage, the hand, sinks relative to the surrounding bodies, the walls of the room. But we may also imprison the imps and the two test bodies in a closed box and pull this box downwards with the hand. The observers in the box then observe nothing which will allow them to establish the motion of the box. They can simply note the fact that the weight of all bodies in the box decreases at the same rate. If the hand is now moved so rapidly that the objects cannot follow but fall freely, the observers in the box will notice to their astonishment that the objects which were just before quite heavy now fly upwards. They acquire a negative weight, or rather, gravitation no longer acts downwards but upwards. Moreover, in spite of their different weights the two bodies fall equally fast upwards. The people in the box can account for these observations in two ways: either they think that the gravitational field continues to act unaltered but that the box is accelerated in the direction of the field, or they assume that the masses which previously exerted an attractive force below the box have disappeared, and that in their place new masses have appeared above the box reversing the direction of action of gravitation. Is there then any means of distinguishing by experiments within the box between these two possibilities?

We are bound to answer that physics knows of *no* such means.

Actually, the effect of gravitation can in no way be distinguished from the effect of acceleration; each is fully equivalent to the other. This is essentially because of the circumstance that all bodies fall equally fast. If this were not the case we could at once distinguish whether an accelerated motion of bodies of different weight is produced by the attraction of outside masses or is an illusion arising from the acceleration of the observer's point of support. In the first case the bodies of different weight move with different velocities, whereas in the latter case the relative acceleration of all freely moving bodies with respect to the observer is equally great and they fall equally fast in spite of their different weights.

This *principle of equivalence* of Einstein is an example of a kind of theorem which we have particularly emphasized in this book, one that asserts that a certain physical statement cannot be established or that two concepts cannot be distinguished. Physics refuses to accept such concepts and statements and replaces them by new ones. Only ascertainable facts have physical reality.

Classical mechanics distinguishes between the motion of a body that is left to itself subject to no forces (inertial motion) and the motion of a body under the action of gravitation. The former is rectilinear and uniform in an inertial system; the latter occurs in curvilinear paths and is nonuniform. According to the principle of equivalence this distinction must be dropped. For by merely passing over to an accelerated system of reference we can transform the rectilinear uniform motion of inertia into a curved, accelerated motion which cannot be distinguished from one produced by gravitation. The converse holds too, at least for limited portions of the motion, as will be explained more fully later. From now on we shall call every motion of a body on which no forces of an electrical, magnetic, or other origin act, but which is only under the influence of gravitating masses, an inertial motion. This term is thus to have a more general significance than earlier. The theorem that the inertial motion relative to the inertial system is uniform and rectilinear—the ordinary law of inertia—now comes to an end. Our problem now is to state the law of inertial motion in the generalized sense.

The solution of this problem releases us from absolute space and at the same time furnishes us with a *theory of gravitation* which thereby becomes linked up much more intimately with the principles of mechanics than in Newton's theory.

We shall supplement these remarks by adding a few calculations. We have shown earlier (III, 8, p. 76) that the equations of motion of mechanics, referred to a system S which has the constant acceleration k with respect to the inertial systems, may be written in the form

$$mb = K',$$

where K' denotes the sum of the true force K and the inertial force $-mk$, i.e.,

$$K' = K - mk.$$

Now, if K is the force of gravitation, then $K = mg$, thus

$$K' = m(g-k).$$

By choosing the acceleration k of the system of reference S appropriately we can make the difference $g-k$ assume any arbitrary positive or negative value, or zero. If, in analogy with electrodynamics, we call the force on unit mass the "*intensity of field*" of gravitation, and the space in which it acts the *gravitational field*, we may say that by choosing the accelerated system appropriately we can produce a constant gravitational field, reduce one that is given, or annul, intensify, or reverse it.

It is clear that within a sufficiently small portion of space and during a small interval of time any arbitrary gravitational field may be regarded as approximately constant. Hence we can always find an accelerated system of reference relative to which there is no gravitational field in the limited space-time region.

Is it then possible to eliminate every gravitational field in its whole extent and for all times by merely choosing an appropriate system of reference? May gravitation, in other words, be regarded as "apparent"? This is clearly not the case. The field of the earth, for example, cannot be fully eliminated. For it is directed towards the center, and the acceleration would also have to point towards the center. But this is obviously possible only for a restricted time even if we were to admit (and this we shall have to do) that the system of reference is not rigid but contracts with acceleration about the center. By rotating the system of reference about an axis we get an inertial force directed away from this axis (III, 9, p. 80, formula (31)), namely, the centrifugal force

$$mk = m\frac{4\pi^2 r}{T^2}.$$

This compensates the gravitational field of the earth for a given period T only at a certain distance r—for instance, for the radius of the moon's orbit (supposed circular) if T is the time of the moon's revolution.

Thus there are "true" gravitational fields, yet the sense of this word in the general theory of relativity is different from that in classical mechanics. For we can always annul an arbitrarily small part of the field by choosing the system of reference appropriately. We shall define the concept of gravitational fields more accurately later.

There are, of course, certain gravitational fields which can be fully eliminated by a suitable choice of the system of reference. To find such we need only start from a system of reference in which a part of space is fieldless and then introduce a system of reference which is accelerated in some way. Relative to this there is then a gravitational field, which vanishes as soon as we return to the original system of reference. The centrifugal field $k=\frac{4\pi^2 r}{T^2}$ is of this kind. The question as to the conditions under which a gravitational field can be made to vanish in its whole extent can be answered, of course, only by the finished theory.

3. The Failure of Euclidean Geometry

But before we proceed we must overcome a difficulty which calls for a considerable effort.

We are accustomed to represent motions in the Minkowski world as world lines. The framework of this four-dimensional geometry was furnished by the world lines of light rays or the orbits of inertial masses moving under no forces. In the old theory these world lines were straight with respect to the inertial systems. But if we take the point of view of the general theory of relativity, accelerated systems are equivalent and in them the world lines that were previously straight are now curved (III, 1, p. 55, Fig. 32, *a–f*). In place of these, other world lines become straight. Moreover, this change is also true of the orbits in space. The concepts "straight" and "curved" become relativized, in so far as they refer to the orbits of the light rays and of freely moving bodies.

Through this the whole structure of Euclidean geometry is made to totter. For it rests essentially (see III, 1, p. 54) on the classical law of inertia, which determines straight lines.

It might now be thought that this difficulty could be surmounted by using only rigid measuring rods to define such geometrical elements as straight line, plane, and so forth. But not even that is possible, as Einstein shows in the following way:

We start from a space-time region in which no gravitational field exists during a certain time relative to an appropriately selected system of reference S. Next we consider a body which rotates in this region with a constant velocity of rotation, say a plane circular disk (Fig. 135) rotating on its axis at right angles to its own plane. We introduce a system of reference S' which is rigidly fixed to this disk. A gravitational field directed outwards then exists in S', and it is given by the centrifugal acceleration

$$k = \frac{4\pi^2 r}{T^2}.$$

Now, an observer situated on S' wishes to measure the disk. To do so, he uses a rod of definite length as his unit, which must thus be at rest relative to S'. An observer in the system of reference S uses exactly the same rod for his unit of length, and in this process it must, of course, be at rest relative to S.

We shall now have to assume that the results of the special principle of relativity hold so long as we restrict ourselves to portions of space and time in which the motion can be regarded as uniform. To make this possible, we assume that the unit rod is small compared with the radius of the disk.

If the observer in S' applies his rod in the direction of a radius of the disk, the observer in S will notice that the length of the moving rod relative to S remains unaltered and equal to, say, 1 cm. For the motion of the rod is perpendicular to the direction of its length. If

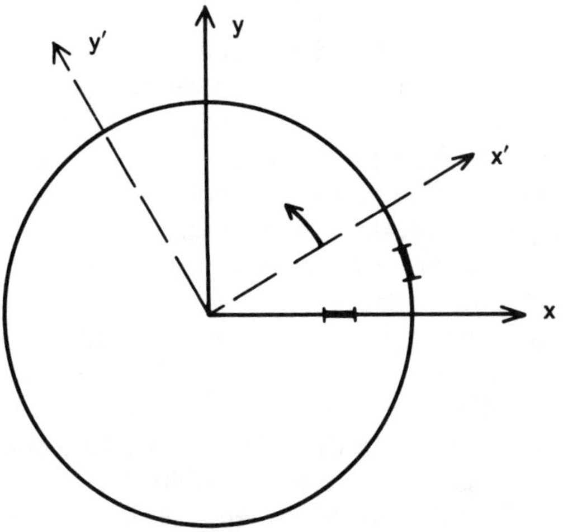

Fig. 135 *A circular disk rotates relative to the system* S (x, y)*; connected with the disk is the system* S′(x′, y′). *Small measuring rods, one radial, one tangential, are indicated.*

the observer in S' applies the rod to the periphery of the disk, then, by the special theory of relativity, it will appear shortened to the observer in S. If it be assumed that 100 small rods have to be applied end to end in order to reach from one end of the diameter to the other, the observer in S would require $\pi = 3.14 \ldots \times 100$, or, about 314 rods at rest relative to S to measure out the periphery; but the observer in S' would not find this number of rods sufficient. For the rods that are at rest in S' appear shortened as seen from S, and it requires more than 314 of them to go completely around the periphery.

Accordingly the observer in S' would assert that the ratio of the circumference of the circle to the diameter is not 3.14 . . . , but more. This ratio increases with increasing radius r, for the gravitational field is proportional to r. Thus we have a contradiction to Euclidean geometry.

A corresponding result holds for the measurements of time. We imagine synchronized clocks at rest in the system S and other clocks of the same construction on a radius of the rotating disk at rest relative to S'. Then a comparison of the clocks on the disk with the resting ones shows that the clocks on the disk are going slower than those in S and that the difference increases with the distance from the center. Only the clock at the center of the disk is synchronous with the clocks in S since it has no velocity relative to S.

But we see further that the clocks on the disk cannot even be synchronized, since they run more slowly as compared with those in S, the larger the distance from the center. It is thus impossible to arrive at a reasonable definition of time with the help of clocks which are at rest relative to the system of reference if this system of reference is rotating, i.e., if it is being accelerated, or, what is the same thing according to the principle of equivalence, if a gravitational field exists in it. In a gravitational field a rod is longer or shorter or a clock goes more quickly or more slowly according to the position at which the measuring apparatus is situated.

This implies that the foundation of the space-time world, on which rest all the arguments we have made so far, collapses. We are again compelled to generalize the concepts of space and time, but now in a much more radical way far exceeding the previous efforts in range.

It is clearly meaningless to define coordinates and time x, y, z, t in the ordinary way. For then the fundamental geometrical concepts—straight line, plane, circle, and so forth—are regarded as immediately given, and the validity of Euclidean geometry in space or of Minkowski's generalization to the space-time world is assumed.

Hence the problem arises of describing the four-dimensional world and its laws without basing it on a definite geometry, a priori.

It seems now as if the ground beneath us is giving way. Everything is tottering, straight is curved, and curved is straight. But the difficulty of this undertaking did not intimidate Einstein. Mathematicians had already accomplished important preparatory work. Gauss (1827) had sketched out the theory of curved surfaces in the form of a general two-dimensional geometry, and Riemann (1854) had extended this doctrine to continuous manifolds of any number of dimensions. The theory was further developed by Christoffel, Ricci and Levi-Cività, among others. We cannot here show how these mathematical instruments are applied, although a deeper understanding of the general principle of relativity is impossible without them. The reader must not, therefore, expect complete elucidation of Einstein's ideas from the following discussion. He will find only pictures and analogies, which are always poor substitutes for exact concepts. But if these indications stimulate the reader to further study, their purpose will have been fulfilled.

4. Geometry on Curved Surfaces

The problem of formulating geometry without the framework of straight lines and their Euclidean network of axioms and theorems being given a priori is by no means so strange as may appear at first sight. Let us suppose that a surveyor has to measure out a hilly piece of land covered by a dense wood and then sketch a map of it. From each point he can see only a limited part of the surroundings. Theodolites are useless to him; he has essentially to resort to the measuring chain. This enables him to measure out small triangles or quadrangles, whose corners are fixed by thrusting poles into the ground; by linking such *directly* measurable figures with one another he can gradually advance to more distant parts of the wood which are not directly visible.

Expressed abstractly, the surveyor may apply the methods of ordinary Euclidean geometry to small regions. But these methods cannot be applied to the piece of land as a whole. It can be investigated geometrically only step by step, by proceeding from one place to the next. Furthermore, Euclidean geometry is not strictly valid in hilly territory; there are *no* straight lines on such a surface at all. Short pieces of the measuring chain may be regarded as straight, but there is no straight connecting line along the ground from valley to valley and from hill to hill. Euclidean geometry thus, in a certain sense, holds only in small or infinitesimal regions, while in larger areas a more general doctrine of space, or rather of surface, holds.

If the surveyor wishes to proceed systematically he will first cover the ground in the wood with a network of lines which are marked by poles or specified trees. He requires two families of lines which intersect (Fig. 136). The lines will be chosen as smoothly and continuously curved as possible, and in each family they will bear consecutive numbers. We take x as the symbol for any member of the one family, and y as that for any member of the other.

Each point of intersection has, then, two numbers x, y, say $x=3$, $y=5$. Intervening points may be characterized by fractional values

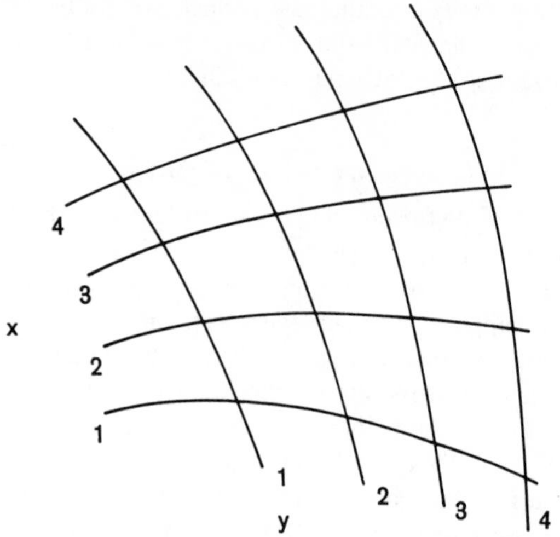

Fig. 136 *Two intersecting systems of curves serving as curvilinear coordinates on a surface.*

of x and y. This method of fixing the points of a curved surface was first applied by Gauss; x and y are, therefore, called *Gaussian coordinates.*

The essential feature involved is that the numbers x and y denote neither lengths nor angles nor other measurable geometrical quantities, but are merely numbers, as in the American system of numbering streets and houses.

The task of introducing a measure into this numbering of the points on the land falls to the surveyor. His measuring chain comprises about the region of one mesh of the network of Gaussian coordinates.

The surveyor will now proceed to measure out mesh for mesh. Each of these may be regarded as a small parallelogram and is defined when the lengths of two adjacent sides and an angle are known. The surveyor has to measure these and plot them in his map for each mesh. When this has been done for all meshes, he clearly has a complete knowledge of the geometry of the land in his map.

In place of the three data for each mesh (two sides and one angle), it is usual to apply a different method in determining the measure, one which has the advantage of greater symmetry.

Let us consider one of the meshes, a parallelogram, whose sides correspond to two consecutive integers (say $x=3$, $x=4$, and $y=7$, $y=8$) (Fig. 137). Let P be any point within this mesh, and s its distance from the corner point O with the smaller numbers. This will be measured out by the measuring chain. We draw the parallels to the net lines through P, and they intersect the net lines in A and B. Further, let C be the foot of the perpendicular dropped from P onto the x-coordinate.

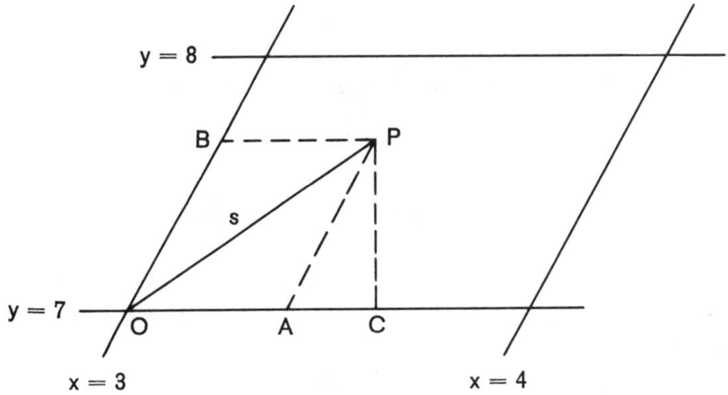

Fig. 137 *The generalized theorem of Pythagoras.*

The points A and B then also have numbers, or Gaussian coordinates, in the net. Point A is determined, say, by measuring the side of the parallelogram on which A lies and the distance AO, and by regarding the ratio of these two lengths as the increase of the x-coordinate of A from O. We shall denote this increase itself by ξ, choosing O as the origin of the Gaussian coordinates. In the same way we determine the Gaussian coordinate η of B as the ratio in which B cuts the corresponding side of the parallelogram. Clearly then ξ and η are the Gaussian coordinates of P relative to O. If x, y are the mesh numbers of O itself, or its Gaussian coordinates with regard to an arbitrarily fixed origin, then ξ, η are small increments of x, y.

The true length of $\overline{OA}$ is, of course, not ξ but, say, $a\xi$ where a is a definite number to be determined by measurement. In the same way the true length of $\overline{OB}$ is not η but by $b\eta$. If we move the point P

about, its Gaussian coordinates change, but the numbers a and b which give the ratio of the Gaussian coordinates to the true lengths remain unchanged.

We next express the distance $\overline{OP}=s$ with the help of the right-angled triangle OPC according to Pythagoras' theorem. We have

$$s^2 = \overline{OP}^2 = \overline{OC}^2+\overline{CP}^2.$$

Now $\overline{OC}=\overline{OA}+\overline{AC}$, thus

$$s^2 = \overline{OA}^2+2\,\overline{OA}\;\overline{AC}+\overline{AC}^2+\overline{CP}^2.$$

On the other hand, in the right-angled triangle APC, we have

$$\overline{AP}^2 = \overline{AC}^2+\overline{CP}^2.$$

Hence

$$s^2 = \overline{OA}^2+2\,\overline{OA}\;\overline{AC}+\overline{AP}^2.$$

Here $\overline{OA}=a\xi$, $\overline{AP}=\overline{OB}=b\eta$. Further, $\overline{AC}$ is the projection of $\overline{AP}=b\eta$, and thus bears a fixed ratio to it, say $\overline{AC}=c\eta$. Hence we get

$$s^2 = a^2\xi^2+2ac\xi\eta+b^2\eta^2.$$

Here a, b, c are fixed ratio numbers. It is usual to designate the three factors of this equation in a different way, namely,

$$s^2 = g_{11}\xi^2+2g_{12}\xi\eta+g_{22}\eta^2. \tag{97}$$

This equation may be called the *generalized Pythagorean theorem* for Gaussian coordinates.

The three quantities g_{11}, g_{12}, g_{22} may serve, just like the sides and angle, to describe positions and distances of points in the parallelogram. We therefore call them the *metrical coefficients* and use the expression *metric of the surface* for the quantity s^2 as given by (97). The metrical coefficients have different values from mesh to mesh, which must be inserted in the map or given as mathematical function of x, y, the Gaussian coordinates of O. If they are known for every mesh, then by formula (97) the true distance of an arbitrary point P within an arbitrary mesh from the origin can be calculated, provided the Gaussian coordinates x, y of O are given.

The metric coefficients thus represent the whole geometry on the surface.

It will be objected that this assertion cannot be right. For the network of Gaussian coordinates was chosen quite arbitrarily and this arbitrariness applies equally to g_{11}, g_{12}, g_{22}. That is quite true. Another network could be chosen, and we should obtain for the distance between the same points O, P an expression built up just like (97) but with different factors g_{11}', g_{12}', g_{22}'. Yet there are, of course, rules for establishing transformation formulae connecting the g_{11}, g_{12}, g_{22} with the g_{11}', g_{12}', g_{22}' similar to those with which we became acquainted earlier.

Every real geometrical fact on the surface must clearly be expressed by such formulae as remain unaltered by a change of the Gaussian coordinates (i.e., are invariant). This makes the geometry of surfaces a theory of invariants of a very general type. The only restriction of the coordinate net is that it must be continuously curved and cover the surface without gaps but so that no point is covered twice.

Now what are the geometrical problems that the surveyor has to solve as soon as he has obtained the metric?

There are no straight lines on the curved surface, but there are *straightest* lines; these are at the same time those which form the shortest connection between two points. Their mathematical name is "geodetic lines," and they are characterized mathematically thus: Divide an arbitrary line on the surface into small, measurable sections of lengths s_1, s_2, s_3 . . .; then the sum

$$s_1+s_2+s_3+\cdots$$

for the geodetic line between two points P_1, P_2 is shorter than any other line between them (Fig. 138). The s_1, s_2, . . . may be determined from the generalized Pythagorean theorem (97) if the g_{11}, g_{12}, g_{22} are known.

On a spherical surface it is known that the "greatest" circles on the sphere are the shortest lines. They are cut out by the planes that pass through the center. On other surfaces, they are often very complicated curves; and yet they are the simplest curves on the surface and form the framework of its geometry, just as straight lines form the framework of Euclidean geometry of the plane.

Geodetic lines are, of course, represented by invariant formulae.

They are real geometric properties of the surface. All the other invariants can be derived from these invariants. But we cannot enter further into this question here.

Another fundamental property of a surface is its *curvature*. It is generally defined with the help of the third dimension of space. The curvature of a sphere, for example, is measured with the help of the sphere's radius, that is, the distance of a surface point from a center lying outside the spherical surface. The surveyor in the woody

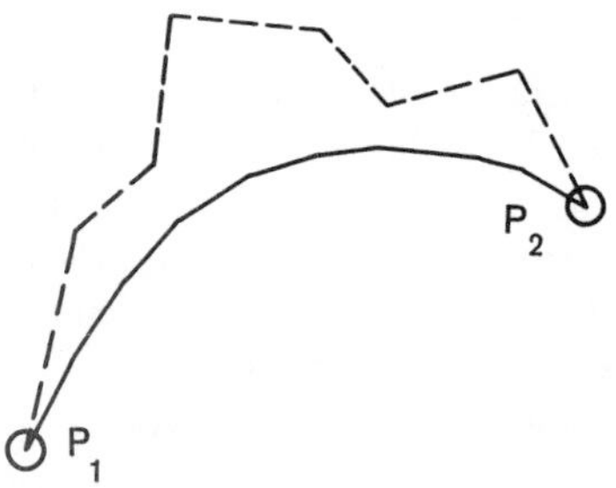

Fig. 138 *A geodetic line compared with another arbitrary line.*

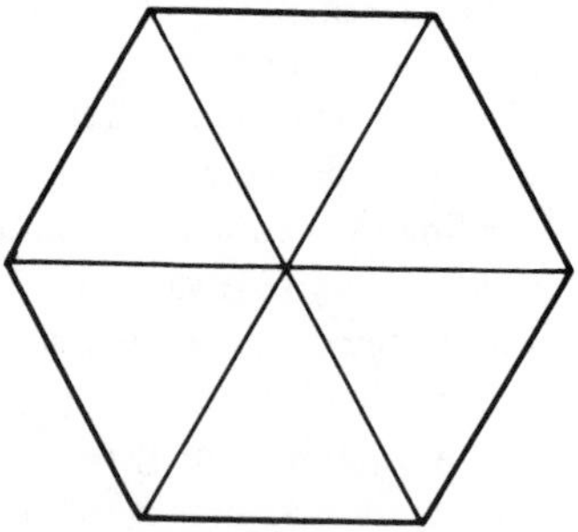

Fig. 139 *Hexagon for determining the intrinsic curvature of a surface.*

regions will not be able to apply this definition of curvature. He cannot move out of his surface, so he has to try to find out the curvature with his measuring chain alone. Gauss showed that this is actually possible. The idea of these considerations can be simply understood in the following way:

The surveyor provides himself with twelve equally long wires and forms with them a regular hexagon together with its radii as shown in Fig. 139. According to a well-known theorem of plane geometry it is possible to have this figure of twelve wires in one plane all stretched tight simultaneously. This is really rather remarkable, for when five of the six equiangular triangles are stretched out, then the last wire must fit into the remaining gap and no adjustment is possible. We learn at school that it does so, but what is learned at school is not usually much reflected on later. Thus it is astonishing that the gap is filled in by a wire of exactly the same length as the other sides.

Actually this works only in the plane. If we attempt the same thing on a curved surface in such a way that the center and the six corners rest on it, the hexagon does not close. On the summits of hills and in the depths of valleys the last wire is too long; on a pass leading from a lower region to another one between two hills (where the surface is saddle-shaped) it is too short. The reader can discover this himself with twelve pieces of wire and some cushions.

The experiment gives us, however, a criterion for finding the curvature of a surface without leaving the surface, for if the hexagon is complete, then the surface is plane; if not, it is curved. We shall not derive the measure of curvature. The indications given are sufficient to show that such a measure can be defined rigorously. It clearly depends on how the metric coefficients change from place to place. As Gauss has proved, the *measure of curvature* can be expressed in terms of the g_{11}, g_{12}, g_{22}, and it is an invariant of the surface, independent of the Gaussian net chosen.

Gauss' theory of surfaces is a method of treating geometry to which we can apply the expression *contact theory*, a term borrowed from physics. It is not the laws of a surface on a large scale that are primarily considered, but rather the differential properties of the surface (properties in the small): the metric coefficients and the invariants formed from them, and above all, the measure of curvature. The form of the surface and its geometrical properties as a whole can then be determined subsequently by calculations which are very similar to the solution of the differential equations of physics. In contrast with this, Euclidean geometry is a typical theory of action at a distance. This is why the new physics, which is entirely built up on the concepts of contact action, on the idea of field, finds the Euclidean scheme insufficient and has to pursue new paths after the manner of Gauss.

5. The Two-dimensional Continuum

Let us suppose that our surveyor is operating with a wire hexagon to establish the curvature of the ground, and that he takes no account of the fact that there is a clear space in the wood in the middle of the hexagon which allows the sun to shine on the ends of the wires that meet there. These wires will stretch a little because of being heated. Hence the six radial wires will be longer than the six outer wires, so that the latter will not connect. If the ground is flat in reality, therefore, the surveyor will believe that he is standing on top of a hill or in the hollow of a valley. If he is conscientious he will repeat the measurement with wires of another material. Under the influence of the sun's heat, these will expand more or less than those used before. This will draw his attention to the error and will lead him to correct it.

Now let us assume that the increase of length produced by the heating is the same for all the available materials of which the wires can be made. The error will then never come to light. Plains will be regarded as mountains and some mountains will be regarded as plains. Or let us imagine that physical forces as yet unknown to us exert some influence on the lengths of rods and wires, but to the same extent in all cases. Then the geometry which the surveyor would determine with his measuring chain and wire polygons would prove to be quite different from the true geometry of the surface. But so long as he operates on his surface and has no possibility of using the third dimension, he will be firmly convinced that he has determined the correct geometry of the surface.

These reflections show us that the concept of geometry in a surface, or, in Gauss' expression, "*geometria intrinsica*," has nothing to do with the form of the surface as it appears to an observer who has the third dimension of space at his disposal. Once the unit of length has been given by a measuring chain, the Gaussian network chosen and the metric determined, the geometry in the surface is fully established *relative* to this system no matter what changes the measures undergo in reality during the process of measurement. These changes do not exist for a creature who is confined to the surface, so long as they affect all substances in the same way. Hence this creature will find curvatures where there are in reality none, and conversely. But this "in reality" becomes meaningless as far as surface creatures are concerned, for they have no conception of a third dimension, just as we human beings have no idea of a fourth dimension of space. It is, therefore, meaningless for these creatures to describe their world as "a surface, embedded in a three-dimensional space"; rather it is a "two-dimensional continuum." This continuum has a definite geometry, definite shortest or geodetic lines, and also a definite "measure of curvature" at every point. But the surface creatures will by no means associate the same idea with the latter phrase as we do with the intuitive concept of the curvature of a surface; rather they will only mean that the wire hexagon remains more or less open or closed and nothing more.

If the reader succeeds in experiencing the feelings of this surface creature and in imagining the world as it appears to this creature, the next stage of abstraction will present no difficulty.

Now exactly the same thing as happened to the surface creature might happen to us as human beings in our three-dimensional world. Perhaps this world is embedded in a four-dimensional space in precisely the same way as a surface is embedded in our three-dimensional space; and unknown forces may change all lengths in certain regions of space without our ever being able to notice this directly. But then it would happen that a spatial polyhedron, constructed after the manner of the six-sided figure, which should close according to ordinary geometry, would turn out to be slightly open.

Have we ever detected anything of this sort? Since olden times Euclidean geometry has always been considered to be exact. Its theorems have even been declared in the critical philosophy of Kant (1781) to be a priori and, as it were, eternal truths. The great mathematicians and physicists, however—above all, Gauss, Riemann, and Helmholtz—have never shared this general belief. Gauss even once undertook an extensively planned geodetic survey to test a theorem of Euclidean geometry, namely, that the sum of the angles in a triangle amounts to two right angles (180°). He measured out the triangle between the three mountains, Brocken, Hoher Hagen, and Inselberg. The result was that the sum of the angles was found to be of the right amount within the limits of error.

On account of this undertaking, Gauss was attacked by philosophers. It was asserted above all that even if he had detected deviations, this would at most have proved that the light rays between the telescopes had been deflected by some perhaps unknown physical causes, but would prove nothing about the validity or nonvalidity of Euclidean geometry.

Now Einstein asserts, as we have already remarked (p. 317), that the geometry of the real world is actually not Euclidean, and he supports this statement by concrete examples. To understand the relation of his doctrine to the preceding investigations about the foundations of geometry we must turn to a short discussion of some fundamental problems on the border of science and philosophy.

6. Mathematics and Reality

The question is: What is the object of geometric concepts at all? Geometry certainly has its origin in the surveyor's art of measurement, that is, in a purely empirical doctrine. The ancients discovered that geometrical theorems could be proved deductively, that only a small number of principles or axioms need be assumed and then the whole system of remaining theorems could be derived from them by mere logic. This discovery had a powerful effect. For geometry became the model of every deductive science, and it became the aim of rigorous thinkers to demonstrate something "*more geometrico*." Now what are the objects with which scientific geometry occupies itself? Philosophers and mathematicians have discussed this question from all points of view and have given a great number of answers. The certainty and incontrovertible correctness of geometric theorems was generally admitted. The only problem was how to arrive at such absolutely certain theorems and what the things were to which they referred.

It is without doubt true that if a person admits the geometric axioms to be correct, he is also compelled to acknowledge all the other theorems in geometry as true. For the sequence of the proofs is compelling for anyone who thinks logically at all. This reduces the question to that of the origin of the axioms. In the axioms we have a small number of statements about points, straight lines, planes, and similar concepts which are to hold exactly. For this reason, unlike most statements of science and of ordinary life, they cannot have their origin in experience; for this always furnishes only approximately correct and more or less probable results. Hence we must look for other sources of knowledge which guarantee that these theorems are absolutely certain. According to Kant (1781), time and space are forms of intuition which are a priori, which precede all experience, and which indeed first make experience possible. According to this view, the objects of geometry must be preconstructed forms of *pure* intuition which are at the base of the judgments that we make about real objects in *empirical* intuition (direct perception). Thus the statement "the edge of this ruler is straight" would come about by comparing the directly perceived edge with the pure intuition of a straight line, without this process, of course, coming into consciousness. The object of geometric science would then be the straight line given in pure intuition, that is, neither a logical concept nor a physical thing but some third kind of notion whose nature can be communicated only by calling attention to the experience connected with the intuition "straight."

We do not presume to pronounce a judgment on this doctrine or on similar philosophical theories. These concern primarily the subjective experience of space, and this lies outside the scope of our book. Here we are dealing with the space and time of physics, that is, of a science which consciously and increasingly clearly turns away from intuition as a source of knowledge and demands more precise criteria.

Fig. 140 *Checking the straightness of an edge with the help of a beam of light.*

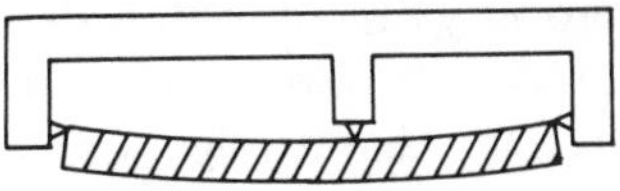

Fig. 141 *Checking the straightness of an edge by rotation.*

We must now take it as a fact that a physicist would never pronounce the statement "the edge of this ruler is straight" on the basis of direct intuition. It is a matter of indifference to him whether there is any such thing as a pure form of intuition of a straight line or not with which the edge of the ruler can be compared. Rather he would make definite *experiments* to test the straightness, just as he would test every other assertion about objects by means of experiments. For instance, he will look along the edge of the ruler, that is, he will ascertain whether a ray of light which touches the initial and the end point of the edge also just glides over all the remaining points of the edge (Fig. 140). Or he will turn the ruler about the end points of the edge and make the point of a pencil touch any arbitrary intermediate point of the edge. If this contact remains unaffected by the rotation, the edge is straight (Fig. 141).

Now, if we subject these processes, which are superior to intuition in so far as they are objective (i.e., can be checked by anybody else), to criticism, we see that they, too, carry us no further into the question of absolute straightness. In the first method it is evidently already assumed that the ray of light follows a straight course. How do we prove that it does? In the second method it is assumed that the points about which the ruler is turned and the point of the pencil are in rigid connection and that the ruler is itself rigid. Suppose that we wish to test the straightness of a rod with circular cross-section, which is lying in a horizontal position and is a little bent owing to its own weight; this bending will then remain unaltered by the rotation, and thus the method of contact will find straightness where it is in reality curved. It is useless to object that these are sources of error which occur in every physical measurement and which are avoided by every expert experimenter. What we are concerned with is to show that straightness or any other geometrical property cannot be directly proved empirically, but is only relative to definite geometrical properties of the apparatus used in the measurement (straightness of the ray of light, rigidity of parts of the instrument). If we divest the operations actually performed of all additional features of thought, memory, or previous knowledge, nothing remains except the discovery that if two points of the ruler's edge lie on a ray of light, then so does this point or another; or that if two points of the ruler coincide with two points of a body, the same also holds for this or that third point. Thus, what is really ascertained is coincidences in space or, rather, space-time, the coinciding of two recognizable material points at the same time and at the same place. All the rest is speculation, even such a simple assertion as that the straightness of the ruler can be determined by such experiments on coincidence.

A critical review of exact science teaches us that all our observations resolve finally into such coincidences. Each measurement states that a pointer or a mark coincides with some division on a scale simultaneously with the coincidence of the hands of a clock with some divisions of the dial. No matter whether the measurement concerns lengths, times, forces, masses, electrical currents, chemical affinities, or anything else, all that is actually observable consists of space-time coincidences. In the language of Minkowski these are world points that are marked in the space-time manifold by the intersection of material world lines. Physics is the doctrine of the relations between such marked world points.

Mathematical theory is the logical working out of these relations. However complicated it may be, its ultimate object is always to represent the actually observed coincidences as the logical consequences of certain fundamental assumptions and principles. Some statements about coincidences have the form of geometrical theorems. Geometry as a doctrine that is applicable to the real world has no favored rank above the other branches of physical science. The geometrical concepts depend in the same way on the actual behavior of natural objects as the concepts in other parts of physics. We cannot allocate a special position to geometry.

The fact that Euclidean geometry hitherto was placed above physics was due to the fact that there are light rays which behave with very great accuracy like the straight lines of the conceptual scheme of Euclidean geometry, and that there are approximately rigid bodies which satisfy with considerable accuracy the Euclidean axioms of congruence. The statement that geometry is exactly valid cannot be credited as having any sense from a physical point of view.

The objects of the geometry which is actually applied to the world of things are thus these things themselves regarded from a definite point of view. A straight line is by definition a ray of light, or an inertial orbit, or the totality of the points of a body regarded as rigid which do not move when the body is turned about two fixed points, or some other physical something. Whether the straight line so defined has the properties which the geometry of Euclid asserts can be determined only from experience. An example of such a property of Euclidean geometry is the theorem which Gauss tested empirically about the sum of the angles in a triangle. We must acknowledge such experiments as thoroughly justified. Another characteristic property of two-dimensional geometry was given by the automatic closing of the wire hexagon (p. 326). Only experience can teach whether a definite choice of physical objects chosen to represent the straight line, the unit of length, and so forth, has this property or not. In the former case Euclidean geometry is applicable relative to these definitions, in the latter it is not.

Now Einstein asserts that all previous definitions of the fundamental concepts of the space-time continuum by means of rigid measuring rods, clocks, rays of light, or inertial orbits in small limited regions certainly obey the laws of Euclidean geometry or of Minkowski's world, respectively, but not on a large scale. That this was not discovered earlier is due to the smallness of the deviations. The problem arises how to deal with this new situation. There are obviously two ways: Either we may give up defining the straight line by means of the ray of light, length by means of a rigid body, and so forth, and look for other realizations of the fundamental Euclidean concepts in order to retain the Euclidean system which expresses their logical relationships; or we may abandon Euclidean geometry itself and endeavor to set up a more general doctrine of space.

It is clear to anyone who is not a stranger to science that the first way does not seriously come into consideration. Nevertheless it cannot be proved that it is impossible. Here it is not logic that decides, but scientific judgment and tact. There is no logical path from fact to theory. Power of imagination, intuition, fantasy are here, as everywhere, the sources of creative achievement, and the criterion of correctness is represented by the power of predicting phenomena that have not yet been investigated or discovered. Let the reader try seriously to assume that a ray of light in empty cosmic space is not the "straightest" thing there is, and let him work out the consequences of this hypothesis. Then he will understand why Einstein pursued the other path.

Since Euclidean geometry failed, Einstein could have fallen back on some other definite non-Euclidean geometry. There are systems of this sort worked out by Lobatschewski (1829), Bolyai (1832), Riemann (1854), Helmholtz (1866), and others, and these systems were evolved chiefly to test whether definite axioms of Euclid were logically necessary consequences of the others; whether logical contradictions would appear if they were replaced by other axioms. If we would choose a special non-Euclidean geometry of this kind to represent the physical world we should simply be substituting one evil for another. Einstein went back to the physical phenomena, namely, the concept of space-time coincidence or the event represented by a world point.

7. The Metric of the Space-Time Continuum

The totality of marked world points is what is actually ascertainable. In itself the four-dimensional space-time continuum is structureless. It is the mutual relations of the world points disclosed by experiment that impresses a geometry with a definite metric on it. Thus, in the real world we are confronted with the same circumstances as those we found in considering surface geometry. Hence the mathematical treatment will follow the same method.

First we shall introduce Gaussian coordinates into the four-dimensional world. We construct a network of marked world points. We consider space to be filled with matter in arbitrary motion, which may be deformed in any way but is to maintain its continuous connection; it is to be a sort of "mollusk," as Einstein expresses it. In it we draw three families of intersecting lines which we number, and we distinguish these families by the letters x, y, z. In the corners of the meshes of the resulting network we imagine clocks to be placed, which go at any arbitrary rate but are arranged so that the difference of the data t of adjacent clocks is small. Thus the whole is a nonrigid system of reference, "a mollusk of reference." In the four-dimensional world we obtain, therefore, a system of Gaussian coordinates consisting of a net of four numbered families of surfaces x, y, z, t.

All moving rigid systems of reference are, of course, special kinds of these general deformable systems. But from our general point of view it is meaningless to introduce rigidity as something given a priori. The separation of time from space is also arbitrary. For, since the rate of the clocks can be assumed arbitrarily but continuously variable, space as the totality of all "simultaneous" world points is not a physical reality. If different Gaussian coordinates are chosen, other world points become simultaneous.

But those things which do not alter when we pass from one system of Gaussian coordinates to another are the points of intersection of the real world lines, the marked world points, space-time coincidences. All really ascertainable facts of physics are qualitative relations between the positions of these world points and thus remain unaffected by a change of Gaussian coordinates.

Such a transformation of the Gaussian coordinates of the

space-time continuum consists of a transition from one system of reference to another one that is arbitrarily deformed and in motion. The postulate that we use only those laws of nature which can really be ascertained has as a consequence that these laws are invariant with respect to *arbitrary transformations of the Gaussian coordinates* from x, y, z, t into x', y', z', t'. This postulate clearly contains the general principle of relativity, for among the transformations of x, y, z, t are those which represent the transition from one three-dimensional system of reference to another moving arbitrarily. Formally, however, it goes beyond this because it also includes arbitrary deformations of the scales of space and time.

In this way we have reached solid foundations for a general treatment of the space-time continuum in a perfectly relativistic way. Our next step will be to link up this mathematical method with the physical reflections which we made earlier and which culminated in the enunciation of the principle of equivalence.

We are now in the same position with respect to the four-dimensional world as the surveyor in the forest after he had marked out his coordinate network but had not yet begun to measure it with his measuring chain. We must look around for a four-dimensional measuring chain.

Such a chain is furnished by the *principle of equivalence*. We know that by choosing the system of reference appropriately we can always make sure that no gravitational field reigns in any sufficiently small part of the world. There are an infinite number of such systems of reference which move rectilinearly and uniformly with respect to each other and for which the laws of the special theory of relativity hold. Measuring rods and clocks behave as expressed by Lorentz transformations: light rays and inertial motion (see p. 236 and p. 239) are straight world lines. Within this small region of the world the quantity

$$F = s^2 = \xi^2 + \eta^2 + \zeta^2 - c^2\tau^2$$

is an invariant with a direct physical meaning. For if the line connecting the origin O (which is assumed to be in the interior of the small region) with the world point $P\ (\xi, \eta, \zeta, \tau)$ is a spacelike world line, then s is the distance $\overline{OP}$ in that system of reference in which the two points are simultaneous. But if the world line $\overline{OP}$ is timelike, then $s=ic\tau$, where τ is the time difference of the events O and P in the coordinate system in which both occur at the same point. Earlier we have called s the four-dimensional distance (VI, 11, p. 307). It is directly measurable by means of measuring rods and clocks, and so, if the imaginary coordinate $\varphi=ic\tau$ is introduced, it has, formally, the character of a Euclidean distance in the four-dimensional space:

$$s = \sqrt{F} = \sqrt{\xi^2+\eta^2+\zeta^2+\varphi^2}.$$

The fact of the validity of the special theory of relativity in small regions corresponds exactly to the fact that Euclidean geometry can be applied to sufficiently small parts of a curved surface. But Euclidean geometry and the special theory of relativity need *not* hold in large regions. There need be no straight world lines at all but only straightest or geodetic lines.

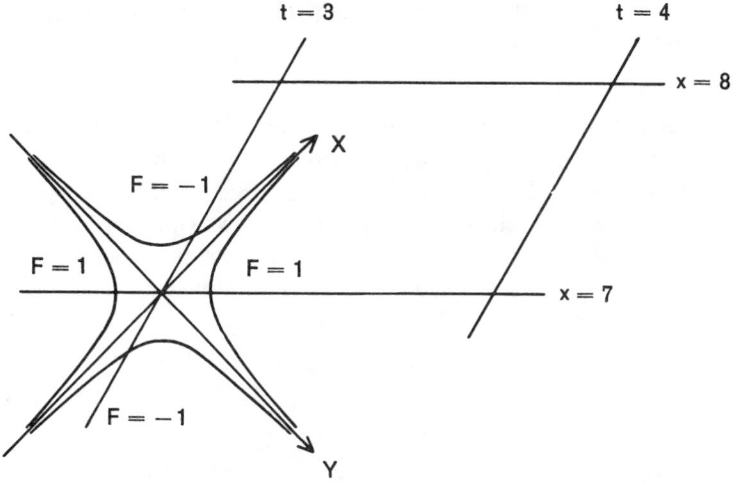

Fig. 142 *Metric in the neighborhood of the world point x=7, t=3.*

The further treatment of the four-dimensional world runs parallel with the theory of surfaces. First we must measure out the meshes of any arbitrary net of Gaussian coordinates with the help of the four-dimensional distance s. We interpret the process in a two-dimensional xt-plane (Fig. 142). Let a mesh of the coordinate net be bounded by the lines $x=3$, $x=4$ and $t=7$, $t=8$ (compare Fig. 137, p. 323). The rays of light that start from the corner $x=3$, $t=7$ correspond to two world lines which intersect and which we can draw as straight lines within a small region. The hyperbolic calibration curves $F=\pm 1$ lie between these light lines. They correspond to the circle which, in ordinary geometry, contains the points which are at the same distance 1.

Then the application of formula (97) of the theory of surfaces leads to the expression

$$s^2 = g_{11}\xi^2 + 2g_{12}\xi\varphi + g_{22}\varphi^2$$

for the invariant s, where ξ and $\varphi = ic\tau$ are the Gaussian coordinates of any point P of the mesh under consideration.

If we insert $\varphi = ic\tau$ we get

$$s^2 = g_{11}\xi^2 + 2icg_{12}\xi\tau - c^2 g_{22}\tau^2,$$

or, if we change the notation (writing now g_{12} instead of icg_{12} and g_{22} instead of $-c^2 g_{22}$):

$$s^2 = g_{11}\xi^2 + 2g_{12}\xi\tau + g_{22}\tau^2.$$

The quantities g_{11}, g_{12}, g_{22} are called coefficients of the metric and may be interpreted directly physically. Thus, for example, for $\tau=0$, $s=\sqrt{g_{11}}\,\xi$, that is, $\sqrt{g_{11}}$ denotes the true length of the spatial side of the mesh in the system of reference in which it is at rest.

In the four-dimensional world the invariant distance s between two neighbor points whose relative Gaussian coordinates are ξ, η, ζ, τ are represented by an expression of the form

$$\left.\begin{aligned} s^2 = {} & g_{11}\xi^2 + g_{22}\eta^2 + g_{33}\zeta^2 + g_{44}\tau^2 \\ & + 2g_{12}\xi\eta + 2g_{13}\xi\zeta + 2g_{14}\xi\tau \\ & + 2g_{23}\eta\zeta + 2g_{24}\eta\tau + 2g_{34}\zeta\tau. \end{aligned}\right\} \qquad (98)$$

This formula may be called *the generalized Pythagorean theorem for the four-dimensional world.*

The *metric coefficients* $g_{11}, \ldots g_{34}$ will have different values from mesh to mesh of the coordinate net; that means, they depend on the coordinates and time value x, y, z, t of the point O. Moreover, they will have other values for another choice of the Gaussian coordinates, and the new values will be connected with the original values by definite formulae of transformation.

8. The Fundamental Laws of the New Mechanics

According to the general principle of relativity, the laws of nature are represented by invariants for arbitrary transformations of the Gaussian coordinates, just as the geometric properties of a surface are invariant for arbitrary transformations of the curvilinear coordinates. The framework in the theory of surfaces consisted of geodetic lines. In just the same way geodetic lines are constructed in the four-dimensional world (i.e., such lines as form the shortest connection between two world points), and in this process the distance between two neighboring points is to be measured by the invariant s.

Now what do the geodetic lines signify? In such regions as are free of gravitation and for an appropriate choice of the system of reference, they are clearly straight lines with respect to this system. But the world lines are either spacelike ($s^2>0$) or timelike ($s^2<0$) or light lines ($s=0$). If we introduce a different system of Gaussian coordinates, the same world lines now become curved but, of course, remain geodetic lines.

From this it follows that the geodetic lines must correspond exactly to those physical phenomena which are represented in ordinary geometry and mechanics by straight lines—namely, rays of light and motions of inertia. Thus we have found the required formulation for the *generalized law of inertia*, which comprises the phenomena of inertia and gravitation in one expression.

If the metric coefficients $g_{11}, \ldots g_{34}$ relative to an arbitrary Gaussian coordinate system are known for every point of the net, the geodetic lines can be obtained by mere calculation. If there is no gravitational field present in a certain region relative to the coordinate system under consideration, then

$$\left.\begin{array}{l} g_{11} = g_{22} = g_{33} = 1, \qquad g_{44} = -c^2, \\ g_{12} = g_{13} = g_{14} = g_{23} = g_{24} = g_{34} = 0; \end{array}\right\} \tag{99}$$

for then the general expression of the distance (98) must reduce to $s^2=\xi^2+\eta^2+\zeta^2-c^2\tau^2$. Deviations of the g from these values thus denote the state that is called gravitational field in ordinary mechanics; then inertial motions are nonuniform and curved—ordinary mechanics regards the Newtonian force of attraction as the cause of this. The ten quantities g have thus a double function: (1) they define the metric, the units of length and time; (2) they represent the gravitational field of ordinary mechanics. The *metrical field* and the *gravitational field* are different aspects of the same thing, both represented by the ten quantities g.

Einstein's theory is thus an amalgamation of geometry and physics, a synthesis of the laws of Pythagoras and Newton. It achieves this by a critical examination of the concepts of space and time in conjunction with the old and well-established experience that gravitational acceleration is independent of the mass of the moving body.

But the new formulation of the law of inertia is only the first step of the theory. We have found in the g's the means of describing mathematically the geometrical-mechanical state of the world relative to an arbitrary Gaussian-coordinate system. Now the essential problem of the theory comes to light. It is as follows:

Laws are to be found according to which the metrical field (the g's) can be determined for every point of the space-time continuum relative to any Gaussian coordinate system.

Concerning these laws we know at present the following:

1. They must be invariant with respect to an arbitrary change of the Gaussian coordinates.
2. They must be fully determined by the distribution of the material bodies.

To these there will be added a formal condition which Einstein has taken over from the ordinary Newtonian theory of gravitation. If we represent Newtonian mechanics as a theory of pseudo action by contact by means of differential equations, then, like all field laws of physics, these are of the second order. This suggests the postulate that the new laws of gravitation, which are differential equations in the g's, should also be at most of the second order.

From these postulates Einstein has succeeded in deriving the equations of the metric or gravitational field. Hilbert, Klein, Weyl, Eddington, and other mathematicians have joined their efforts in investigating thoroughly and illuminating the formal structure of Einstein's formulae. We cannot give these laws and the arguments on which they are founded here because this is impossible without the application of higher mathematics. A few indications will suffice.

We know from the theory of surfaces that curvature is an invariant with respect to any arbitrary change of the Gaussian coordinates and that it may be determined from measurements in the surface. (The reader will remember the use of the hexagon of wires.) Moreover, it is a differential expression of the second order.

In an exactly analogous way invariants may be found for the four-dimensional world which are direct generalizations of the curvature invariant of the theory of surfaces. The procedure can be described as follows: Consider all the geodetic world lines which touch a two-dimensional surface of the four-dimensional world in a point P. These geodetic lines themselves stretch out another surface which may be called a geodetic surface. Now, if we draw a hexagon within this surface such that its sides and radii have the same four-dimensional length, this hexagon will not in general be closed; thus the geodetic surface is curved. If we choose a geodetic surface through the point P with a different orientation in the four-dimensional space (touching another surface), the curvature alters. The totality of all the curvatures of the geodetic surfaces through a point furnishes a number of independent invariants. If these are zero, the geodetic surfaces are such that in a properly chosen Gaussian coordinate system they are planes; then the four-dimensional space is Euclidean. The deviations of the invariants from zero thus determine the gravitational fields and must depend on the distribution of the material bodies. But according to the special theory of relativity (VI, 8, formula (83), p. 280), the mass of a body is equal to the energy divided by the square of the velocity of light. The distribution of matter is thus determined by certain energy-momentum invariants. It is these to which the curvature invariants are set proportional. The factor of proportionality corresponds to the gravitational constant (III, 3, p. 63) of Newton's theory. The formulae thus obtained are the equations of the metric field. If the space-time distribution of energy and momentum are given, the g's can be calculated, and they in their turn determine the motion of the material bodies and the distribution of their energy. The whole is a highly complicated system of differential equations. But the mathematical complexity is offset by an enormous conceptual advance, which consists of its general invariance. For this is the expression of the complete relativity of all events. Absolute space has finally vanished from the laws of physics.

There is a feature of terminology here which worries nonmathematicians. We are accustomed to call the invariants of three-dimensional space that are analogous to surface curvature (or even those of four-dimensional space itself) the *measures of curvature.* We say of space-time regions in which they differ from zero that they are "curved." A person untrained in mathematical speech usually becomes indignant at this. He says that he can understand something *in* space being curved but that it is sheer nonsense to imagine space itself curved. Well, no one demands that it be imagined; can invisible light be imagined, or inaudible tones? If it be admitted that our senses fail us in these things and that the methods of physics reach further, we must make up our minds to allow the same privilege to the doctrine of space and time. For intuition perceives only what develops as a mental process through the joint working of physical, physiological, and psychological phenomena, and is, therefore, actually determined by it. Physics does not, of course, deny that the actual perceptions can be interpreted with considerable accuracy according to the classical laws of Euclid. The deviations which Einstein's theory predicts are so small that only the extraordinary precision of measurement of present-day physics and astronomy can disclose them. Nevertheless they are there, and if the sum of our experiments leads to the result that the space-time continuum is non-Euclidean or "curved," intuition must give way to the judgment based on the integration of all our knowledge.

9. Mechanical Inferences and Confirmations

The first task of the new physics is to show that classical mechanics and physics is correct to a high degree of approximation, for otherwise it would be impossible to understand how two centuries of untiring and careful research could have been satisfied with it. The next problem is, then, to find out the deviations which are characteristic of the new theory and which suggest experimental tests.

How is it that classical mechanics suffices to describe all terrestrial phenomena and almost all phenomena of cosmic motions? For example, what takes the place of the concept of absolute space and the concept of absolute time without which, according to Newtonian principles, even the simplest facts like the behavior of the Foucault pendulum, inertial and centrifugal forces, and so forth, cannot be explained?

In principle we have already answered these questions at the beginning of our discussion about the general principle of relativity. We there (VII, 1, p. 309) agreed to take as the basis of relativistic dynamics the assumption first suggested, that distant masses as real causes must now replace what was previously taken as a fictitious cause of physical phenomena, absolute space. The cosmos as a whole, the multitude of stars, produces at every point and at every moment a definite metric or gravitational field. How this is constituted on a large scale can be found only by speculations of a cosmological kind such as we shall later briefly discuss (VII, 12, p. 361). On a small scale, however, the metric field must be "Euclidean" if the system of reference be appropriately chosen; that is, the inertial orbits and rays of light must be straight world lines. Now, compared with the cosmos, even the dimensions of our planetary system are small, and hence the Newtonian laws hold in them with respect to an appropriate coordinate system, apart from local deviations produced by the sun and the planetary masses, deviations which correspond to the attractions of the Newtonian theory. Astronomy teaches us that such a system of reference, in which the action of the masses of the fixed stars within the region of our planetary system leads to the Euclidean metric, is at rest relative to (or in uniform rectilinear motion with respect to) the totality of cosmic masses, and that the fixed stars produce only extremely small and irregular forces which in the average cancel each other. An *explanation* of this astronomic fact can be given only by applying the new dynamic principles to the whole cosmos, which we shall talk about at greater length in a concluding section. For the present we are dealing with the mechanics and physics of the region within the planetary system. There all results of Newtonian mechanics remain almost unaltered. But we must bear in mind that the vibration plane of Foucault's pendulum remains fixed, not with respect to absolute space, but with respect to the system of distant masses, that is, that centrifugal forces are due not to absolute rotations but to rotations with respect to distant masses. Furthermore, we are quite free to refer the laws of physics not to the ordinary system of coordinates, in which the metric field is Euclidean and a gravitational field in the ordinary sense does not exist (except for the local fields of the sun and the planets), but to a system moving and deformed in any way whatsoever; only in this case gravitational fields at once appear and geometry loses its Euclidean character. The general form of all physical laws remains always the same, except that the values of the quantities $g_{11}, g_{22}, \ldots g_{34}$ which determine the metric field or the gravitational field, are different in every system of reference. This invariance of the laws alone contains the difference between the new and the old dynamics; here, too, we were able to use systems of reference moving arbitrarily (or deformed), but then the physical laws did not retain their form. Rather, there were "simplest" forms of the physical laws, the Newtonian ones, which were assumed in definite systems of coordinates at rest in absolute space. In the general theory of relativity there are no such simplest or "normal" forms of the laws; at the most, the numerical values of the quantities $g_{11}, \ldots g_{34}$, which occur in all physical laws, might be particularly simple within limited spaces or be only slightly different from such simple values. Thus, the ordinary geometrical and mechanical formulae hold in a system of reference which would be Euclidean within the small space of the planetary system if there were no sun and no planets; in this system the $g_{11}, \ldots g_{34}$ would have the simple values of (99). In reality, however, the $g_{11}, \ldots g_{34}$ do not have exactly these values but differ from them in the vicinity of the planetary masses, as we shall explain later. Any other (say rotating) system of reference in which the $g_{11}, \ldots g_{34}$ do not have the simple values of (99) (provided the planetary masses were disregarded as sources of the metric field) is thus, in principle, fully equivalent to the first. Thus we may return to Ptolemy's point of view of a "motionless earth." This would mean that we use a system of reference rigidly fixed to the earth in which all stars are performing a rotational motion with the same angular velocity around the earth's axis. It is not sufficient simply to transform the usual metric (99) to this rotating system; one has to show that the transformed metric can be regarded as produced, according to Einstein's field equations, by rotating distant masses. This has been done by Thirring. He calculated the field due to a rotating, hollow, thick-walled sphere and proved that inside the cavity it behaved as if there were centrifugal and other inertial forces usually attributed to absolute space.

Thus from Einstein's point of view Ptolemy and Copernicus are equally right. What point of view is chosen is a matter of expediency. For the mechanics of the planetary system the view of Copernicus is certainly the more convenient. But it is meaningless to call the gravitational fields that occur when a different system of reference is chosen "fictitious" in contrast with the "real" fields produced by near masses: it is just as meaningless as the question of the "real" length of a rod (VI, 5, p. 253) in the special theory of relativity. A gravitational field is neither "real" nor "fictitious" in itself. It has no meaning at all independent of the choice of coordinates, just as in the case of the length of a rod. Nor are the fields distinguished by the fact that some are directly produced by masses while others are not; in the one case it is particularly the near masses that produce an effect; in the other it is the distant masses of the cosmos.

Arguments of "common sense" have been advanced against this doctrine, among them the following: If a railway train encounters an obstacle and becomes shattered, this event can be described in two ways. First, we may choose the earth (which is here regarded as at rest relative to the cosmic masses) as system of reference, and make the (negative) acceleration of the train responsible for the destruction. Or, second, we may choose a coordinate system rigidly attached to the train, in which case, at the moment of collision, the whole world makes a jerk relative to this system; everywhere we get a very strong gravitational field parallel to the original motion and this field causes the destruction of the train. Why does the church tower in the neighboring village not tumble down, too? Why, in light of the claim that the following two statements are to be equivalent, do the consequences of the jerk and of the gravitational field associated with it make themselves evident one-sidedly through the destruction of the train: that the world is at rest and the train is suddenly retarded—or that the train is at rest and the world is retarded? The answer is this: the church tower does not fall down because, during the retardation, its relative position to the distant cosmic masses is not changed at all. The jerk, which, as seen from the train, the whole world experiences, affects all bodies equally from the most distant stars to the church tower. All these bodies fall freely in the gravitational field which is present during the retardation, with the exception of the train, which is prevented by the retarding forces from falling freely. But with respect to internal events (such as the equilibrium of the church tower) freely falling bodies behave just like bodies that are poised freely and are withdrawn from all influences. Thus, no disturbances of the equilibrium occur, and the church tower does not tumble down. The train, however, is prevented from falling freely. This gives rise to forces and stresses which lead to its consequent destruction.

To appeal to "common sense" in these difficult questions is altogether precarious. There are supporters of the theory of a substantial ether who reject the theory of relativity because it is too far removed from our power of intuition and too abstract. Some of these have finally come to recognize the special principle of relativity now that experiments have indisputably decided in favor of it, but they still struggle against the principle of general relativity because it is contrary to common sense. To these Einstein makes the following reply: According to the special theory of relativity, the train in uniform motion is certainly a system of reference which is equivalent to the earth. Will the common sense of the engine driver admit this? He will object that he has to heat and oil not the "surroundings" but the engine, and that it must, therefore, be the motion of the engine which shows the effect of his work. This argument can be answered by referring to an electric engine which by hitting an obstacle will be smashed in the same way as a steam engine, though the driver has to do some oiling but has no part in the production of the energy which is supplied by a power station fixed on the earth. Common sense often has the tendency to lead us astray.

We now resume our consideration of celestial mechanics from the point of view of Einstein, turning our attention to the local gravitational fields which superimpose themselves on the cosmic field as a result of the existence of planetary masses.

We can give but a short summary of these researches, since they concern essentially mathematical consequences of the field equations.

The simplest problem is to determine the motion of a planet about the sun. Here we do best to start from the Gaussian system already mentioned, in which the gravitational field is Euclidean and no gravitational field in the ordinary sense would be present in the region of the solar system if the sun and the planet were absent; this system is characterized by the $g_{11}, \dots g_{34}$ having the values (99) (p. 339) if the sun's action is disregarded. It is merely a question of determining deviations from these values effected by the sun's mass. The corrected field has to be independent of time and of spherical symmetry around the sun's center. Einstein himself gave an approximate solution of the field equations; later Schwarzschild found that in this case an exact solution exists which leads to rather simple expressions for the $g_{11}, \dots g_{34}$. From these one can calculate the planetary orbits as geodetic lines. Their curvature, which is regarded in Newton's theory as due to the attractive force, appears now as a consequence of the curvature of the space-time world of which they are the straightest lines.

The planetary orbits determined in this way approximate with a high degree of accuracy those in Newton's theory. This result is not at all trivial if we bear in mind the totally different standpoint of the two theories. In the case of Newton we start from absolute space (which is philosophically unsatisfactory) and a deflecting force which is invented *ad hoc* and has the strange and unexplained property that it is proportional to the inertial mass; in the case of Einstein we start from a general principle, which is philosophically satisfactory, and develop the theory as the simplest possible representation of this idea. Even if Einstein's theory were to achieve no more than to express the results of Newtonian mechanics in conformity with the general principle of relativity, those who are seeking the simplest harmony in the laws of nature would prefer it.

People of this type, however, are rare, and if Einstein's theory had not gone beyond that, it would have been appreciated only by a few theoretical physicists and astronomers. General interest in a theory always depends on its ability to explain things previously inexplicable or to predict phenomena not yet observed. The general theory of relativity has been successful in both these directions.

10. Predictions of the New Mechanics and Their Confirmations

We said that the motion of a planet about the sun regarded—according to Einstein and Schwarzschild—as a geodetic line in four-dimensional space-time, turns out to be very nearly the same as that predicted from Newton's theory. The degree of approximation is very high; still a precise calculation reveals slight deviations, and the difference increases with the gravitational field. Therefore the best opportunity to discover these deviations would be with a planet near the sun. Now in dealing with Newton's celestial mechanics (III, 4, p. 66), we have already seen that the only certain case of failure is precisely that of the planet nearest the sun, Mercury. There is left an unexplained *motion of Mercury's perihelion* of 43 seconds of arc per century. But this is just the amount required by Einstein's theory. The confirmation of this result of Einstein's mechanics was therefore actually anticipated by Leverrier's calculation. This result is of extraordinary importance, for no new arbitrary constant enters into Einstein's formula, and the "anomaly" of Mercury is just as necessary a consequence of the theory as the result that Kepler's laws are valid for the planets far removed from the sun.

For other planets, the movement of the perihelion due to the relativistic effect is very small. For the earth and Venus there is a rough confirmation of the theory within the limits of observational errors. The tiny effect of these two planets can only be separated with great difficulty and low accuracy from the other causes of perturbation which produce a much bigger rotation of the perihelion of the orbits. The following table shows the motion of the perihelion of the three planets nearest the sun according to Einstein's theory and to observation:

	Motion of the Perihelion in Seconds of Arc per Century	
	Theoretical	*Observed*
Mercury	43.03±0.03	43.11±0.45
Venus	8.63	8.4 ±4.8
Earth	3.8	5.0 ±1.2

The table shows the accuracy with which the theory is confirmed by the observations. For the earth the observation is relatively exact but the agreement with the theory is not very satisfactory; for Venus the observation is less accurate, but the agreement is better.

The anomaly of the perihelion of Mercury is so far the only confirmation of the general theory of relativity in the domain of mechanics.

Einstein and others have of course looked out for other effects which might be observable. In section 9 of Chapter III and in the first section of this chapter we have discussed centrifugal forces. According to Newton they indicate the movement in absolute space. According to Mach and Einstein they indicate the movement relative to the distant masses of the stars. If the latter explanation is correct, the force on a body due to a large mass in its neighborhood ought to be different according to whether the latter is at rest or rotating. Applied to our planetary system, this leads to a perturbation of the motion of the planets due to the rotation of the sun around its axis (one revolution per 21 days). This perturbation turns out again to be a movement of the perihelion. Unfortunately an estimate of the size of this effect shows that for all planets it is much too small to be observed. In fact, as we stated before, there is no unaccounted-for deviation in the planetary motions. Concerning the satellites of the planets, the situation is the same; the effect of the rotation of the planets is too small to give an observable effect.

But now there has appeared a new possibility to confirm the motion of the perihelion—both that due to the difference of the law of force according to Newton and Einstein, and that due to the rotation of the central body—namely, with the help of an artificial satellite of the earth.

We give a table for both effects.

Radius of the Earth = 6367 *km.*

Mean Distance of Satellite (*in km.*)		*Ratio of small to large semi-axis of elliptic orbit*	*Rotation of Perihelion (in seconds of arc per century)*	
From the center of the earth	*From the surface of the earth*		*Through deviations from Newton's law*	*Through the rotation of the earth*
6770	400	0.99995	1450	−43
10000	3650	0.969	586.6	

The orbits chosen are almost circles, since the ratio of the semiaxes is very nearly unity. The negative sign of the figure due to the earth's rotation means that this small rotation is in the opposite direction of the larger effect and must be subtracted from it.

We see that for artificial satellites both effects may become observable. However, the difficulties are large. Satellites near the earth's surface suffer friction in the uppermost layers of the atmosphere and are thus retarded and fall after a certain interval of time. Furthermore, there are perturbations, not only those due to the sun, the moon, and planets, but also those due to the deviation of the earth surface from an exact sphere, the effects of which cannot be calculated with great accuracy.

So far we have considered only the movement of one body around another, neglecting the motion of the latter. This assumption is justified if the central body is very heavy compared with the other (as is the case for the pairs sun planet or planet moon). But it is easy to eliminate this restriction; the two-body problem can be solved exactly in Newtonian mechanics, with the result that both bodies perform Kepler motions around their center of gravity.

For more than two bodies there exist no simple exact solutions of Newton's equations of motion and one has to apply perturbation theory (see III, 4, p. 64). Now the question arises whether at least the Newtonian equations of motion for a system of many bodies can be obtained as a first approximation from Einstein's theory, and what deviations are to be expected. One has to show that the total field due to the moving bodies is, according to Einstein's equations, in the first approximation nothing but the superimposed Newtonian fields due to the single masses, and that the law of the geodetic lines reduces to the Newtonian equations of motion in this field. This proof is not difficult and has been given by Einstein himself. But Einstein was not content with this result.

There are still two fundamental assumptions in this theory: [1] the idea of general relativity which leads (in conjunction with the postulate of simplicity) to the field equations and, by solving these, to the metric-gravitational field; [2] the hypothesis that the free motions of particles are represented by geodetic lines of this metric. Postulate [1] roughly corresponds to Newton's law of force (inverse square of the distance); postulate [2], to his equations of motion (acceleration proportional to the force). Einstein, in his later years, regarded this duality as unsatisfactory and attempted to remove it. Although the interpretation of the orbital motions as geodetic lines of a non-Euclidean geometry (assumption [2]) was the starting point of his considerations on general relativity, he came to the conviction that it was superfluous and actually contained in the field equations (assumption [1]).

The idea is that the field produced by a body in turn reacts on the body and thus determines its world line. This is mathematically a very complicated problem the nature of which we cannot even indicate. Einstein attacked it together with his collaborators Infeld and Hoffmann; their first investigations were so extended that only summaries could be published. Later Infeld succeeded in simplifying the calculations considerably. Another method was developed by the Russian physicist Fock. Thus we have now a very satisfactory relativistic mechanics consisting simply of the general invariant field equations, which accounts for all facts known at present about the motion of celestial bodies and predicts numerous phenomena very likely open to observation in the near future.

11. Optical Deductions and Confirmations

So far there have been found, besides these astronomic deductions, only a few optical phenomena which do not escape observation because of the smallness of their effects.

One is the *red shift of the spectral lines* of the light which comes from stars of great mass. At their surfaces there is a strong gravitational field. This affects the metric and causes a clock to go more slowly there than on the earth where the gravitational field is smaller. Now we have such clocks in the atoms and molecules of luminescent gases. Their mechanism of vibration is certainly the same wherever the molecule happens to be, and thus the time of vibration is the same in those systems of reference in which the same gravitational field, say the field zero, is present.

If the time of vibration in the fieldless region of space is T, then $s=icT$ is the corresponding invariant distance of the world points corresponding to two successive extreme points of the vibration relative to the system of reference in which the atom is at rest. In a relatively accelerated system of reference in which there is a gravitational field, the same $s=icT$ is given by formula (98), in which ξ, η, ζ characterize the differences of the space coordinates of the atom at the instances when the observation of the vibration begins and when it ends, and where τ is the corresponding time interval, where all these quantities are measured in the chosen system of reference. If we take the origin of the space coordinates in the center of the atom, we can put $\xi=\eta=\zeta=0$ and have

$$s^2 = -c^2T^2 = g_{44}\tau^2.$$

Thus

$$\tau = T\frac{c}{\sqrt{-g_{44}}}.$$

Now it is only in a fieldless space that we have $g_{44}=-c^2$ (see formula (99), p. 339), thus $\tau=T$. But in the gravitational field g_{44} is different from $-c^2$, say $g_{44}=-c^2(1-\gamma)$. Thus the time of vibration is altered to

$$\tau = T\frac{1}{\sqrt{1-\gamma}}$$

or, if the deviation γ is small, approximately (see note on p. 216)

$$\tau = T\left(1+\frac{\gamma}{2}\right). \qquad (100)$$

This is the difference in the beating of two clocks situated at different places for which the difference of the gravitational field given by g_{44} has the relative value γ.

Whether γ is positive or negative can be found by considering a simple case in which the question can be answered directly with the help of the principle of equivalence. This is the case for a constant gravitational field such as occurs at the immediate surface of a celestial body. The action of such a field g may be replaced by an acceleration on the part of the observer of the same value g, and directed opposite to the attraction. If l is the distance of the observer from the surface of the star, a light wave from the surface will take the time $t=\frac{l}{c}$ to reach the observer, and what he finds will be the same as if he had during that time moved in the outward direction with the acceleration g. When the light wave reaches him he would have the velocity $v=gt=\frac{gl}{c}$ in the direction of the light's motion; hence, by Doppler's principle (formula (41), p. 124) he observes the diminished frequency

$$\nu' = \nu\left(1-\frac{v}{c}\right) = \nu\left(1-\frac{gl}{c^2}\right). \qquad (101)$$

The formula can also be directly proved from the principle of equivalence (VII, 2) with the help of the idea of light quanta mentioned before. Light of frequency ν can be regarded according to quantum theory as consisting of quanta of energy $\epsilon = h\nu$. These have an inertial mass $m = \frac{\epsilon}{c^2} = \frac{h\nu}{c^2}$, and this is, according to the principle of equivalence, equal to its gravitational mass. When light quanta $h\nu$ have travelled the distance l against the field of gravitation g, their energy has decreased by glm. Hence at the end of this journey the energy of a quantum $\epsilon' = h\nu'$ is only

$$h\nu' = h\nu - gl\frac{h\nu}{c^2} = h\nu\left(1 - \frac{gl}{c^2}\right).$$

If the factor h is omitted on both sides, we obtain again the formula (101). The time of vibration $\tau = \frac{1}{\nu'}$, observed in the gravitational field is related to the time of vibration $T = \frac{1}{\nu}$ determined in the field-less space in the following way:

$$\tau = T\frac{1}{1 - \frac{gl}{c^2}}$$

or, approximately,

$$\tau = T\left(1 + \frac{gl}{c^2}\right). \qquad (101a)$$

The physical content of this formula is this: Given two equally constructed, synchronous clocks initially at rest relative to each other; if one of them is exposed to a gravitational field for a certain period of time, they will no longer be synchronized, but the clock which was in the field will be retarded.

By comparing (101*a*) with (100), one sees that in this case of a constant field $\gamma = \frac{2gl}{c^2}$. Now according to formula (15) (II, 14, p. 48) Gx is the potential energy of a body in a constant gravitational field G when lifted through a distance x; here, according to (13) (III, 12, p. 42), $G = mg$, where m is the mass. Hence the difference of potential energy per unit mass of two bodies in the distance l is lg. If we call this φ, we obtain $\gamma = \frac{2\varphi}{c^2}$.

Now the Newtonian notion of potential energy is alien to Einstein's theory, but as Newtonian mechanics is an approximation to the Einsteinian, one can take this quantity φ over and then show that the formula

$$\gamma = \frac{2\varphi}{c^2}$$

holds for any gravitational field and that γ is positive if the light travels against the direction of the field.

The formula can be applied to light coming from the sun or a star, where it has to escape a very strong field of attraction, while arriving on the earth it finds only a very faint accelerating field. Hence all spectral lines of the stars ought to be displaced a little towards the red end of the spectrum. Although the predicted effect is very small, its existence has been confirmed by observation, at least qualitatively. A perfect quantitative agreement has not yet been achieved because neither the masses nor the radii of the fixed stars suitable for such measurement are known with sufficient accuracy. As far as results have been obtained they are in fair agreement with Einstein's formula.

The red shift for the sun is difficult to observe as it is rather small and superimposed on other effects of a similar appearance. It has been measured by several astronomers for diverse points across the sun's surface, and it was found that at the inner part of the sun's disk it is much smaller than the value predicted by the theory but increases towards the rim of the disk where it reaches the theoretical value. This can be easily understood if one takes into account the physical state of the gaseous substances forming the outer layer of the sun. These are not in static equilibrium, but in turbulent motion; hot, bright masses of gas are rising violently from the interior while cooler and darker masses sink towards the interior. Thus the Doppler effect (IV, 8) will produce an additional violet shift which reduces the red shift predicted by Einstein.

At this stage we can fill in a gap which was left earlier (VI, 5, p. 262), namely, the complete explanation of the "clock paradox." We assume two observers A and B, of which one, A, was at rest in an inertial system (of the special theory of relativity), while the other,

B, set out on a journey. On B's return A's clock, by (76), (p. 257), is in advance of B's by the amount $\frac{1}{2}\beta^2 t_0$, where t_0 is the total time of the journey as measured in the system A. This formula of course holds only approximately, yet it suffices for our purpose as long as we use corresponding approximations in all our calculations.

Now we may also regard B as at rest. Observer A then makes a journey in the reverse direction. But of course we cannot simply infer that B's clock must now be in advance of A's by exactly the same amount, for B is not at rest in an inertial system but is experiencing accelerations.

From the standpoint of the general theory of relativity we must rather take into account that when the system of reference is altered, definite gravitational fields must be introduced during the times of acceleration.

In the first case under consideration A is at rest in a region of space in which the metric is Euclidean and in which gravitational fields are absent. In the second, B is at rest in a system of reference in which, during the short periods of departure, turning, and arrival at A, gravitational fields occur in which A falls freely while B is held fixed by external forces. Of these three gravitational fields the first and the last have no influence on the relative rates of the clocks of A and B, since they are at the same point at the moments of departure and return and since a difference of rate occurs in a gravitational field, by (101), only when there is a distance l between the clocks. But a difference of rate occurs when A reverses his direction. If t is the time taken to reverse during which a gravitational field arises (B being supposed at rest), then A's clock, which is at a distance l and in the gravitational field g, is in advance of B's clock. This time difference is given to a sufficient degree of approximation by (101a) (p. 353), namely, by $\frac{gl}{c^2}t$. During the times, however, when A is moving uniformly and the special principle of relativity must be applied, A's clock must, conversely, be behind B's clock by the amount $\frac{\beta^2}{2}t_0$. Thus, on the whole A's clock will be in advance of B's by

$$\frac{gl}{c^2}t-\frac{\beta^2}{2}t_0$$

on A's return.

Now it can be shown that this value agrees exactly with the result of the first point of view in which A was regarded at rest, namely, that the advance of A's clock with respect to B's is equal to $\frac{1}{2}\beta^2 t_0$.

For since the moving observer, in reversing his velocity v, assumes the velocity $-v$, his total change of velocity is $2v$. We get his acceleration by dividing this by t, the time taken to effect this change. This gives $g=\frac{2v}{t}$ as his acceleration. On the other hand, at the moment of turning back, half the duration t_0 of the journey is over. The distance between the two observers is then $l=v\frac{t_0}{2}$.

From this it follows that

$$gl = v^2 \frac{t_0}{t}$$

and

$$\frac{gl}{c^2}t - \frac{\beta^2}{2}t_0 = \frac{v^2}{c^2}t_0 - \frac{\beta^2}{2}t_0 = \frac{\beta^2}{2}t_0,$$

which concludes the proof.

Thus the clock paradox is due to a false application of the special theory of relativity, namely, to a case in which the methods of the general theory should be applied.

A similar error lies at the root of the following objection, which is continually being brought forward, although the explanation is very simple.

According to the general theory of relativity, a coordinate system which is rotating with respect to the fixed stars (i.e., which is rigidly connected with the earth) is fully equivalent to a system which is at rest with respect to the fixed stars. In such a system, however, the fixed stars themselves acquire enormous velocities. If r is the distance of a star, its velocity becomes $v=\frac{2\pi r}{T}$, where T denotes the duration of a day. This becomes equal to the velocity of light c if $r=\frac{cT}{2\pi}$. If r is measured in terms of the astronomic unit of length, the light year*, we must divide this by $c\times 365$, T being set equal to

* A light year is the distance which light traverses with the velocity 300,000 km./sec. in one year (365 days).

1 day. As soon as the distance exceeds $1/2\pi \times 365$ light years, the velocity becomes greater than c. But even the nearest stars are several light years distant from the sun. On the other hand the theory of relativity (VI, 6, p. 266) asserts that the velocity of material bodies must always be less than that of light. Here there seems to be a glaring contradiction.

This, however, arises only because the law $v < c$ is entirely restricted to the special theory of relativity. In the general theory it has to be formulated in the following more elaborate way. As we know, it is always possible to choose a system of reference such that Minkowski's world geometry holds in the immediate neighborhood of any arbitrary point—that is, so that the geometry is Euclidean, that there is no gravitational field, and that $g_{11}, \ldots g_{34}$ have the values of (99) on page 339. With respect to this system and in this narrow space the velocity of light $c = 3 \times 10^{10}$ cm./sec. is the upper limit for all velocities.

As soon as these conditions are not fulfilled, however—if gravitational fields are present—the velocity either of material bodies or of light can assume any numerical value. For the light lines in the world are determined by $F = s^2 = 0$, or, if we restrict our attention to the xt-plane, by

$$s^2 = g_{11}\xi^2 + 2g_{14}\xi\tau + g_{44}\tau^2 = 0.$$

We can calculate $\frac{\xi}{\tau}$ from this quadratic equation, and this is the velocity of light. For example, if $g_{14} = 0$, we get from $g_{11}\xi^2 + g_{44}\tau^2 = 0$ the value $\frac{\xi}{\tau} = \sqrt{-\frac{g_{44}}{g_{11}}}$ as the velocity of light, and this depends on just how great g_{11} and g_{44} happen to be. The velocity of a material body has only to be smaller than $\sqrt{-\frac{g_{44}}{g_{11}}}$.

If we take the earth as the system of reference, we have the centrifugal field (III, 9, p. 80) $\frac{4\pi^2 r}{T^2}$ which assumes enormous values at great distances. Hence the g's have values that differ greatly from the Euclidean values of (99). Therefore the velocity of light is much greater for some directions of the light ray than its ordinary value c, and other bodies can also attain much greater velocities.

In any arbitrary Gaussian coordinate system not only does the velocity of light become different, but the light rays no longer remain straight. A second optical test of the general principle of relativity depends on this curvature of the light rays. The world lines of light are geodetic lines just like the inertial orbits of material bodies, and hence, like the latter, will in general be curved. But on account of the great velocity of light, the deflection of its rays is much less than the curvature of the celestial orbits. From the principle of equivalence, we can see why this deflection should come about. For, in an accelerated system of reference, every rectilinear and uniform motion is curved and nonuniform, so that the same must hold for any arbitrary gravitational field.

A ray of light which, coming from a fixed star, passes close by the sun will thus be attracted to it and will describe a somewhat concave

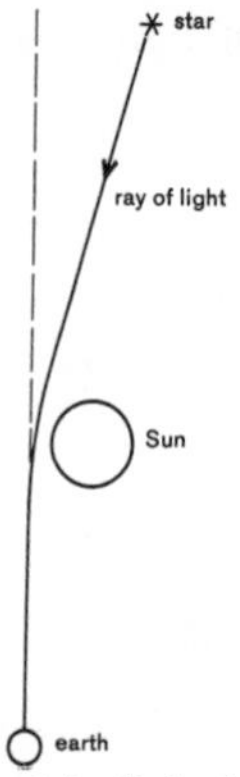

Fig. 143 *Deflection of the light from a star by the sun.*

orbit with respect to the sun (Fig. 143). The observer on the earth will assign to the star a position in the direction of the ray that strikes his eye, and hence the star will appear to be displaced a small amount outwards. This deflection might be calculated from Newton's theory of attraction, in which the ray of light may be treated, say, as a comet which approaches the sun with the velocity of light; since the hyperbolic orbit of a comet, like the elliptic orbit of a planet, does not depend on its mass (because of the equality of inertial and gravitational mass), it does not matter what mass one ascribes to the "light particles." It is of historic interest that this idea was carried out as early as 1801 by the German mathematician and surveyor Soldner. We then get a formula similar to that of Einstein, but giving only half the value for the deflection. This is due to the circumstance that according to Einstein's theory the gravitational field in the neighborhood of the sun is more intense than in Newton's theory. It is just this minute difference (which escaped Einstein's attention when he made his first provisional publication of the theory) that constitutes a particularly sharp criterion of the correctness of the general theory of relativity.

The deflection of the apparent positions of the fixed stars in the neighborhood of the sun can be observed only during a total eclipse of the sun, since otherwise the bright radiation of the sun renders invisible the stars in its vicinity.

The first test of Einstein's prediction was made, mainly through the initiative of the British astronomer Eddington, on May 29, 1919. Two British expeditions were sent out to observe a total eclipse of the sun—one to the West Coast of Africa, the other to north Brazil—and they returned with a number of photographs of the stars surrounding the sun. The result obtained by measuring the plates was announced November 6, 1919, and proclaimed as the triumph of Einstein's theory. The displacement predicted by Einstein, which is to amount to 1.75 seconds of arc, was confirmed.

Since that time measurements of the deflection of the light by the sun have been made at a number of total eclipses. Though the measurements are difficult, there is no doubt about the existence of an effect very close to the predicted magnitude; in any case the value obtained from Newtonian mechanics (first given by Soldner and then by Einstein in his preliminary publication), which is only half of the relativistic value, is certainly not right. But an exact agreement between theory and measurements has not yet been obtained. There are some recent observations which give a deflection 10 per cent larger than the theoretical value. Whether this deviation is due to errors of observation or indicates a genuine insufficiency of Einstein's theory must be left to future research. But there is no doubt that this theory is nearer the truth than the classical one or any other theory so far proposed.

Now the question arises whether the refinement of modern methods of measurement will not permit the observation of the optical relativistic effects in the gravitational field of the earth. This has recently been achieved for the shift of frequency of spectral lines with the help of a surprising discovery of the German physicist Mössbauer. To explain this we have to digress a little to atomic and nuclear physics. An atomic nucleus which has emitted a γ-quantum of a certain frequency can absorb just the same quantum. This is the resonance effect well known from acoustics, optics and electric vibrations. Hence an atom which is a source of γ-rays of a definite frequency can be used as receiver of the same radiation. But there exist two causes which disturb the resonance: the Doppler effect and the recoil effect.

The atoms in gases and liquids are in perpetual fast motion, representing the thermal agitation. In consequence of the Doppler effect the frequency of radiation emitted by the atoms is enlarged or diminished according to the direction of the atomic motion in relation to the direction of emission (see p. 121). In the same way the atomic movement puts the frequency of absorbing atoms out of tune. Hence the emitted γ-quanta have not exactly the same frequency as the receiving atoms and are not absorbed. The same holds also for solids. These consist in regular arrangements of atoms bound together with strong forces (so-called crystal lattices), but even here the atoms vibrate around equilibrium positions and thus produce the Doppler effect.

The second disturbance of the resonance is caused by the recoil suffered by the atoms (or better: atomic nuclei) when emitting γ-quanta, as it changes the frequency ν of the γ-quanta. The recoil is due to the momentum $p=\epsilon/c$ which is carried by the quanta of energy $\epsilon=h\nu$ (see p. 292). According to the theorem of conservation of momentum the emitting atom obtains the same momentum in the opposite direction; thus it gets a velocity $v=p/m$ and an energy $e=\frac{1}{2}mv^2$. This energy is supplied by the γ-quantum, hence its frequency is diminished and is only $\nu'=(\epsilon-e)/h$ instead of $\nu=\epsilon/h$ for an immobile atom. The receiving atom which, if fixed, can absorb the frequency ν needs, if movable, a larger frequency $\nu''=(\epsilon+e)/h$ because it has, after the absorption, to carry the additional momentum of the γ-quantum, again because of the Doppler effect; in addition to the energy ϵ absorbed it obtains the kinetic energy e, and both are supplied from the absorbed quantum.*

* As e is small compared with ϵ we have calculated p from ϵ, i.e., from ν, and not from ν' or ν'', respectively. A more correct calculation would give a result differing only by an amount of higher order, too small to be observed.

Both effects—and this is Mössbauer's discovery—can be eliminated: the Doppler effect vanishes if the emitting and the absorbing atoms are cooled down to temperatures so low that the thermal agitation has practically ceased. The recoil can be circumvented by building-in the atoms into solids where they are bound with very strong forces to their neighbor atoms. Then the momentum of the γ-quantum will not be transferred to a single atom, but to the whole crystal of mass M. As M is large compared with the atomic masses, the velocity $V=p/M$ and the energy $E=\frac{1}{2}MV^2=p^2/2M$ are extremely small and not observable. Therefore the emitting atoms produce actually the unchanged frequency ν, and the absorbing atoms can absorb the same frequency, provided both emitter and receiver are at rest. If they have a relative motion u the Doppler effect comes into play again and the absorption is diminished. This decrease is measurable if the Doppler displacement is not too small compared with the natural width $\delta\nu$ of the spectral line. This width $\delta\nu$ one can understand from the standpoint of wave theory: The energy emitted produces a damping of the vibrating atoms. Vibration of the frequency ν means an undisturbed wave train without beginning and end. Every actual vibration, e.g., that of the atoms, which has a certain beginning and an ending through the damping, can be composed of endless vibrations of different frequencies. Therefore one finds neighbor frequencies near to the given frequency ν. Hence the line has a certain width called its natural width.

The precision of the method is amazingly great. One can discover even velocities of 1/1000 mm./sec.

Thus the displacement of spectral lines through the terrestrial gravitational field has been measured (Pound and Rebka, Harvard, 1960; Cranshaw, Schiffer, and Whitehead, Harwell, 1960). The relation of gravity on earth to the corresponding quantity produced by the sun is 1:3000. The emitter was put on the top, the absorber on the foot of a tower of 22 m. height. In this way only a small part, $1:3\times 10^5$, was exploited. A γ-line of iron (Fe 57) was used with the relative width of 5×10^{-13}. The quantity to be measured was about $\frac{1}{100}$ of this. The result of the measurement was $\frac{\Delta\nu}{\nu}=5.1\times 10^{-15}$ with an accuracy of about 10 per cent and is in good agreement with the theoretical value of 4.9×10^{-15}.

Before these experiments were performed, other methods were much discussed; it was suggested not to use visible light nor ultra-short γ-waves, but much longer waves, namely those electromagnetic waves in the 1*cm* region which are called "radar." The technique of these rays has made great progress during the last years; it is now possible to keep the frequency of a radar emitter constant to a high degree and to measure it and its alterations with an extreme precision. One method would consist in putting the emitter and receiver on the foot and on the top of a high mountain. But this would need an accuracy of measurement not yet achieved. Another method would consist in putting the emitter in an artificial satellite, the receiver on the surface of the earth. Because of the greater difference in height the accuracy required is much less and not far from that available at present.

12. Cosmology

The idea first expressed by Ernst Mach, that the inertial forces are due to the total system of fixed stars, suggests the application of the theory of general relativity to the whole universe. This step was actually made by Einstein in 1917, and from that time dates the modern development of cosmology and cosmogony, the sciences of the structure and genesis of the cosmos. This development is still in full swing and rich in important results, though far from final conclusions. To give an account of this enormous field of research and speculation would require another book of the same extent as this one. But as the cosmological investigations are regarded by many as the most important part of Einstein's work they cannot be entirely omitted. Therefore we shall give a short outline of the present situation.

Speculations about the universe have been made since time immemorial. The ancients believed the stars fixed on a crystal sphere; the question of what was beyond was not discussed. On the other hand Aristotle regarded time as infinite. In medieval times Thomas Aquinas taught that this opinion of Aristotle's could neither be confirmed nor refuted; the creation of cosmos and time could only be based on belief. Beginning of the world in time and its finite extension was a settled doctrine in scholasticism. The idea that the cosmos could be regarded as infinite was expressed by the medieval thinker Nicolas of Cusa (1401–1464). Newton included the infinity of space and time in his fundamental principles (see p. 57) and speculated on the question of whether or not the stars were finite in number and filled only a finite part of infinite space. He came to the conclusion that the number of stars must be infinite and spread rather uniformly through space, for a finite number would collapse in consequence of their mutual attraction. Later it turned out that this argument led to mathematical difficulties of so severe a kind that even modifications of the Newtonian law of gravitation for large distances were contemplated. There is another objection to the assumption of a finite number of stars which is just the opposite of Newton's—namely, that such a system would disperse and thus vanish by dilution. The stars have considerable velocities, and these are distributed at random in all directions. Thus the system resembles the molecules of a gas and it is obvious that a gas not enclosed in rigid walls will expand and diffuse itself. The kinetic theory of gases teaches us that in order to avoid this diffusion it is not sufficient to replace the walls by mutual attractions which, as in Newton's law, are inversely proportional to the square of the distance. Therefore, it seems difficult to understand why the stellar system still exists in spite of this tendency of expansion. However, this argument has also lost its power since modern research has just revealed the actual existence of an expansion of the universe of this type, as we shall discuss presently.

There are other criteria which can be used with regard to the problem of whether the stellar system is finite or infinite. In the latter case, it has been said that the whole heaven would shine with a bright light, for, though the intensity of the light reaching us from a single star decreases with the square of the distance, the number of stars in successive spherical shells of equal thickness increases in the same rate, and if these shells would extend to infinity while the density of stars remains approximately constant the eye would see brightness in every direction. This argument was thought to be weakened by the remark that space is not quite empty: there are atoms and dust particles everywhere, mostly of an extremely small density, but sometimes concentrated in clouds, and these absorb and scatter the light passing through, thus dimming and obscuring the stars. But this argument would hold only if there were a beginning of the world in time. Otherwise there would be thermal equilibrium and the dust clouds would be just as hot and brilliant as the stars. Now this leads to the problem of the "heat death of the world," i.e., the irreversible levelling of thermal differences taught by thermodynamics. But we shall not enter into a discussion of this difficult question.

All these considerations led to no definite answer. Thus it was a decisive step when Einstein began to investigate the problem from the standpoint of his theory. He first tried to answer the traditional problem: How can matter be homogeneously dispersed in space without acquiring an outward motion and dispersing itself? But he was disappointed when he struck the same difficulties as other scholars before who had treated the problem with classical methods. As we said above, these difficulties were so severe that one was driven to trying the rather radical remedy of changing Newton's law of force for large distances. In a similar way Einstein suggested a modification of his law of gravitation, preserving, of course, the principle of general invariance, but changing his field equations in such a way that the alterations were imperceptible for the planetary system and conspicuous only for cosmic distances. He made use of the fact that his space was non-Euclidean and curved. The curvature is expressed by ten quantities

$$R_{11}, R_{22}, R_{33}, R_{44}, R_{23}, R_{31}, R_{12}, R_{14}, R_{24}, R_{34},$$

which are of the same geometrical type as the ten metric coefficients $g_{11}, g_{12}, \ldots g_{23}, \ldots g_{34}$; Einstein now replaced the $R_{11}, \ldots R_{34}$ by $R_{11}+\lambda g_{11}, \ldots R_{34}+\lambda g_{34}$, where λ is a universal constant, and assumed that these combinations are determined by the distribution of the masses in the same way as the R's before. The result was what he had hoped for—there existed a static solution (i.e., a solution independent of time) of the new equations which corresponds to a uniform density of masses (stars) in a space possessing remarkable properties: it is non-Euclidean, finite, but without boundaries.

We have to make some comments about this strange statement, that a space may be finite yet have no bounds or borders. Consider the two-dimensional case: there is no difficulty in imagining a finite but unbounded surface, e.g., a sphere. Einstein asserts that the three-dimensional space behaves in the same way; in particular, for a uniform distribution of mass it is the three-dimensional analogue of a spherical surface. A geodetic line on a sphere is a greatest circle, hence closed. The same should hold for the geodetic lines of our world, which are represented by rays of light or orbits of free particles (not disturbed by local masses). Hence a light signal or a body shot out in one direction should return from the opposite direction, after, of course, a very long time. But there are other consequences which are not completely out of reach of actual experience. In a European observatory a certain star may be photographed; at the antipodes, say at Sydney in Australia, a star in the opposite direction may be observed. Then it is conceivable than in a spherical universe both observers actually see one and the same star—just as on the surface of the earth a wireless message from the antipodes, which follows the curvature of the globe, can reach us from opposite directions. It is even conceivable that the identity of the two star images can be established by some feature of the spectrum. And if ordinary light fails us for these large journeys, there is the modern method of radar astronomy which reaches much farther out into space. Though these are only reflections on future possibilities, they show that a closed, finite, and boundless space is a possibility accessible to empirical investigations.

Concerning Einstein's statical model, it turned out that the radius of curvature of the three-dimensional spherical surface is connected to the value of the constant λ, and that both depend on the total amount of matter in the universe. The larger the total mass, the smaller the radius; if matter is dilated more and more, the curvature diminishes.

This rather simple model of the universe appeared at that time quite satisfactory, since it was in agreement with the known facts. For the observations indicated only small and irregular movements and a rather uniform distribution of the stars. But the new ideas initiated by Einstein stimulated research, and soon the aspect of the problem was thoroughly changed.

In the same year 1917, when Einstein made known his statical model of the cosmos (with the λ-term), the Dutch astronomer De Sitter published another model which was also a solution of Einstein's field equations (with the λ-term); it had the property that it existed in some way even for an "empty" world, free of matter, and if there were masses it was not static: there was a kind of cosmic repulsion between the masses which tended to drive them apart and to dilute the system. The tendency of expansion was, of course, noticeable only for very large distances. De Sitter began to look for data about the movements of very distant objects. He found in the literature a few, not very reliable records concerning the motion of the so-called spiral nebulae. These are actually enormous accumulations of stars similar to the galactic system to which our sun belongs, but so far away that most of them appear as nebulous patches, while others can be partly resolved into stars. They are now often called galaxies. At that time knowledge of these objects was scanty. But in the cases where the radial velocities could be derived from the Doppler effect (see p. 121), they showed a red shift remarkably large as compared with that of closer objects, stars in our own galaxy. These indications led to further theoretical investigations and to new and better measurements of the distances and the velocities of spiral nebulae. About 1929 the American astronomer Hubble demonstrated the existence of a strange correlation between distance and speed of the nebulae: they all move outwards, away from us, and with a velocity which increases proportional to the distance; or, in other words, the system of the spiral nebulae is expanding—just as the primitive comparison of this system with a gas had suggested to earlier thinkers. Now if one regards the expansion to have been the same in the past as it is today, one is led to the idea that the whole system must have had a beginning when all matter was condensed in a small "supernucleus," and one can calculate the time interval since this "beginning of the world" and the present instant. The result obtained from Hubble's data was 2000 to 3000 millions of years.

Meanwhile the relativistic cosmology initiated by Einstein and De Sitter began to ripen in the hands of Friedmann, Lemaître, Tolman, Robertson, and others. A series of new possible models of the world were discovered between the extreme cases found by Einstein and De Sitter, and the question arose which of them fitted the empirical facts best, in particular those facts established by Hubble. Today there are many ramifications and refinements of the theory and there has been so enormous an increase of observational material that it is difficult to judge the actual situation. Earlier ideas which seemed to be most fertile have turned out to be too narrow or even wrong. There exist nonstatic solutions of Einstein's equation which have the characteristic property of his statical model of 1917 of being finite and closed; they correspond in two dimensions to the surface of a uniformly expanding sphere, like a rubber ball being blown up. But this very finiteness and closedness of the universe which was so stirring at its first announcement has turned out not to be such a compelling idea, for there are other nonstatic solutions in which the world is infinite and "plane." One can even say that the classical model of an expanding gas, the particles of which obey Newton's law, is quite sufficient in ordinary Euclidean space to represent essential features of the observations. Such a theory could have been applied to the expanding universe 100 or 150 years ago. However, the idea of a nonstatical system of stars was foreign to these periods, and there is hardly any mention of it in the literature; only Boltzmann, one of the founders of the kinetic theory of gases and of the statistical theory of matter in general, hinted in 1895 on the possibility of expanding systems of stars, but without going into the matter seriously. As a matter of fact, such a classical treatment has to be modified, namely, for very distant and therefore very fast objects. There Newtonian mechanics breaks down and has to be replaced by the mechanical laws of special relativity. The English astronomer Milne has constructed a theory of the expanding universe from this standpoint, using only special relativity and the principle of homogeneity, which says that the aspect of the world is in general the same for any observer wherever he is placed. Milne was so convinced of the strength of his principles that he regarded them as logically compelling. He, like Eddington before him, believed that he could derive the structure of the world from a priori principles without appeal to experience—yet both advocated quite different and contradictory "a priori" foundations of their system. Neither system has proved to be fruitful for the development of science.

The λ-term which Einstein introduced in 1917 and which initiated the whole cosmological development had a tortuous fate. Weyl and Eddington interpreted it as a universal cosmic length and erected on this idea a theory surrounded by much philosophy. Later, when the wide choice of possible theories intermediate between the models of Einstein and De Sitter became clear, the λ-term appeared to be rather superfluous, and Einstein himself recommended its omission. But he and other cosmologists seem to have overlooked the fact that the λ-term was quite indispensable for reconciling the age of the universe, calculated by extrapolating backwards from Hubble's data, with the maximum age of single meteorites, stars, and star systems derived from quite different and independent observations (e.g., determining the age of meteorites found on the earth by analyzing the content of radioactive elements and their decay products, the known periods of decay then providing a kind of atomistic clock applicable to the cosmic time scale). Both ages, that of the world as a whole and that of special objects, turned out to be of the same order, some 1000 million years; but one needed the λ-term to make the world as a whole older than the special objects mentioned. The situation changed again when careful new investigations, made from 1952 onwards, of the cosmic distances revealed that these were actually larger than those accepted by Hubble. The λ-term could again be discarded without getting into trouble concerning the age of the world as computed by Hubble's formula and the radioactive measurements of the age of meteorites and other celestial objects. Derived from modern measurements, the age of the world is some 10000 million years.

The idea of a definite "beginning of the world" was so strange that efforts were made to avoid it and to replace it by a steady state. This is obviously not possible without making the assumption that matter is permanently created out of nothing—for how could the stars have an outward movement without leaving behind them a more and more dilated region? But this is not the case; there is strong evidence for a rather uniform average distribution of stars in the whole of space accessible to the largest telescopes.

Therefore theories have been suggested which assume the world to be in a steady state by creation of matter. It is a fact that so-called "new stars," "novae" and "supernovae," appear. They are usually explained as explosions of existing stars of low luminosity. But Jordan has suggested that they may be really new, that gravitational energy may be transformed into actual matter. His theory is a generalization of Einstein's; following a suggestion made by Dirac, he assumed the gravitational constant of general relativity (a generalization of Newton's constant of gravitation; see p. 341) to be actually not constant but variable and coordinated as an eleventh field quantity to the ten components of the metric field. But in spite of great efforts, no tangible result has been derived. The same holds for a theory proposed by Hoyle, Bondi, and Gold, who assume the production of single hydrogen atoms out of nothing everywhere in space, and modify Einstein's equations accordingly. Fortunately—for the inventors of this theory—it turns out that the number of atoms created is so small (of the order of 1 atom in a cube of 100 m. side length during a century) that it is far below any possible observation.

The reader may get the impression that modern cosmology has strayed from the sound empirical road into a wilderness where statements can be made without fear of observational check. Indeed, this can be said of the theories just sketched, particularly as the mixed feeling of admiration and slight disgust which they produce is enhanced by the almost fanatical assurance with which they are advertised by their authors. Unfortunately, but rather naturally, this state of affairs has been used by different ideologies to claim one of these theories as a confirmation of their dogma and to anathematize the other. There are theologians who welcome the cosmology which introduces a beginning of the world because it can be interpreted as an act of divine creation. As geology and palaeontology have already taught us, the time scales of the Bible have to be multiplied by a large factor; one has only to enlarge this factor again to interpret the biblical tale of the creation as a symbolic presentation of what science teaches. On the other hand, materialists and atheists prefer the steady state universe of the Hoyle type which avoids an act of creation and the embarrassing question: What was before this act? The *Soviet Encyclopedia* takes a rather vague position by considering the expansion as a phenomenon of limited extent in an otherwise stationary universe.

Views of this kind, preached as dogma, are foreign to the spirit of science, and each of them can be refuted by showing that it does not take all aspects into account. Those who welcome the idea of a "beginning" forget that all one can assuredly say is that this is a state of high density of matter quite distinct from the distribution of isolated stars known to us; one may doubt that in this state the notions of space and time are applicable, because these notions are intimately related to the dispersed system of stars. The "beginning" refers only to our ability to describe the state of things in terms of accustomed concepts. Whether there was a creation from nothing is not a scientific question, but a matter of belief and beyond experience, as the old philosophers and theologians like Thomas Aquinas knew. To the atheists who dislike the "beginning" because it may be interpreted as a creation must be said that the beginning of the world as known to us may be the end of another development of matter, though it will be practically impossible ever to learn anything about it because all traces have been destroyed in the turmoil of collapse and reconstruction. Finally, the official declaration of the *Soviet Encyclopedia*, with its rejection of the relativistic cosmology which is branded as "idealistic," excludes itself from any rational treatment of cosmology at all.

Eccentricities and fantasies must not obscure the fact that Einstein's ideas have opened a new aspect of the universe and given a new impetus to the old science of astronomy comparable to that of Copernicus.

13. The Unified Field Theory

We have mentioned that Einstein's law connecting energy and mass, $E=mc^2$ (p. 280), which sprang from the theory of special relativity, has its most important applications in the domain of elementary particles, nuclei, and electrons. The masses of these represent enormous concentrations of energy in very small regions of space.

Hence one should presume that they will produce considerable local curvatures of space and corresponding gravitational fields. Can these fields explain the cohesive forces which keep the particles together against the repulsion of the electric charges which they carry?

An estimate of the two forces, electric repulsion and gravitational attraction, is discouraging. Consider two electrons in the distance r; both forces have the form $\frac{\text{const.}}{r^2}$, and the value of the constant is e^2 for the electric force (formula (46), p. 153) and km^2 for the gravitational force (formula (26) p. 64), where k is the gravitational constant. Their ratio is $\frac{(e/m)^2}{k}$; and as $\frac{e}{m}\sim 5\times 10^{17}$ electrostatic units per gram $=5\times 10^{17}$ cm.$^{3/2}$/sec. gm.$^{1/2}$ (we have used the dimension of charge given on p. 153) and $k\sim 7\times 10^{-8}$ cm.^{3}/sec.^{2} gm., one obtains the enormous (dimensionless) number $\sim 3\times 10^{42}$. This seems to indicate that gravitation is much too small to explain the cohesion of the electron.

In spite of this, Einstein began, a short time after finishing the general theory of relativity, to work on a unified field theory, which was to combine the laws of electromagnetism and gravitation in one system of formulae, always hoping that he could obtain in this way not only a formal unification but an explanation of the existence of the elementary particles and their strange behavior which is commonly described with the help of quantum theory.

It is impossible to give an idea of quantum theory in the frame of this book (see p. 291). We must content ourselves with reporting that this theory, started by Max Planck in 1900, owes much of its fundamental development to Einstein himself. In fact, the same volume (1905) of the German periodical *Annalen der Physik* which contains Einstein's first and main paper on relativity, also contains his most important paper on quantum theory, which transformed it from a strange hypothesis into statements open to experimental research. Quantum theory is an extension of classical mechanics in another direction than relativity; while the latter modified our concepts of space and time, quantum theory altered our attitude to the concept of causality. Classical theory uses differential equations which are of a deterministic character, since they allow us to predict the future from present observations in a rigorous way. The laws of quantum theory are of a statistical character and allow us only to predict probabilities for future events. Einstein himself has made the most important steps in this direction. But when later these ideas were systematized in so-called *quantum mechanics*, he did not follow these developments. He believed that they gave only an incomplete description of nature and that they ought to be reduced to laws of the deterministic type.

This conviction was the driving force for his untiring efforts to evolve a unified theory in which he expected to give not only an explanation of the existence of the proton and the electron but also all the results usually described by quantum mechanics. He published many versions of unified theories, all based on generalizations of his theory of the metric field. Other prominent scientists—Weyl, Eddington, Schrödinger—worked in the same direction. But when Einstein died (1956) his aim seemed to be as far away as ever.

Most of the physicists were skeptical about this work. Their main argument came from the fact that other elementary particles (neutron, different mesons, hyperons, neutrinos, etc.) have been found which are unstable and disintegrate like radioactive atoms into other particles. Quantum theory associates with each of these a type of field (wave fields of matter, according to de Broglie), and a unified field theory as envisaged by Einstein ought to embrace these all.

Thus the problem has shifted and must be answered from a much more general standpoint. Still, to Einstein is due the merit of having insisted on the importance of the problem, namely, to find all-embracing laws which unify the whole of the physical world.

Modern research follows Einstein on this way to unification, though not in his conviction that the unified laws ought to be of the classical, deterministic character. There are two directions of research. One accepts only special relativity and tries to construct the universal law from the facts of observation, in particular from the most general symmetry properties of the interaction of different elementary particles as revealed by experiment. The main feature of the new laws is their "nonlinearity," which means that different terms of the equations are of different degree in the unknown quantities. (To illustrate this, it may be said that in classical physics all laws of wave propagation—as, for example, Maxwell's equations—are linear, a statement which implies that waves can penetrate one another without interaction. Nonlinearity occurs in hydrodynamics, which describes the motion of liquids; two currents cannot be superposed without mutual disturbance.) One has no other guide for guessing the right laws than the ideal of simplicity. Such is the approach to the problem chosen, for instance, by Heisenberg.

The other direction of research tends to establish laws which are invariant in regard to general transformations, and follows therefore Einstein's general relativity in its procedure. Now Einstein's laws of gravitation are nonlinear, and therefore this way of introducing nonlinearity (as has been tried, for example, by Green) seems to be the most natural one.

Whether one or the other of these methods will lead to the anticipated "world law" must be left to future research.

14. Conclusion

The outline of modern physics and its astronomic applications given in the concluding sections shows that Einstein's ideas have lost none of their power since their first enunciation more than half a century ago. They have given the physical sciences the impetus which has liberated them from outdated philosophical doctrine and made them one of the decisive factors in the modern world of man.

INDEX

A CATALOG OF SELECTED

DOVER BOOKS

IN ALL FIELDS OF INTEREST

A CATALOG OF SELECTED DOVER BOOKS IN ALL FIELDS OF INTEREST

STICKLEY CRAFTSMAN FURNITURE CATALOGS, Gustav Stickley and L. & J. G. Stickley. Beautiful, functional furniture in two authentic catalogs from 1910. 594 illustrations, including 277 photos, show settles, rockers, armchairs, reclining chairs, bookcases, desks, tables. 183pp. 6½ x 9¼. 23838-5 Pa. $9.95

AMERICAN LOCOMOTIVES IN HISTORIC PHOTOGRAPHS: 1858 to 1949, Ron Ziel (ed.). A rare collection of 126 meticulously detailed official photographs, called "builder portraits," of American locomotives that majestically chronicle the rise of steam locomotive power in America. Introduction. Detailed captions. xi + 129pp. 9 x 12. 27393-8 Pa. $12.95

AMERICA'S LIGHTHOUSES: An Illustrated History, Francis Ross Holland, Jr. Delightfully written, profusely illustrated fact-filled survey of over 200 American lighthouses since 1716. History, anecdotes, technological advances, more. 240pp. 8 x 10¾. 25576-X Pa. $12.95

TOWARDS A NEW ARCHITECTURE, Le Corbusier. Pioneering manifesto by founder of "International School." Technical and aesthetic theories, views of industry, economics, relation of form to function, "mass-production split" and much more. Profusely illustrated. 320pp. 6⅛ x 9¼. (USO) 25023-7 Pa. $9.95

HOW THE OTHER HALF LIVES, Jacob Riis. Famous journalistic record, exposing poverty and degradation of New York slums around 1900, by major social reformer. 100 striking and influential photographs. 233pp. 10 x 7⅞. 22012-5 Pa. $10.95

FRUIT KEY AND TWIG KEY TO TREES AND SHRUBS, William M. Harlow. One of the handiest and most widely used identification aids. Fruit key covers 120 deciduous and evergreen species; twig key 160 deciduous species. Easily used. Over 300 photographs. 126pp. 5⅜ x 8½. 20511-8 Pa. $3.95

COMMON BIRD SONGS, Dr. Donald J. Borror. Songs of 60 most common U.S. birds: robins, sparrows, cardinals, bluejays, finches, more–arranged in order of increasing complexity. Up to 9 variations of songs of each species. Cassette and manual 99911-4 $8.95

ORCHIDS AS HOUSE PLANTS, Rebecca Tyson Northen. Grow cattleyas and many other kinds of orchids–in a window, in a case, or under artificial light. 63 illustrations. 148pp. 5⅜ x 8½. 23261-1 Pa. $4.95

MONSTER MAZES, Dave Phillips. Masterful mazes at four levels of difficulty. Avoid deadly perils and evil creatures to find magical treasures. Solutions for all 32 exciting illustrated puzzles. 48pp. 8¼ x 11. 26005-4 Pa. $2.95

MOZART'S DON GIOVANNI (DOVER OPERA LIBRETTO SERIES), Wolfgang Amadeus Mozart. Introduced and translated by Ellen H. Bleiler. Standard Italian libretto, with complete English translation. Convenient and thoroughly portable–an ideal companion for reading along with a recording or the performance itself. Introduction. List of characters. Plot summary. 121pp. 5¼ x 8½. 24944-1 Pa. $2.95

TECHNICAL MANUAL AND DICTIONARY OF CLASSICAL BALLET, Gail Grant. Defines, explains, comments on steps, movements, poses and concepts. 15-page pictorial section. Basic book for student, viewer. 127pp. 5⅜ x 8½. 21843-0 Pa. $4.95

BRASS INSTRUMENTS: Their History and Development, Anthony Baines. Authoritative, updated survey of the evolution of trumpets, trombones, bugles, cornets, French horns, tubas and other brass wind instruments. Over 140 illustrations and 48 music examples. Corrected and updated by author. New preface. Bibliography. 320pp. 5⅜ x 8½. 27574-4 Pa. $9.95

HOLLYWOOD GLAMOR PORTRAITS, John Kobal (ed.). 145 photos from 1926-49. Harlow, Gable, Bogart, Bacall; 94 stars in all. Full background on photographers, technical aspects. 160pp. 8⅜ x 11¼. 23352-9 Pa. $11.95

MAX AND MORITZ, Wilhelm Busch. Great humor classic in both German and English. Also 10 other works: "Cat and Mouse," "Plisch and Plumm," etc. 216pp. 5⅜ x 8½. 20181-3 Pa. $6.95

THE RAVEN AND OTHER FAVORITE POEMS, Edgar Allan Poe. Over 40 of the author's most memorable poems: "The Bells," "Ulalume," "Israfel," "To Helen," "The Conqueror Worm," "Eldorado," "Annabel Lee," many more. Alphabetic lists of titles and first lines. 64pp. 5 3/16 x 8¼. 26685-0 Pa. $1.00

PERSONAL MEMOIRS OF U. S. GRANT, Ulysses Simpson Grant. Intelligent, deeply moving firsthand account of Civil War campaigns, considered by many the finest military memoirs ever written. Includes letters, historic photographs, maps and more. 528pp. 6⅛ x 9¼. 28587-1 Pa. $11.95

AMULETS AND SUPERSTITIONS, E. A. Wallis Budge. Comprehensive discourse on origin, powers of amulets in many ancient cultures: Arab, Persian Babylonian, Assyrian, Egyptian, Gnostic, Hebrew, Phoenician, Syriac, etc. Covers cross, swastika, crucifix, seals, rings, stones, etc. 584pp. 5⅜ x 8½. 23573-4 Pa. $12.95

RUSSIAN STORIES/PYCCKNE PACCKA3bl: A Dual-Language Book, edited by Gleb Struve. Twelve tales by such masters as Chekhov, Tolstoy, Dostoevsky, Pushkin, others. Excellent word-for-word English translations on facing pages, plus teaching and study aids, Russian/English vocabulary, biographical/critical introductions, more. 416pp. 5⅜ x 8½. 26244-8 Pa. $8.95

PHILADELPHIA THEN AND NOW: 60 Sites Photographed in the Past and Present, Kenneth Finkel and Susan Oyama. Rare photographs of City Hall, Logan Square, Independence Hall, Betsy Ross House, other landmarks juxtaposed with contemporary views. Captures changing face of historic city. Introduction. Captions. 128pp. 8¼ x 11. 25790-8 Pa. $9.95

AIA ARCHITECTURAL GUIDE TO NASSAU AND SUFFOLK COUNTIES, LONG ISLAND, The American Institute of Architects, Long Island Chapter, and the Society for the Preservation of Long Island Antiquities. Comprehensive, well-researched and generously illustrated volume brings to life over three centuries of Long Island's great architectural heritage. More than 240 photographs with authoritative, extensively detailed captions. 176pp. 8¼ x 11. 26946-9 Pa. $14.95

NORTH AMERICAN INDIAN LIFE: Customs and Traditions of 23 Tribes, Elsie Clews Parsons (ed.). 27 fictionalized essays by noted anthropologists examine religion, customs, government, additional facets of life among the Winnebago, Crow, Zuni, Eskimo, other tribes. 480pp. 6⅛ x 9¼. 27377-6 Pa. $10.95

THE BEST TALES OF HOFFMANN, E. T. A. Hoffmann. 10 of Hoffmann's most important stories: "Nutcracker and the King of Mice," "The Golden Flowerpot," etc. 458pp. 5⅜ x 8½. 21793-0 Pa. $9.95

FROM FETISH TO GOD IN ANCIENT EGYPT, E. A. Wallis Budge. Rich detailed survey of Egyptian conception of "God" and gods, magic, cult of animals, Osiris, more. Also, superb English translations of hymns and legends. 240 illustrations. 545pp. 5⅜ x 8½. 25803-3 Pa. $11.95

FRENCH STORIES/CONTES FRANÇAIS: A Dual-Language Book, Wallace Fowlie. Ten stories by French masters, Voltaire to Camus: "Micromegas" by Voltaire; "The Atheist's Mass" by Balzac; "Minuet" by de Maupassant; "The Guest" by Camus, six more. Excellent English translations on facing pages. Also French-English vocabulary list, exercises, more. 352pp. 5⅜ x 8½. 26443-2 Pa. $8.95

CHICAGO AT THE TURN OF THE CENTURY IN PHOTOGRAPHS: 122 Historic Views from the Collections of the Chicago Historical Society, Larry A. Viskochil. Rare large-format prints offer detailed views of City Hall, State Street, the Loop, Hull House, Union Station, many other landmarks, circa 1904-1913. Introduction. Captions. Maps. 144pp. 9⅜ x 12¼. 24656-6 Pa. $12.95

OLD BROOKLYN IN EARLY PHOTOGRAPHS, 1865-1929, William Lee Younger. Luna Park, Gravesend race track, construction of Grand Army Plaza, moving of Hotel Brighton, etc. 157 previously unpublished photographs. 165pp. 8⅞ x 11¾. 23587-4 Pa. $13.95

THE MYTHS OF THE NORTH AMERICAN INDIANS, Lewis Spence. Rich anthology of the myths and legends of the Algonquins, Iroquois, Pawnees and Sioux, prefaced by an extensive historical and ethnological commentary. 36 illustrations. 480pp. 5⅜ x 8½. 25967-6 Pa. $8.95

AN ENCYCLOPEDIA OF BATTLES: Accounts of Over 1,560 Battles from 1479 B.C. to the Present, David Eggenberger. Essential details of every major battle in recorded history from the first battle of Megiddo in 1479 B.C. to Grenada in 1984. List of Battle Maps. New Appendix covering the years 1967-1984. Index. 99 illustrations. 544pp. 6½ x 9¼. 24913-1 Pa. $14.95

SAILING ALONE AROUND THE WORLD, Captain Joshua Slocum. First man to sail around the world, alone, in small boat. One of great feats of seamanship told in delightful manner. 67 illustrations. 294pp. 5⅜ x 8½. 20326-3 Pa. $5.95

ANARCHISM AND OTHER ESSAYS, Emma Goldman. Powerful, penetrating, prophetic essays on direct action, role of minorities, prison reform, puritan hypocrisy, violence, etc. 271pp. 5⅜ x 8½. 22484-8 Pa. $6.95

MYTHS OF THE HINDUS AND BUDDHISTS, Ananda K. Coomaraswamy and Sister Nivedita. Great stories of the epics; deeds of Krishna, Shiva, taken from puranas, Vedas, folk tales; etc. 32 illustrations. 400pp. 5⅜ x 8½. 21759-0 Pa. $10.95

BEYOND PSYCHOLOGY, Otto Rank. Fear of death, desire of immortality, nature of sexuality, social organization, creativity, according to Rankian system. 291pp. 5⅜ x 8½. 20485-5 Pa. $8.95

A THEOLOGICO-POLITICAL TREATISE, Benedict Spinoza. Also contains unfinished Political Treatise. Great classic on religious liberty, theory of government on common consent. R. Elwes translation. Total of 421pp. 5⅜ x 8½. 20249-6 Pa. $9.95

PERSPECTIVE FOR ARTISTS, Rex Vicat Cole. Depth, perspective of sky and sea, shadows, much more, not usually covered. 391 diagrams, 81 reproductions of drawings and paintings. 279pp. 5⅜ x 8½. 22487-2 Pa. $6.95

DRAWING THE LIVING FIGURE, Joseph Sheppard. Innovative approach to artistic anatomy focuses on specifics of surface anatomy, rather than muscles and bones. Over 170 drawings of live models in front, back and side views, and in widely varying poses. Accompanying diagrams. 177 illustrations. Introduction. Index. 144pp. 8⅜ x11¼. 26723-7 Pa. $8.95

GOTHIC AND OLD ENGLISH ALPHABETS: 100 Complete Fonts, Dan X. Solo. Add power, elegance to posters, signs, other graphics with 100 stunning copyright-free alphabets: Blackstone, Dolbey, Germania, 97 more–including many lower-case, numerals, punctuation marks. 104pp. 8⅛ x 11. 24695-7 Pa. $8.95

HOW TO DO BEADWORK, Mary White. Fundamental book on craft from simple projects to five-bead chains and woven works. 106 illustrations. 142pp. 5⅜ x 8. 20697-1 Pa. $4.95

THE BOOK OF WOOD CARVING, Charles Marshall Sayers. Finest book for beginners discusses fundamentals and offers 34 designs. "Absolutely first rate . . . well thought out and well executed."–E. J. Tangerman. 118pp. 7¾ x 10⅝. 23654-4 Pa. $6.95

ILLUSTRATED CATALOG OF CIVIL WAR MILITARY GOODS: Union Army Weapons, Insignia, Uniform Accessories, and Other Equipment, Schuyler, Hartley, and Graham. Rare, profusely illustrated 1846 catalog includes Union Army uniform and dress regulations, arms and ammunition, coats, insignia, flags, swords, rifles, etc. 226 illustrations. 160pp. 9 x 12. 24939-5 Pa. $10.95

WOMEN'S FASHIONS OF THE EARLY 1900s: An Unabridged Republication of "New York Fashions, 1909," National Cloak & Suit Co. Rare catalog of mail-order fashions documents women's and children's clothing styles shortly after the turn of the century. Captions offer full descriptions, prices. Invaluable resource for fashion, costume historians. Approximately 725 illustrations. 128pp. 8⅜ x 11¼. 27276-1 Pa. $11.95

THE 1912 AND 1915 GUSTAV STICKLEY FURNITURE CATALOGS, Gustav Stickley. With over 200 detailed illustrations and descriptions, these two catalogs are essential reading and reference materials and identification guides for Stickley furniture. Captions cite materials, dimensions and prices. 112pp. 6½ x 9¼. 26676-1 Pa. $9.95

EARLY AMERICAN LOCOMOTIVES, John H. White, Jr. Finest locomotive engravings from early 19th century: historical (1804–74), main-line (after 1870), special, foreign, etc. 147 plates. 142pp. 11⅜ x 8¼. 22772-3 Pa. $10.95

THE TALL SHIPS OF TODAY IN PHOTOGRAPHS, Frank O. Braynard. Lavishly illustrated tribute to nearly 100 majestic contemporary sailing vessels: Amerigo Vespucci, Clearwater, Constitution, Eagle, Mayflower, Sea Cloud, Victory, many more. Authoritative captions provide statistics, background on each ship. 190 black-and-white photographs and illustrations. Introduction. 128pp. 8⅞ x 11¾. 27163-3 Pa. $13.95

EARLY NINETEENTH-CENTURY CRAFTS AND TRADES, Peter Stockham (ed.). Extremely rare 1807 volume describes to youngsters the crafts and trades of the day: brickmaker, weaver, dressmaker, bookbinder, ropemaker, saddler, many more. Quaint prose, charming illustrations for each craft. 20 black-and-white line illustrations. 192pp. 4⅝ x 6. 27293-1 Pa. $4.95

VICTORIAN FASHIONS AND COSTUMES FROM HARPER'S BAZAR, 1867–1898, Stella Blum (ed.). Day costumes, evening wear, sports clothes, shoes, hats, other accessories in over 1,000 detailed engravings. 320pp. 9⅜ x 12¼. 22990-4 Pa. $14.95

GUSTAV STICKLEY, THE CRAFTSMAN, Mary Ann Smith. Superb study surveys broad scope of Stickley's achievement, especially in architecture. Design philosophy, rise and fall of the Craftsman empire, descriptions and floor plans for many Craftsman houses, more. 86 black-and-white halftones. 31 line illustrations. Introduction 208pp. 6½ x 9¼. 27210-9 Pa. $9.95

THE LONG ISLAND RAIL ROAD IN EARLY PHOTOGRAPHS, Ron Ziel. Over 220 rare photos, informative text document origin (1844) and development of rail service on Long Island. Vintage views of early trains, locomotives, stations, passengers, crews, much more. Captions. 8⅞ x 11¾. 26301-0 Pa. $13.95

THE BOOK OF OLD SHIPS: From Egyptian Galleys to Clipper Ships, Henry B. Culver. Superb, authoritative history of sailing vessels, with 80 magnificent line illustrations. Galley, bark, caravel, longship, whaler, many more. Detailed, informative text on each vessel by noted naval historian. Introduction. 256pp. 5⅜ x 8½. 27332-6 Pa. $7.95

TEN BOOKS ON ARCHITECTURE, Vitruvius. The most important book ever written on architecture. Early Roman aesthetics, technology, classical orders, site selection, all other aspects. Morgan translation. 331pp. 5⅜ x 8½. 20645-9 Pa. $8.95

THE HUMAN FIGURE IN MOTION, Eadweard Muybridge. More than 4,500 stopped-action photos, in action series, showing undraped men, women, children jumping, lying down, throwing, sitting, wrestling, carrying, etc. 390pp. 7⅞ x 10⅝. 20204-6 Clothbd. $25.95

TREES OF THE EASTERN AND CENTRAL UNITED STATES AND CANADA, William M. Harlow. Best one-volume guide to 140 trees. Full descriptions, woodlore, range, etc. Over 600 illustrations. Handy size. 288pp. 4½ x 6⅜. 20395-6 Pa. $5.95

SONGS OF WESTERN BIRDS, Dr. Donald J. Borror. Complete song and call repertoire of 60 western species, including flycatchers, juncoes, cactus wrens, many more–includes fully illustrated booklet. Cassette and manual 99913-0 $8.95

GROWING AND USING HERBS AND SPICES, Milo Miloradovich. Versatile handbook provides all the information needed for cultivation and use of all the herbs and spices available in North America. 4 illustrations. Index. Glossary. 236pp. 5⅜ x 8½. 25058-X Pa. $6.95

BIG BOOK OF MAZES AND LABYRINTHS, Walter Shepherd. 50 mazes and labyrinths in all–classical, solid, ripple, and more–in one great volume. Perfect inexpensive puzzler for clever youngsters. Full solutions. 112pp. 8⅛ x 11. 22951-3 Pa. $4.95

PIANO TUNING, J. Cree Fischer. Clearest, best book for beginner, amateur. Simple repairs, raising dropped notes, tuning by easy method of flattened fifths. No previous skills needed. 4 illustrations. 201pp. 5⅜ x 8½. 23267-0 Pa. $6.95

A SOURCE BOOK IN THEATRICAL HISTORY, A. M. Nagler. Contemporary observers on acting, directing, make-up, costuming, stage props, machinery, scene design, from Ancient Greece to Chekhov. 611pp. 5⅜ x 8½. 20515-0 Pa. $12.95

THE COMPLETE NONSENSE OF EDWARD LEAR, Edward Lear. All nonsense limericks, zany alphabets, Owl and Pussycat, songs, nonsense botany, etc., illustrated by Lear. Total of 320pp. 5⅜ x 8½. (USO) 20167-8 Pa. $6.95

VICTORIAN PARLOUR POETRY: An Annotated Anthology, Michael R. Turner. 117 gems by Longfellow, Tennyson, Browning, many lesser-known poets. "The Village Blacksmith," "Curfew Must Not Ring Tonight," "Only a Baby Small," dozens more, often difficult to find elsewhere. Index of poets, titles, first lines. xxiii + 325pp. 5⅝ x 8¼. 27044-0 Pa. $8.95

DUBLINERS, James Joyce. Fifteen stories offer vivid, tightly focused observations of the lives of Dublin's poorer classes. At least one, "The Dead," is considered a masterpiece. Reprinted complete and unabridged from standard edition. 160pp. 5$\frac{3}{16}$ x 8¼. 26870-5 Pa. $1.00

THE HAUNTED MONASTERY and THE CHINESE MAZE MURDERS, Robert van Gulik. Two full novels by van Gulik, set in 7th-century China, continue adventures of Judge Dee and his companions. An evil Taoist monastery, seemingly supernatural events; overgrown topiary maze hides strange crimes. 27 illustrations. 328pp. 5⅜ x 8½. 23502-5 Pa. $8.95

THE BOOK OF THE SACRED MAGIC OF ABRAMELIN THE MAGE, translated by S. MacGregor Mathers. Medieval manuscript of ceremonial magic. Basic document in Aleister Crowley, Golden Dawn groups. 268pp. 5⅜ x 8½. 23211-5 Pa. $8.95

NEW RUSSIAN-ENGLISH AND ENGLISH-RUSSIAN DICTIONARY, M. A. O'Brien. This is a remarkably handy Russian dictionary, containing a surprising amount of information, including over 70,000 entries. 366pp. 4½ x 6⅛. 20208-9 Pa. $9.95

HISTORIC HOMES OF THE AMERICAN PRESIDENTS, Second, Revised Edition, Irvin Haas. A traveler's guide to American Presidential homes, most open to the public, depicting and describing homes occupied by every American President from George Washington to George Bush. With visiting hours, admission charges, travel routes. 175 photographs. Index. 160pp. 8¼ x 11. 26751-2 Pa. $11.95

NEW YORK IN THE FORTIES, Andreas Feininger. 162 brilliant photographs by the well-known photographer, formerly with *Life* magazine. Commuters, shoppers, Times Square at night, much else from city at its peak. Captions by John von Hartz. 181pp. 9¼ x 10¾. 23585-8 Pa. $12.95

INDIAN SIGN LANGUAGE, William Tomkins. Over 525 signs developed by Sioux and other tribes. Written instructions and diagrams. Also 290 pictographs. 111pp. 6⅛ x 9¼. 22029-X Pa. $3.95

ANATOMY: A Complete Guide for Artists, Joseph Sheppard. A master of figure drawing shows artists how to render human anatomy convincingly. Over 460 illustrations. 224pp. 8⅜ x 11¼. 27279-6 Pa. $10.95

MEDIEVAL CALLIGRAPHY: Its History and Technique, Marc Drogin. Spirited history, comprehensive instruction manual covers 13 styles (ca. 4th century thru 15th). Excellent photographs; directions for duplicating medieval techniques with modern tools. 224pp. 8⅜ x 11¼. 26142-5 Pa. $11.95

DRIED FLOWERS: How to Prepare Them, Sarah Whitlock and Martha Rankin. Complete instructions on how to use silica gel, meal and borax, perlite aggregate, sand and borax, glycerine and water to create attractive permanent flower arrangements. 12 illustrations. 32pp. 5⅜ x 8½. 21802-3 Pa. $1.00

EASY-TO-MAKE BIRD FEEDERS FOR WOODWORKERS, Scott D. Campbell. Detailed, simple-to-use guide for designing, constructing, caring for and using feeders. Text, illustrations for 12 classic and contemporary designs. 96pp. 5⅜ x 8½. 25847-5 Pa. $2.95

SCOTTISH WONDER TALES FROM MYTH AND LEGEND, Donald A. Mackenzie. 16 lively tales tell of giants rumbling down mountainsides, of a magic wand that turns stone pillars into warriors, of gods and goddesses, evil hags, powerful forces and more. 240pp. 5⅜ x 8½. 29677-6 Pa. $6.95

THE HISTORY OF UNDERCLOTHES, C. Willett Cunnington and Phyllis Cunnington. Fascinating, well-documented survey covering six centuries of English undergarments, enhanced with over 100 illustrations: 12th-century laced-up bodice, footed long drawers (1795), 19th-century bustles, 19th-century corsets for men, Victorian "bust improvers," much more. 272pp. 5⅝ x 8¼. 27124-2 Pa. $9.95

ARTS AND CRAFTS FURNITURE: The Complete Brooks Catalog of 1912, Brooks Manufacturing Co. Photos and detailed descriptions of more than 150 now very collectible furniture designs from the Arts and Crafts movement depict davenports, settees, buffets, desks, tables, chairs, bedsteads, dressers and more, all built of solid, quarter-sawed oak. Invaluable for students and enthusiasts of antiques, Americana and the decorative arts. 80pp. 6½ x 9¼. 27471-3 Pa. $7.95

HOW WE INVENTED THE AIRPLANE: An Illustrated History, Orville Wright. Fascinating firsthand account covers early experiments, construction of planes and motors, first flights, much more. Introduction and commentary by Fred C. Kelly. 76 photographs. 96pp. 8¼ x 11. 25662-6 Pa. $8.95

THE ARTS OF THE SAILOR: Knotting, Splicing and Ropework, Hervey Garrett Smith. Indispensable shipboard reference covers tools, basic knots and useful hitches; handsewing and canvas work, more. Over 100 illustrations. Delightful reading for sea lovers. 256pp. 5⅜ x 8½. 26440-8 Pa. $7.95

FRANK LLOYD WRIGHT'S FALLINGWATER: The House and Its History, Second, Revised Edition, Donald Hoffmann. A total revision–both in text and illustrations–of the standard document on Fallingwater, the boldest, most personal architectural statement of Wright's mature years, updated with valuable new material from the recently opened Frank Lloyd Wright Archives. "Fascinating"–*The New York Times.* 116 illustrations. 128pp. 9¼ x 10¾. 27430-6 Pa. $11.95

AUTOBIOGRAPHY: The Story of My Experiments with Truth, Mohandas K. Gandhi. Boyhood, legal studies, purification, the growth of the Satyagraha (nonviolent protest) movement. Critical, inspiring work of the man responsible for the freedom of India. 480pp. 5⅜ x 8½. (USO) 24593-4 Pa. $8.95

CELTIC MYTHS AND LEGENDS, T. W. Rolleston. Masterful retelling of Irish and Welsh stories and tales. Cuchulain, King Arthur, Deirdre, the Grail, many more. First paperback edition. 58 full-page illustrations. 512pp. 5⅜ x 8½. 26507-2 Pa. $9.95

THE PRINCIPLES OF PSYCHOLOGY, William James. Famous long course complete, unabridged. Stream of thought, time perception, memory, experimental methods; great work decades ahead of its time. 94 figures. 1,391pp. 5⅜ x 8½. 2-vol. set.
Vol. I: 20381-6 Pa. $12.95
Vol. II: 20382-4 Pa. $12.95

THE WORLD AS WILL AND REPRESENTATION, Arthur Schopenhauer. Definitive English translation of Schopenhauer's life work, correcting more than 1,000 errors, omissions in earlier translations. Translated by E. F. J. Payne. Total of 1,269pp. 5⅜ x 8½. 2-vol. set. Vol. 1: 21761-2 Pa. $11.95
Vol. 2: 21762-0 Pa. $11.95

MAGIC AND MYSTERY IN TIBET, Madame Alexandra David-Neel. Experiences among lamas, magicians, sages, sorcerers, Bonpa wizards. A true psychic discovery. 32 illustrations. 321pp. 5⅜ x 8½. (USO) 22682-4 Pa. $8.95

THE EGYPTIAN BOOK OF THE DEAD, E. A. Wallis Budge. Complete reproduction of Ani's papyrus, finest ever found. Full hieroglyphic text, interlinear transliteration, word-for-word translation, smooth translation. 533pp. 6½ x 9¼.
21866-X Pa. $10.95

MATHEMATICS FOR THE NONMATHEMATICIAN, Morris Kline. Detailed, college-level treatment of mathematics in cultural and historical context, with numerous exercises. Recommended Reading Lists. Tables. Numerous figures. 641pp. 5⅜ x 8½.
24823-2 Pa. $11.95

THEORY OF WING SECTIONS: Including a Summary of Airfoil Data, Ira H. Abbott and A. E. von Doenhoff. Concise compilation of subsonic aerodynamic characteristics of NACA wing sections, plus description of theory. 350pp. of tables. 693pp. 5⅜ x 8½. 60586-8 Pa. $14.95

THE RIME OF THE ANCIENT MARINER, Gustave Doré, S. T. Coleridge. Doré's finest work; 34 plates capture moods, subtleties of poem. Flawless full-size reproductions printed on facing pages with authoritative text of poem. "Beautiful. Simply beautiful."–*Publisher's Weekly*. 77pp. 9¼ x 12. 22305-1 Pa. $6.95

NORTH AMERICAN INDIAN DESIGNS FOR ARTISTS AND CRAFTSPEOPLE, Eva Wilson. Over 360 authentic copyright-free designs adapted from Navajo blankets, Hopi pottery, Sioux buffalo hides, more. Geometrics, symbolic figures, plant and animal motifs, etc. 128pp. 8⅜ x 11. (EUK) 25341-4 Pa. $8.95

SCULPTURE: Principles and Practice, Louis Slobodkin. Step-by-step approach to clay, plaster, metals, stone; classical and modern. 253 drawings, photos. 255pp. 8⅛ x 11.
22960-2 Pa. $10.95

PHOTOGRAPHIC SKETCHBOOK OF THE CIVIL WAR, Alexander Gardner. 100 photos taken on field during the Civil War. Famous shots of Manassas Harper's Ferry, Lincoln, Richmond, slave pens, etc. 244pp. 10⅝ x 8¼. 22731-6 Pa. $9.95

FIVE ACRES AND INDEPENDENCE, Maurice G. Kains. Great back-to-the-land classic explains basics of self-sufficient farming. The one book to get. 95 illustrations. 397pp. 5⅜ x 8½. 20974-1 Pa. $7.95

SONGS OF EASTERN BIRDS, Dr. Donald J. Borror. Songs and calls of 60 species most common to eastern U.S.: warblers, woodpeckers, flycatchers, thrushes, larks, many more in high-quality recording. Cassette and manual 99912-2 $8.95

A MODERN HERBAL, Margaret Grieve. Much the fullest, most exact, most useful compilation of herbal material. Gigantic alphabetical encyclopedia, from aconite to zedoary, gives botanical information, medical properties, folklore, economic uses, much else. Indispensable to serious reader. 161 illustrations. 888pp. 6½ x 9¼. 2-vol. set. (USO)
Vol. I: 22798-7 Pa. $9.95
Vol. II: 22799-5 Pa. $9.95

HIDDEN TREASURE MAZE BOOK, Dave Phillips. Solve 34 challenging mazes accompanied by heroic tales of adventure. Evil dragons, people-eating plants, bloodthirsty giants, many more dangerous adversaries lurk at every twist and turn. 34 mazes, stories, solutions. 48pp. 8¼ x 11. 24566-7 Pa. $2.95

LETTERS OF W. A. MOZART, Wolfgang A. Mozart. Remarkable letters show bawdy wit, humor, imagination, musical insights, contemporary musical world; includes some letters from Leopold Mozart. 276pp. 5⅜ x 8½. 22859-2 Pa. $7.95

BASIC PRINCIPLES OF CLASSICAL BALLET, Agrippina Vaganova. Great Russian theoretician, teacher explains methods for teaching classical ballet. 118 illustrations. 175pp. 5⅜ x 8½. 22036-2 Pa. $5.95

THE JUMPING FROG, Mark Twain. Revenge edition. The original story of The Celebrated Jumping Frog of Calaveras County, a hapless French translation, and Twain's hilarious "retranslation" from the French. 12 illustrations. 66pp. 5⅜ x 8½.
22686-7 Pa. $3.95

BEST REMEMBERED POEMS, Martin Gardner (ed.). The 126 poems in this superb collection of 19th- and 20th-century British and American verse range from Shelley's "To a Skylark" to the impassioned "Renascence" of Edna St. Vincent Millay and to Edward Lear's whimsical "The Owl and the Pussycat." 224pp. 5⅜ x 8½.
27165-X Pa. $4.95

COMPLETE SONNETS, William Shakespeare. Over 150 exquisite poems deal with love, friendship, the tyranny of time, beauty's evanescence, death and other themes in language of remarkable power, precision and beauty. Glossary of archaic terms. 80pp. 5 3/16 x 8¼. 26686-9 Pa. $1.00

BODIES IN A BOOKSHOP, R. T. Campbell. Challenging mystery of blackmail and murder with ingenious plot and superbly drawn characters. In the best tradition of British suspense fiction. 192pp. 5⅜ x 8½. 24720-1 Pa. $6.95

THE WIT AND HUMOR OF OSCAR WILDE, Alvin Redman (ed.). More than 1,000 ripostes, paradoxes, wisecracks: Work is the curse of the drinking classes; I can resist everything except temptation; etc. 258pp. 5⅜ x 8½. 20602-5 Pa. $5.95

SHAKESPEARE LEXICON AND QUOTATION DICTIONARY, Alexander Schmidt. Full definitions, locations, shades of meaning in every word in plays and poems. More than 50,000 exact quotations. 1,485pp. 6½ x 9¼. 2-vol. set.
Vol. 1: 22726-X Pa. $16.95
Vol. 2: 22727-8 Pa. $16.95

SELECTED POEMS, Emily Dickinson. Over 100 best-known, best-loved poems by one of America's foremost poets, reprinted from authoritative early editions. No comparable edition at this price. Index of first lines. 64pp. 5$\frac{3}{16}$ x 8¼.
26466-1 Pa. $1.00

CELEBRATED CASES OF JUDGE DEE (DEE GOONG AN), translated by Robert van Gulik. Authentic 18th-century Chinese detective novel; Dee and associates solve three interlocked cases. Led to van Gulik's own stories with same characters. Extensive introduction. 9 illustrations. 237pp. 5⅜ x 8½. 23337-5 Pa. $6.95

THE MALLEUS MALEFICARUM OF KRAMER AND SPRENGER, translated by Montague Summers. Full text of most important witchhunter's "bible," used by both Catholics and Protestants. 278pp. 6⅝ x 10. 22802-9 Pa. $12.95

SPANISH STORIES/CUENTOS ESPAÑOLES: A Dual-Language Book, Angel Flores (ed.). Unique format offers 13 great stories in Spanish by Cervantes, Borges, others. Faithful English translations on facing pages. 352pp. 5⅜ x 8½.
25399-6 Pa. $8.95

THE CHICAGO WORLD'S FAIR OF 1893: A Photographic Record, Stanley Appelbaum (ed.). 128 rare photos show 200 buildings, Beaux-Arts architecture, Midway, original Ferris Wheel, Edison's kinetoscope, more. Architectural emphasis; full text. 116pp. 8¼ x 11. 23990-X Pa. $9.95

OLD QUEENS, N.Y., IN EARLY PHOTOGRAPHS, Vincent F. Seyfried and William Asadorian. Over 160 rare photographs of Maspeth, Jamaica, Jackson Heights, and other areas. Vintage views of DeWitt Clinton mansion, 1939 World's Fair and more. Captions. 192pp. 8⅞ x 11. 26358-4 Pa. $12.95

CAPTURED BY THE INDIANS: 15 Firsthand Accounts, 1750-1870, Frederick Drimmer. Astounding true historical accounts of grisly torture, bloody conflicts, relentless pursuits, miraculous escapes and more, by people who lived to tell the tale. 384pp. 5⅜ x 8½. 24901-8 Pa. $8.95

THE WORLD'S GREAT SPEECHES, Lewis Copeland and Lawrence W. Lamm (eds.). Vast collection of 278 speeches of Greeks to 1970. Powerful and effective models; unique look at history. 842pp. 5⅜ x 8½. 20468-5 Pa. $14.95

THE BOOK OF THE SWORD, Sir Richard F. Burton. Great Victorian scholar/adventurer's eloquent, erudite history of the "queen of weapons"–from prehistory to early Roman Empire. Evolution and development of early swords, variations (sabre, broadsword, cutlass, scimitar, etc.), much more. 336pp. 6⅛ x 9¼.
25434-8 Pa. $9.95

THE INFLUENCE OF SEA POWER UPON HISTORY, 1660–1783, A. T. Mahan. Influential classic of naval history and tactics still used as text in war colleges. First paperback edition. 4 maps. 24 battle plans. 640pp. 5⅜ x 8½. 25509-3 Pa. $12.95

THE STORY OF THE TITANIC AS TOLD BY ITS SURVIVORS, Jack Winocour (ed.). What it was really like. Panic, despair, shocking inefficiency, and a little heroism. More thrilling than any fictional account. 26 illustrations. 320pp. 5⅜ x 8½. 20610-6 Pa. $8.95

FAIRY AND FOLK TALES OF THE IRISH PEASANTRY, William Butler Yeats (ed.). Treasury of 64 tales from the twilight world of Celtic myth and legend: "The Soul Cages," "The Kildare Pooka," "King O'Toole and his Goose," many more. Introduction and Notes by W. B. Yeats. 352pp. 5⅜ x 8½. 26941-8 Pa. $8.95

BUDDHIST MAHAYANA TEXTS, E. B. Cowell and Others (eds.). Superb, accurate translations of basic documents in Mahayana Buddhism, highly important in history of religions. The Buddha-karita of Asvaghosha, Larger Sukhavativyuha, more. 448pp. 5⅜ x 8½. 25552-2 Pa. $9.95

ONE TWO THREE . . . INFINITY: Facts and Speculations of Science, George Gamow. Great physicist's fascinating, readable overview of contemporary science: number theory, relativity, fourth dimension, entropy, genes, atomic structure, much more. 128 illustrations. Index. 352pp. 5⅜ x 8½. 25664-2 Pa. $8.95

ENGINEERING IN HISTORY, Richard Shelton Kirby, et al. Broad, nontechnical survey of history's major technological advances: birth of Greek science, industrial revolution, electricity and applied science, 20th-century automation, much more. 181 illustrations. ". . . excellent . . ."–*Isis*. Bibliography. vii + 530pp. 5⅜ x 8¼. 26412-2 Pa. $14.95

DALÍ ON MODERN ART: The Cuckolds of Antiquated Modern Art, Salvador Dalí. Influential painter skewers modern art and its practitioners. Outrageous evaluations of Picasso, Cézanne, Turner, more. 15 renderings of paintings discussed. 44 calligraphic decorations by Dalí. 96pp. 5⅜ x 8½. (USO) 29220-7 Pa. $4.95

ANTIQUE PLAYING CARDS: A Pictorial History, Henry René D'Allemagne. Over 900 elaborate, decorative images from rare playing cards (14th–20th centuries): Bacchus, death, dancing dogs, hunting scenes, royal coats of arms, players cheating, much more. 96pp. 9¼ x 12¼. 29265-7 Pa. $11.95

MAKING FURNITURE MASTERPIECES: 30 Projects with Measured Drawings, Franklin H. Gottshall. Step-by-step instructions, illustrations for constructing handsome, useful pieces, among them a Sheraton desk, Chippendale chair, Spanish desk, Queen Anne table and a William and Mary dressing mirror. 224pp. 8⅛ x 11¼. 29338-6 Pa. $13.95

THE FOSSIL BOOK: A Record of Prehistoric Life, Patricia V. Rich et al. Profusely illustrated definitive guide covers everything from single-celled organisms and dinosaurs to birds and mammals and the interplay between climate and man. Over 1,500 illustrations. 760pp. 7½ x 10⅛. 29371-8 Pa. $29.95

Prices subject to change without notice.

Available at your book dealer or write for free catalog to Dept. GI, Dover Publications, Inc., 31 East 2nd St., Mineola, N.Y. 11501. Dover publishes more than 500 books each year on science, elementary and advanced mathematics, biology, music, art, literary history, social sciences and other areas.